M O D U L A R
Mathematics
for GCSE

Brian Gaulter and Leslye Buchanan

Oxford University Press

Oxford University Press, Walton Street, Oxford OX2 6DP

Oxford   New York   Toronto
Delhi   Bombay   Calcutta   Madras   Karachi
Kuala Lumpur   Singapore   Hong Kong   Tokyo
Nairobi   Dar es Salaam   Cape Town
Melbourne   Auckland   Madrid

and associated companies in
Berlin   Ibadan

Oxford is a trademark of the Oxford University Press

© B. Gaulter and L. Buchanan 1991

First published 1991

Reprinted 1991 (with minor corrections)
Reprinted 1992 (with further corrections)
Reprinted 1992

A CIP catalogue record for this book is available from the British
Library.

ISBN 0 19 914399 4

Typeset and designed by Gecko Limited, Bicester, Oxon

Printed and bound in Great Britain by
BPCC Hazells Ltd
Member of BPCC Ltd

CONTENTS

Introduction   v
Acknowledgements   x

Money management module

1 Percentages
   1.1 Finding the percentage of
       an amount   1
   1.2 Increasing an amount by a
       given percentage   2
   1.3 Decreasing an amount by a
       given percentage   3
   1.4 Expressing one quantity as a
       percentage of another   4

2 Ratio and Proportion
   2.1 Ratio   6
   2.2 Proportion   7
   2.3 Division in a given
       ratio   8

3 Wages and Salaries
   3.1 Basic pay   10
   3.2 Overtime rates   10
   3.3 Commission   12
   3.4 Piecework   13
   3.5 Deductions from pay   14
   *3.6 National Insurance   16

4 Value-added Tax
   4.1 Prices exclusive of
       VAT   18
   4.2 Prices inclusive of
       VAT   19

5 Profit and Loss
   5.1 Calculations involving one
       price   21
   5.2 Calculations involving more
       than one price   23

6 Savings
   6.1 Simple interest   25
   6.2 Compound interest   26
   6.3 Personal savings and
       investments   29

7 Banking
   7.1 Cheques   30
   7.2 Current accounts   31
   7.3 Other accounts   33
   7.4 Banking services   34

8 Borrowing and Spending
   8.1 True interest rates   36
   8.2 Buying on credit   38
   8.3 Credit cards   42
   8.4 Discrimination in
       spending   45

9 Travel
   9.1 Foreign currency   48
   9.2 Time   50
   9.3 Timetables   52

10 Household Costs
   10.1 Mortgages   54
   10.2 The cost of renting   57
   10.3 The Community
        Charge   58
   10.4 Household bills   59
   *10.5 Redecoration   65
   *10.6 Repairs and
         improvements   73

11 Insurance
   11.1 House buildings and
        contents insurance   74
   11.2 Life insurance   77
   11.3 Car insurance   80

12 Further percentages
   12.1 Inverse percentages   84
   12.2 Compound
        percentages   85
   12.3 Depreciation   86

Statistics module

13 Statistical Terms   87

14 Sampling, Surveys,
   Questionnaires
   14.1 Surveys   88
   14.2 Censuses   88
   14.3 Samples   88
   14.4 Sampling methods   88
   14.5 Bias   89
   14.6 Questionnaires   90
   14.7 Pilot surveys   91

15 Classification and Tabulation of
   Data
   15.1 Tabulation   92
   15.2 Tally charts   92
   15.3 Frequency tables   93

16 Pictorial Representation of Data
   16.1 Pictograms   94
   16.2 Bar charts   95
   16.3 Pie charts   98
   16.4 Line graphs   100
   16.5 Histograms   101

17 Interpretation of Statistical
   Diagrams
   17.1 Reading and interpreting
        diagrams   107
   17.2 Drawing inferences from
        diagrams   113
   17.3 Dangers of visual
        representation   115
   17.4 Interpretation of statistical
        information   118

18 Averages
   18.1 The arithmetic mean   119
   18.2 The mode   120
   18.3 The median   121
   18.4 The use of mean, mode,
        and median   122
   18.5 The mean of a grouped
        distribution   123
   *18.6 Moving averages   127
   *18.7 Weighted averages   129

19 Cumulative Frequency
   19.1 The cumulative frequency
        curve (or ogive)   131
   19.2 The median   132
   19.3 The interquartile
        range   133
   19.4 Percentiles   133

20 Dispersion
   20.1 Range   136
   20.2 The interquartile
        range   137
   *20.3 The standard
         deviation   137

21 Probability
   21.1 Introduction   140
   21.2 Probability from theory
        and experiment   141
   21.3 Simple probabilities   142
   21.4 Possibility space   142
   21.5 Simple laws of addition
        and multiplication   144
   21.6 Tree diagrams   146

Core mathematics module

22 Basic Numeracy
22.1 Numbers 150
22.2 Directed numbers 152
22.3 Powers, roots and reciprocals 156
22.4 Factors and multiples 158
22.5 Highest common factor and lowest common multiple 160

23 Using a Calculator
23.1 Approximations 162
23.2 Estimation 163
23.3 Standard form 164
23.4 The keys of a calculator 167

24 Fractions
24.1 Types of fraction 169
24.2 Equivalent fractions 169
24.3 Operations involving fractions 171
24.4 The conversion of fractions to decimal fractions 174
24.5 Percentages 175

25 Ratio and Proportion
25.1 Ratio 177
25.2 Division in a given ratio 178
25.3 Direct proportion 179
25.4 Inverse proportion 180
25.5 Measures of rate 181
25.6 Models and map scales 186

26 Measurement
26.1 Metric and imperial units 191
26.2 Conversion between metric and imperial units 194
26.3 Perimeters of polygons 195
26.4 Area 197
26.5 The circumference and area of a circle 205
26.6 Composite areas 208
26.7 Volume 210
26.8 Areas and volumes of similar shapes 214

27 Algebra
27.1 The basics of algebra 218
27.2 Brackets 224
27.3 Common factors 226
27.4 The addition and subtraction of fractions 227
27.5 Equations 229
27.6 Formulae 234

28 Geometry
28.1 Lines and angles 239
28.2 Symmetry 243
28.3 Triangles 247
28.4 Quadrilaterals 253
28.5 Polygons and angles 256
28.6 Circles and angles 257
28.7 The use of drawing instruments 260

29 Trigonometry
29.1 Pythagoras' theorem 265
29.2 Trigonometrical ratios 269
29.3 Bearings 276
*29.4 Angles of elevation and depression 278
*29.5 Three-dimensional problems 281

30 Graphs
30.1 Graphs and curves 285
30.2 The interpretation of graphs 286
30.3 Graph plotting 289
30.4 Conversion graphs 293
30.5 Travel graphs 295
30.6 Cartesian coordinates 298
30.7 Straight-line graphs 298
30.8 Graphs of simple curves 303
*30.9 Graphs of quadratic functions 306
*30.10 The gradient of a curve 308

***31 Further Algebra**
31.1 Simultaneous equations 310
31.2 Removing brackets 315
31.3 Quadratic factors 316
31.4 Quadratic equations 320
31.5 The transformation of formulae 325

***32 Sets**
32.1 Introduction 328
32.2 Union and intersection 330
32.3 Venn diagrams 333

***33 Vectors**
33.1 Vector representation 339
33.2 The modulus of a vector 341
33.3 Combining vectors 342
33.4 Vector geometry 346

***34 Matrices**
34.1 Matrix representation 348
34.2 Addition and subtraction of matrices 350
34.3 Multiplication by a scalar 352
34.4 Matrix multiplication 354
34.5 Identity and inverse matrices 360
34.6 Matrix transformations 362

***35 Iteration** 367

Surveying module

36 Making a Small Survey
36.1 The field book 370
36.2 Triangulation 370
36.3 Offsets 371

37 Plane Areas, including Circles
37.1 The area of a trapezium 375
37.2 The areas of other polygons 376
37.3 The area of a sector 377

***38 Volumes**
38.1 The volume of a pyramid 379
38.2 The volume of a cone 379
*38.3 The volume of a sphere 380

39 The Surface Areas of Bodies
39.1 The surface areas of bodies with straight edges 382
39.2 The surface area of a cylinder 383
*39.3 The surface area of a sphere 385

40 Areas and Volumes of Irregular Shapes
40.1 The trapezium rule 386
40.2 Simpson's rule 390
40.3 Volumes using the trapezium rule and Simpson's rule 392

41 Ordnance Survey Maps
41.1 Triangulation points 395
41.2 Contour lines 395
41.3 Gradients 395
41.4 Cross-sections 399

Solutions to Exercises 402

Index 420

Introduction

Modular syllabuses for GCSE Mathematics

This book is designed for the 'mature' student who hopes to obtain a GCSE certificate in Mathematics or Mathematical Studies. A mature student is anyone over the age of 16.

The book is specifically designed for the GCSE Modular Mathematics Syllabus as examined by the Southern Examining Group (SEG). In this syllabus, three different sections of mathematics are studied in turn, with an examination at the end of each module or section. The Modular Mathematics course has been created to offer a new opportunity in mathematics for mature students, and the modules contain mathematics relevant to everyday life, particularly the Money Management and Statistics modules. Mature students usually have considerable skills in mathematics, of which they are often not aware. The authors' intention is for the students to build upon these skills, while gaining an appreciation of their use in GCSE mathematics.

The modular course offered by the SEG leads to one of two certificates:

- in **Mathematics**, by studying these three modules:

    c Money Management
    d Statistics
    a Core

- in **Mathematical Studies**, by studying these two modules:

    f Money Management
    g Statistics
    and one of these:
    h Surveying
    i Vocational Studies: the application of Money Management and Statistics and spatial skills into a vocational area (e.g. Technical services, Tourism, etc.).
    n Pre-Nursing

The text of the book has been divided into four parts to help as many students as possible without making the book too long. The parts are:

- Money Management
- Statistics
- Core Mathematics
- Surveying

The work to be studied for each of the SEG modules is given in the grid below:

Module		Sections of the book to be studied
Core	a	Core
Money Management	c	Money Management (excluding 10.4, 10.5, 10.6)
Statistics	d	Statistics (excluding 18.7)
Money Management	f	Money Management
Statistics	g	Statistics
Surveying	h	Core sections 26, 28 and 29 and Surveying
Vocational Mathematics	i	Core sections 22–26 and 28.1 and those parts relating to the actual vocational area studied
Pre-Nursing	n	Core sections 22–26, 28.1 and 30.7

There are two differentiated levels of entry for the examination in each module. Higher (Grades A, B, C, D), and General (Grades C, D, E, F). Material which is only relevant at the Higher level is indicated with an asterisk (*).

The book contains all the topics in most GCSE Mathematics syllabuses.

The SEG syllabuses A, B, E are covered by:

- Core (excluding sections 32, 34, 35)
- Money Management (excluding sections 3.6, 4.2, 7.2, 7.3, 7.4, 8.1, 10.3, 10.5, 10.6, 11.1, 11.3)
- Statistics (excluding sections 14, 17, 18.5, 18.6, 18.7, 19)

Examination technique

The use of calculators is encouraged throughout the course, except in the Aural (Mental Arithmetic) examination, which is part of the Core section. For this reason, in part of the Core module, the text does not encourage the use of calculators. However, as stated in the National Curriculum, mental arithmetic should be used in all cases when the student can readily manage without a calculator.

The accuracy required in GCSE syllabuses is three

significant figures (at higher level entry) and two significant figures (at general level entry). Where appropriate, other levels of accuracy can be required and in Probability fractional answers are usually expected. Students should use at least four figures in their working for calculations in questions requiring answers to three significant figures. The answer can then be rounded to the third figure. Similarly for questions requiring two significant figure answers, at least three figures should be used for calculations throughout. There are many occasions (e.g. when using iteration) when it is more convenient to use the memory function of the calculator and retain all figures until the end, rather than keep rounding off at each stage.

Students who show working and show step-by-step solutions gain credit in examinations even when the final answer is incorrect, whereas students who obtain an incorrect answer and have not written down any working are unable to gain any credit.

It is essential that students consider the sense of their answers. Decimal points are frequently misplaced in error, and, for instance, electricity bills of £17,411 should be checked to see whether 17411 pence (or £174.11) was the real answer!

Coursework: Techniques and topics

Coursework is included in GCSE Mathematics and Mathematical Studies to assess skills which cannot be assessed by means of a written examination. These skills include practical skills, the ability to produce imaginative and creative work arising from mathematical ideas, and the conducting of individual and co-operative enquiry and experiment. Coursework cannot, therefore, include a selection of short-answer-type calculations or any other questions of a similar nature which appear in written examinations. Where possible, the coursework should include work of a practical nature and should also involve a group of students working together.

Coursework should be an investigation into a problem which contains mathematics. SEG Coursework should be 'open-ended', i.e. the problems should be capable of allowing more able students to produce better work at a greater depth and will often not have a 'correct' answer. This reflects everyday life when the way in which mathematics is used is more complex than a simple numerical calculation – there is a 'real' problem. For example, suppose John and Anne were thinking of

going on holiday to the South of France with their children. Naturally, before they made a final decision, they would work out the price of the journey. This would include calculations, and hence they would be applying their knowledge of mathematics. Before John and Anne could make the calculations they would need to collect a considerable amount of information. Such information doesn't just land on a desk the way it does in an examination paper. John and Anne would need additional skills before they got to the point of using their mathematics. In fact, the time spent in obtaining the correct and suitable information will often be far greater than the time spent on the mere calculation. Coursework is used to assess the skill of candidates in finding and using the relevant information.

In our example, John and Anne may need to find the distance to a suitable port, possibly Dover or Portsmouth, the cost of the crossing, which varies at different times of the day, the distance from the French port to the South of France, the cost of petrol in both countries, and the petrol consumption of the car. A more accurate estimate of the cost might include tolls charged on the French motorways, meals and overnight accommodation on the journey, together with insurance. Such an estimate for the cost, correctly calculated, will be typical of the sort of work students are required to do in their coursework. A greater depth of investigation into a problem will produce a higher grade.

To obtain a grade of B or C for this type of coursework it is essential to use **comparisons**.

John and Anne could decide to price alternative methods of travelling in France – for example, using a coach or using a plane with or without hiring a car. At a more basic level, most couples would consider the use of different ferry routes.

There may not be a single 'correct answer' to John and Anne's problem, so coursework doesn't have a 'correct answer' either.

Think of the Motorail, for instance. Not everybody uses it though some do. Why? Because the extra cost and possible benefits or disadvantages are assessed individually and different couples come to different conclusions. These factors, sensibly argued, could be a part of a coursework exercise on foreign travel.

In the best coursework there could be a series of 'answers' showing how the solution to a problem depends on circumstances.

In real life, most problems involving mathematics are

solved in a similar way, which is:

- the identification of the problem,
- the gathering of the information needed to solve the problem,
- the mathematical calculation,
- the conclusion.

In SEG examinations, the coursework is assessed under four headings: **comprehension of task, planning, carrying out the task**, and **communication**. These are similar to the four headings above. Note that the mathematical part is only a small part of the assessment (a quarter), and in real life it could occupy only a small part of the time spent on the problem (possibly even less than a quarter). It is vital to discuss logically the initially perceived problem and the conclusion.

Naturally, the complexity of the problem is reflected in the examiner's assessment. The mathematical techniques used must be appropriate and relevant to the problem. However, to obtain a higher grade students must show an appropriately higher level of competence and understanding. Coursework submitted as part of a higher level entry would gain benefit from including techniques which only occur on the higher level syllabus. For example, in Statistics, standard deviation, moving averages and the mean of grouped data could be included where appropriate.

It must be stressed that the actual topic for the coursework is the choice of the student, in consultation with his or her tutor, teacher, or lecturer. It must, however, be capable of assessment under the four headings and must enable the student to produce work which reflects his or her ability.

A student capable of achieving grade A cannot achieve grade A on coursework if the task given is too restrictive for the necessary skills to be shown. For example: 'investigate the colours of cars in a car park' would produce coursework of grade E or F standard, with D as the maximum grade obtainable.

Additional coursework, which simply shows the same skills as have previously been shown, does not gain credit. **Quality, not quantity** is the important criterion. Two pieces of coursework for a module which show the same skills would gain no more credit than a single piece of coursework. Two pieces of coursework which complement each other and show different skills would gain more credit, and, particularly in the Core module, would often be better than a single piece of coursework.

Suitable coursework tasks

Module c or f: Money management

Appropriate coursework could be in the form of an extended piece of work undertaken as an individual study. Suitable topics would be:

1. **An investigation of a major financial transaction.** This could be on one of many varied topics, such as **buying a house**. This might include consideration of:

   (a) the type of house to be purchased, with reasons
   (b) the part of the country in which to live, including the consequences of the choice, and considered opinions of these consequences
   (c) the cost of purchase, mortgage, stamp duty and legal fees
   (d) the advantages/disadvantages of buying a house compared with renting it.

   The addition of selected literature on houses for sale would be beneficial. The insertion of hundreds of unselected advertisements would not.

2. **A costing project.** One example of a costing project would be **redecorating a room**. This could include:

   (a) measuring the room and producing a diagram
   (b) the cost of carpet, choice of carpet and reasons for selecting the type of carpet
   (c) the cost of painting/wallpapering the walls, and the advantages of each
   (d) the cost of curtains, different materials, etc.
   (e) the cost of other relevant items, including perhaps duvet covers, lampshades, etc., together with a conclusion and suggestions for amendments.

3. **A banking project**
   Students should choose such a project with caution. Simple extraction and copying of material available in leaflet form will not gain credit.

Topics which could be considered include:

A financial transaction

- buying a house
- buying a car by cash or credit/obtaining a car by contract hire
- renting electrical goods or buying them
- choosing a holiday
- buying a villa abroad/renting a villa

Costing projects

- redecorating/refurbishing a room
- redecoration of a house
- building an extension
- keeping a horse
- running a car
- using a car/train/bus to travel to work
- getting married
- having a baby

Banking projects

- comparison of different banks
- comparison of banks and building societies
- consideration of banking services available from the different banks

Module d or g: Statistics

Appropriate coursework could include an extended assignment on a suitable topic, for example, **the popularity of different national newspapers**. The data could be found as a group exercise by 2 or 3 students. The write-up and conclusions must, however, be individual tasks. This could include:

(*a*) obtaining from books of national statistical data the required statistics
(*b*) creating a questionnaire and carrying out a survey (or surveys) to discover whether your acquaintances at college, or work, or in your neighbourhood, have the same preferences as those found in (*a*)
(*c*) using your questionnaire, or creating a new one, to discover reasons for this difference
(*d*) producing a final result, with information illustrated in a variety of appropriate statistical ways
(*e*) suggesting ways in which the results could be extended.

Clearly, if your intuition is correct, you can create a single questionnaire which records people's preferences, and also identifies the reasons for their choice.

Topics which could be considered as a basis for a statistical investigation include:

- types of employment
- the average weekly hours of full-time workers
- the number of people who own shares
- the ownership of freezers, washing machines, cars, etc.
- types of savings, e.g. in banks, building societies, etc.
- the number of visits people make to the doctor
- the number of people who have regular dental check-ups
- the types of letter people receive, e.g. bills, cards, bank statements, etc.
- the time people spend on certain activities during a typical week
- the amount of time people spend listening to the radio
- the amount of time people spend watching the television
- the popularity of national newspapers
- the popularity of selected magazines
- the use of recreational facilities in the countryside
- the money spent on certain items as a percentage of the family income
- the percentage of people who own or rent flats, houses or bungalows
- the length of time people live at a particular address
- holiday locations
- the popularity of various sporting activities
- the popularity of organisations for young people, e.g. youth clubs.

Module a: Core

Coursework can include an investigation of a mathematical problem or the construction of a plan of a piece of land as detailed in module *h* (Surveying). Investigations could be:

- the practical use of mathematics in the real world, e.g. why are soup tins cylinders? how do you use trigonometry to measure the height of a church steeple?
- an integral part of the teaching process, e.g. how do you find where two straight lines ($y = mx + c$) meet without drawing the lines?
- an application of known mathematics into a new field, e.g. when is the volume of a box greatest for a given surface area? at what speed is the running cost of a car least?
- a perceived mathematical problem, e.g. the number of squares on a chess board, number

patterns and their extensions into more general cases.

Module h: Surveying

Coursework can either be an investigation of a mathematical problem based upon geometry or trigonometry or the construction of a plan of a piece of land.

The area used for the survey should be large enough for the student to have a choice of suitable methods (triangulation or offsets only). The area should also have irregular boundaries and preferably should contain items which can be inserted on the plan (e.g. swings, trees, paths, lake, etc.). The final coursework should include a detailed plan to a suitable scale together with the method used, why this method was used, and ways in which the work could have been improved. Students can extend their coursework by redesigning the layout of the area, calculating the volume of tarmac used in the construction of the paths, etc.

Presentation of coursework

All such coursework needs a suitable writing up. This should satisfy the following guidelines.

There are three parts to any coursework, which often deserve approximately equal allocations of time. These are the planning, the work itself, and the writing up. The writing up should not be left to the end, as sections like the introduction can be written up when the planning is complete.

The following is only a guide on what to include in your coursework. Although your work is unlikely to fit this format exactly, try to include as many suggestions as possible, as this is *vital* to your grade.

Remember that a good investigation can be let down by a poor write-up.

Planning

Once you have chosen a topic, you should spend some time thinking about:

- the starting point for the work, e.g. a newspaper article, a set of statistics or a task which you are going to investigate
- what further information you will need
- how you are intending to find this information
- what you expect this information to tell you, and how it will relate back to your starting point.

Having a plan like this will mean that you have a good idea of how much work and time the coursework will involve before you even start. This should allow you to organise your time efficiently and not waste it on unnecessary tasks. As the work progresses, you may have to change your plan in the light of your increasing knowledge of the topic. If you keep notes of the plan and any changes made as you proceed, you will have the information to look back on when you write up your project.

Carrying out your project

Try to fit your plan of action into the time available for your coursework. Do not be tempted to leave the work until the last minute, as problems can arise and tasks could take longer than expected. Also, you may need to check your write-up and possibly do further investigations before arriving at a final conclusion. Keep notes on what you do and whether you feel it is successful. This will help when you come to write up your project.

Writing up

Introduction. This could include why you chose to do the particular piece of work and not others that you may have considered. It should state what you expect to find, and include details of any previous projects or results on the same topic (for instance, national statistics with which you wish to compare your results). Include any problems you expected to encounter in carrying out your investigation.

Method. Give details of how you obtained your information, and why you chose to use this method and not any other. State what problems, if any, you had in obtaining the information. State how each part of your investigation relates to your coursework topic.

If appropriate, give details of how you checked the accuracy of your measurements. If you have used diagrams or methods of measurement that have not been taught on your course, say why you have used them and not the standard methods.

Results. Each table or diagram should have a number and a title. The title should make it obvious what the figures refer to without having to read the rest of the report. This section should be totally factual and consist of a commentary linking each table or diagram.

Conclusion. This section should briefly outline your investigation and what you found. You need to make interpretations and judgements about your results and make suggestions as to how your work could have developed further. It should include any generalisations or conclusions that can reasonably be made from your results. Finally, include any benefits that have been gained from doing your investigation.

That's all the advice we can offer you for the moment. We hope you will find our book useful, and wish you every success in the examination.

Brian Gaulter and Leslye Buchanan
Hampshire, January 1991

Acknowledgements

The publisher and authors are grateful to the following organisations for permission to reproduce previously published material:

Abbeylink, Debenhams, Marks and Spencer, and Principles for credit and store cards, and the Association for Payment Clearing Services, for the Cheque Guarantee logo. (p. 42)
Daily Mail, for the InterCity headline (p. 2)
HMSO, for extracts from *Social Trends* and the *Annual Abstract of Statistics* (pp. 285, 287)
Jacobs Suchard Ltd, for the use of Toblerone (p. 212)
Lloyds Bank, for the Access card and cheques and statement forms (pp. 30, 33, 42)
Pedigree Pet Foods, for allowing a mention of Pedigree Chum (p. 5)
Service Publications Ltd, for the map of York, here much simplified (p. 190)
Southern Examining Group, for questions from past examination papers (pp. 283, 338, 369, 394, 400)
Which?, for information on insurance

Every reasonable effort has been made to contact copyright owners, but we apologise for any unknown errors or omissions. The list will be corrected, if necessary, in the next reprint.

1 *P*ercentages

'*Inflation now stands at 7.3%.*' '*Ford have given their workforce a 21.3% rise.*' '*Unemployment in Winchester is less than 2%.*'

Percentages are a part of our everyday lives. They are often quoted in the media, particularly in connection with money matters.

Percentages often help us make comparisons between numbers, but we must know exactly what a 'percentage' is.

1.1 Finding a percentage of an amount

A percentage is a fraction with a particular number divided by 100:

$$20\% \text{ means } \frac{20}{100}$$

A decrease of 20% would be a decrease of $\frac{20}{100}$, which is the same thing as a decrease of $\frac{1}{5}$, a fifth.

Example

Julian reads in the newspaper that the average pocket money for 12-year-olds has increased nationally by 14% in the last year. Julian's 12-year-old daughter Gilly has been given £1.60 per week for the last two years. How much more per week should he provide for a 14% increase?

$$14\% = \frac{14}{100}$$

$$14\% \text{ of } £1.60 = \frac{14}{100} \times £1.60$$

$$= 0.14 \times £1.60$$

Increase in pocket money = 22.4p = 22p (to the nearest 1p)
Note. See section 23.2 for a further explanation of approximations.

MONEY MANAGEMENT

1 Calculate the following percentages to the nearest 1p.

   a 10% of £13.75   **e** 123% of £4.20

   b 50% of £637.24   **f** $12\frac{1}{2}$% of £69.80

   c 7% of £316   **g** 10.80% of £900

   d 15% of £92.72   **h** 34.4% of £128.50

1.2   Increasing an amount by a given percentage

Example

INTERCITY FARES TO RISE BY 21pc

What does this mean in cash terms to the 20 000 long-distance commuters who travel to London every day?

If an InterCity season ticket costs £2632 now, how much will it cost after the rise?

Method 1
To find the new cost of a ticket we can find 21% of £2632 and then add this to the original:

$$21\% \text{ of } £2632 = \frac{21}{100} \times £2632$$

Increase in fare = £552.72

New cost of fare = £2632 + £552.72

= £3184.72

Method 2
Consider the original amount of £2632 as 100%. Increasing it by 21% is the same as finding 121% of the original cost.

$$121\% = \frac{121}{100} = 1.21$$

Therefore the quickest way of increasing the original fare by 21% is to multiply it by 1.21.

$$121\% \text{ of } £2632 = 1.21 \times £2632$$

New cost of fare = £3184.72

Exercise 2

Give all answers to the nearest 1p.

1 Increase the following rail fares by 10%.

    a £9.20   **d** £9.81

    b £3.70   **e** £9.13

    c £5.00

2 Increase the given amount by the required percentage.

    a £72.12 by 50%   **d** £220 by $6\frac{1}{4}$%

    b 95p by 10%   **e** £124.80 by 25%

    c £360 by 120%   **f** £19.99 by $8\frac{1}{2}$%

Note. If you have difficulty in changing $6\frac{1}{4}$% to $\dfrac{6.25}{100}$ see Unit 24.

1.3   Decreasing an amount by a given percentage

Example

The marked price of a sweater is £24.90.
What is its sale price?

Method 1

$$\text{The reduction} = 20\% \text{ of } £24.90$$

$$= \frac{20}{100} \times £24.90$$

$$= £4.98$$

$$\text{The sale price} = £24.90 - £4.98$$

$$= £19.92$$

Method 2

£24.90 is the equivalent of 100% and so decreasing the price by 20% is equivalent to finding 80% of the original price.

$$\frac{80}{100} \times £24.90 = 0.80 \times £24.90$$

$$\text{Sale price} = £19.92$$

Exercise 3

Give all answers to the nearest 1p.

1 Reduce the following marked prices by 20% to find the sale prices:

    a £30.00   **d** 45p

    b £10.50   **e** £12.99

    c £17.60

2 Decrease the given amount by the required percentage:

    a £54.10 by 8%   **d** £27.15 by $12\frac{1}{2}$%

    b 84p by 30%   **e** £99.05 by 40%

    c £128 by 60%   **f** £1.62 by 33%

1.4 Expressing one quantity as a percentage of another

In a survey of insurance companies it was found that the most common type of car accident was one car running into the back of another.

Out of 35 000 claims, 6280 were for this type of accident.

Information of this type is usually quoted as a percentage.

Example 1

Find 6280 as a percentage of 35 000.

First express 6280 as a fraction of 35 000:

$$\frac{6280}{35\,000}$$

Then multiply this fraction by 100 to express it as a percentage:

6280 as a percentage of 35 000

$$= \frac{6280}{35\,000} \times 100\%$$

$$= 17.9\%$$

This means that almost 18% of car accidents are caused by cars running into the backs of other vehicles.

Note. See Unit 23 for a further explanation of significant figures.

Example 2

A shop buys wallpaper from a wholesaler at £6 per roll and sells it to customers at £8.20 per roll.
What is the percentage increase in price?

The **increase** in price is £8.20 − £6.00 = £2.20

This is $\frac{£2.20}{£6.00}$ as a fraction of the **original** price.

$$\text{Percentage increase} = \frac{£2.20}{£6.00} \times 100\%$$

$$= 36.7\%$$

Example 3

By what percentage has the marked price of £4.70 been decreased to give a sale price of £3.80?

The **decrease** in price is expressed as a fraction of the **original** price and then multiplied by 100.

Decrease in price = 90p

$$\text{Percentage decrease} = \frac{90p}{£4.70} \times 100\%$$

$$= \frac{£0.90}{£4.70} \times 100\% \quad \text{(both quantities must be in the same units)}$$

$$= 19.1\%$$

Exercise 4

Give all answers correct to 3 significant figures.

1 Express the first quantity as a percentage of the second:

a 20, 25

b 3, 87

c 140, 80

d 54, 108

e 60p, £1.10

f £16.25, £12.50

2 Find the percentage by which the first amount is increased or decreased to give the second amount:

a £65, £80

b £250, £300

c 20p, 95p

d £499, £399

e £1.23, 67p

f £24.50, £138.20

Exercise 5 (Miscellaneous questions)

1 Restaurants often add a service charge of 10% to your bill. If a meal for two costs £28.60, how much service charge will be added?

2 In a survey of 348 dog owners, it was found that 103 bought Pedigree Chum dog food.
What percent of the market prefer Pedigree Chum?

3 Your employer offers to raise your weekly wage by 10% or £10. Which should you choose and why?

4 In the survey of 35 000 car accidents, 17.7% happened on a Friday.
How many of the accidents occurred on a Friday?

5 In the same survey, 3213 accidents happened in November. What percentage of the total number of car accidents occurred in November?

6 A holiday company offers a 5% discount on all holidays booked before 31 December the previous year. How much will a family pay for a holiday whose advertised price is £996?

7 A wrist watch has a MRRP (maker's recommended retail price) of £32.99, but a jeweller's shop advertises it for £28.99.
By what percentage has the shop reduced the price? (Give your answer to the nearest whole number.)

2 *Ratio* and *Proportion*

2.1 Ratio

Carmen's parents give her a weekly allowance of £3.60.
Her younger brother, Leroy, is given an allowance of £1.20. Carmen receives three times as much allowance as Leroy. Here is another way of saying the same thing:

The ratio of Carmen's allowance to Leroy's allowance is **3 : 1**

Ratios can be written with a colon between the amounts, like this:

First quantity : Second quantity

or as a fraction, like this:

$$\frac{\text{First quantity}}{\text{Second quantity}}$$

Example 1

When Leroy is older, his parents decide that the ratio between his allowance and his sister's should now be 2 : 3.
If Carmen receives £3.90 per week, how much should they give Leroy?

$$\text{Leroy's allowance : Carmen's allowance} = 2 : 3 = \frac{2}{3}$$

$$\text{Leroy should receive } \frac{2}{3} \text{ of Carmen's allowance}$$

$$\text{Leroy's allowance} = \frac{2}{3} \times £3.90$$
$$= £2.60$$

Example 2

A large jar of coffee costs £2.38 and a small jar costs 84p.
Express these prices as a ratio in its lowest terms.

Converting both prices to pence gives the ratio

$$\text{Large jar : small jar} = 238 : 84$$

The ratio can be reduced if 238 and 84 can both be divided by the same number (called a common factor).

The largest factor which is common to 238 and 84 may not be immediately obvious, in which case the reduction to lowest terms can be carried out in stages.

2 is a common factor of 238 and 84. Dividing by 2 reduces the ratio to 119 : 42

Possible factors of 42 are 2, 3, 6 and 7.
Only 7 is also a factor of 119.

Dividing by 7 gives the ratio in its lowest terms

$$= 17 : 6$$

Exercise 6

Write all ratios in their lowest terms.

1 Two brothers have £20 and £24 in their respective savings accounts.
   Express these amounts as a ratio.

2 Miss Morgan has £320 in her current account, £400 in her deposit account, and £800 in her savings account.
   Express these amounts as a ratio.

3 A pound of grapes costs £1.60 and a pound of pears 72p. Write these prices as a ratio.

4 On their birthdays David and Emily are given money in the ratio of their ages.
   When David is 15 years old, Emily is 12.

   a Express the children's ages as a ratio.

   b If David is given £12.50, how much does Emily receive?

   c When the children are three years older, Emily is given £12.50. How much will David receive?

5 On a particular day a newsagent sold 45 copies of the *Star*, 50 copies of the *Sun*, 30 copies of the *Daily Mirror* and 20 copies of the *Daily Mail*.

   a Express the sales figures as a ratio of whole numbers.

   b Express the sales figures in the form $x : y : z : 1$.

   c On the following day the ratio of sales was 2 : 2.25 : 1.75 : 1 and again 20 copies of the *Daily Mail* were sold.
      How many copies of each of the other papers were sold?

2.2 Proportion

Two quantities are in **proportion** if they increase or decrease in the same **ratio**.

For example:
Tomatoes cost 60p per pound.

We expect to pay twice as much for two pounds of tomatoes, i.e. £1.20.
We pay 30p for half a pound.

Example

Shona decides to deposit £45 each month in a savings account. Her monthly salary is £675.

What proportion of her salary does she save?

$$\text{Proportion of salary saved} = \frac{45}{675} = \frac{1}{15}$$

This can also be expressed as a percentage:

$$\frac{1}{15} \times 100 = 6.67\%$$

Exercise 7

1   A boat-building firm loses business and is declared bankrupt. Shareholders receive compensation in the proportion of £1.20 for every £200 worth of shares.
How much is received by someone holding shares worth:
a £450   **b** £25 000   **c** £12 500?

2   The rate of income tax is 25p in the pound. How much tax is paid by a person whose taxable income is:
a £10 140   **b** £7120   **c** £17 610?

3   Two brothers, who live with their parents, each contribute the same proportion of their weekly wage to the household budget. The elder earns £98.55 and contributes £21.90. How much does the younger brother contribute from his earnings of £75.60?

2.3   Division in a given ratio

FAIR SHARES FOR ALL!!

This does not necessarily mean equal shares for all.

For example, if three partners invest different amounts of money in a business, they might expect the profits to be shared in proportion to their investment.

Example 1

Divide £672 between Emily, Faye and Geoff in the ratio 7 : 5 : 9 respectively. How much does each person receive?

Method
 (i) Find the total number of shares.
 (ii) Find the amount of one share.
 (iii) Find the amount each receives.

Calculation
 (i)   Total number of shares = $7 + 5 + 9$ = 21
 (ii)   Amount of one share = $\dfrac{£672}{21}$ = £32

 (iii)   Emily receives $£32 \times 7$ = £224
              Faye receives $£32 \times 5$ = £160
              Geoff receives $£32 \times 9$ = £288

(*Check.* 224 + 160 + 288 = 672.)

Example 2

Three partners, A, B and C, invest money in a small business. The amounts they invest are £10000, £12000 and £6000, respectively.

At the end of the first year of trading the profits from the business are £14350.

They each receive profits in proportion to their investment.
How much does each partner receive?

The investments are in the ratio 10000 : 12000 : 6000

$$= 5 : \quad 6 \quad : \quad 3$$

The total number of shares	= 5 + 6 + 3	= 14
One share of the profits	= £14350 ÷ 14	= £1025
A receives 5 shares	= £1025 × 5	= £5125
B receives 6 shares	= £1025 × 6	= £6150
C receives 3 shares	= £1025 × 3	= £3075

(*Check.* £5125 + £6150 + £3075 = £14350.)

Exercise 8

1   Divide £650 in the ratio 2 : 3.

2   Divide £12000 in the ratio 1 : 3 : 4.

3   Divide £104 in the ratio 6 : 4 : 3.

4   Mrs Chandra shared £4000 among her three children in the ratio 7 : 5 : 4.
How much did each receive?

5   The sum of £1000 is invested in unit trusts for income and for capital growth in the ratio 3 : 5.
How much is invested in each type of trust?

6   Three sisters win a prize in a Mathematics competition for the best Statistics project. They decide to share the prize of £200 according to the amount of work each contributed.
They calculate that this was in the ratio 3 : 3 : 2.

   a   How much should each of the sisters who contributed most receive?

   b   How much should the sister who contributed least receive?

7   X and Y invested money in a home computing business. X put in £6000, but Y could only afford £4000. The profits were divided in the same ratio as their investment.

   a   At the end of the first year the profits were £10530.
How much did each receive?

   b   At the end of the second year X's share of the profits was £9456.
How much was the total profit?

   c   After two years Y increased his investment to £5000. At the end of the year his share of the profits was £8270.
How much did X receive?

8   Four office workers run a pools syndicate and each week pay £4.95, £6.60, £3.30 and £4.95 respectively for their entry. When they win £104616 they divide the winnings in the ratio of their weekly contribution.
How much does each receive?

3 Wages and Salaries

3.1 Basic pay

All employees receive a wage or salary as payment for their labour.

A **wage** is paid weekly and is calculated on a fixed hourly rate.
A **salary** is paid monthly and is calculated on a fixed annual amount.

Many wage earners are required to work a fixed number of hours in a week,
and they are paid for these hours at the basic hourly rate.

Example 1

Simon works in a hairdressers. His basic pay is £3.75 per hour for a 40-hour week.
Calculate his weekly wage.

Weekly wage = Rate of pay × Hours worked

$$= £3.75 \times 40$$

$$= £150$$

Example 2

Layla's gross weekly wage (i.e. her wage before deductions) is £166.95. She works a 35-hour week.
What is her hourly rate of pay?

$$\text{Hourly rate of pay} = \frac{\text{Weekly wage}}{\text{Hours worked}}$$

$$= \frac{£166.95}{35}$$

$$= £4.77$$

Exercise 9

1 Calculate Donna's gross weekly wage if she works for 42 hours per week at a rate of pay of £4.10 per hour.

2 Daniel's yearly salary is £5756. How much is he paid per month?

3 A basic working week is 36 hours and the weekly wage is £231.48. What is the basic hourly rate?

4 An employee's gross monthly pay is £965.20. What is his annual salary?

5 The basic week in a factory is 35 hours. How much is the weekly wage for someone whose basic rate per hour is:
a £7.42   b £4.45   c £12.60?

3.2 Overtime rates

An employee can increase a basic wage by working longer than the basic week,
i.e. by doing overtime.
A higher hourly rate is usually paid for these additional hours.
The most common rates are **time and a half** and **double time**.

Example

Mr Arkwright works a basic 36-hour week for which he is paid a basic rate of £5.84 per hour. In addition, he works 5 hours overtime at time and a half and 3 hours overtime at double time.

Calculate his gross weekly wage.

36 hours basic pay	$= £5.84 \times 36$	$= £210.24$
5 hours overtime at time and a half	$= (£5.84 \times 1.5) \times 5 =$	$£43.80$
3 hours overtime at double time	$= (£5.84 \times 2) \times 3 =$	$£35.04$
	$\therefore$ Gross pay $=$	$£289.08$

Exercise 10

1   A garage mechanic works a basic 37.5-hour week at an hourly rate of £4.90. He is paid overtime at time and a half. How much does he earn in a week in which he does 9 hours overtime?

2   Mr Cooper's basic wage is £6.20 per hour, and he works a basic 5-day, 40-hour week. If he works overtime during the week, he is paid at time and a half. Overtime worked at the weekend is paid at double time.

Calculate his gross wage for a week when he worked five hours overtime during the week and four hours overtime on Saturday.

3   The basic working week in a small factory is 35 hours (i.e. 7 hours per day) and the basic rate of pay is £3.98 per hour. The overtime rate is time and a half from Monday to Friday and double time on Saturdays.
The table below shows the hours worked by five employees. For each employee, calculate the gross weekly pay.

	Mon	Tue	Wed	Thu	Fri	Sat
Andrews A J	7	7	8	9	8	0
Collins F	8	8	9	9	7	5
Hammond C	9	9	9	10	10	0
Jali Y	8	10	11	11	9	4
Longman B H	9	10	10	9	7	6

4   The number of hours worked by an employee is often calculated from a clock card similar to the one shown below. Mr Meyer works a basic eight-hour day, five days a week and his basic hourly rate is £4.15.

Day	In	Out	In	Out	Clock hours	O/T hours
SAT	0730	1200	1230	1400	6	6
SUN	0800	1200				
MON	0730	1200	1300	1700		
TUES	0730	1200	1300	1730		
WED	0800	1230	1300	1730		
THUR	0730	1200	1245	1815		
FRI	0730	1200	1230	1600		

Overtime is paid at the following rates:
Time and a quarter for Monday to Friday
Time and a half for Saturday
Double time for Sunday
Calculate Mr Meyer's gross pay for this week.

5   a   The basic weekly wage of employees in a small firm is £133 for a 38-hour week. What is the basic hourly rate?

    b   All overtime is paid at time and a half. Calculate the number of hours of overtime worked by Ms Wiley during a week when her gross pay was £154.

3.3 Commission

People who are employed as salespersons or representatives and some shop assistants are paid a basic wage plus a percentage of the value of the goods they have sold.

Their basic wage is often small, or non existent, and the **commission** on their sales forms the largest part or all of their gross pay.

Example

A salesman earns a basic salary of £690 per month plus a commission of 5% on all sales over £5000.
Find his gross income for a month in which he sold goods to the value of £9400.

He earns commission on (£9400 − £5000) worth of sales

$$\text{Commission} = 5\% \text{ of } (£9400 - £5000)$$
$$= 0.05 \times £4400$$
$$= £220$$

$$\text{Gross salary} = \text{Basic salary} + \text{Commission}$$
$$= £690 + £220$$
$$= £910$$

Exercise 11

1 An estate agent charges a commission of $1\frac{1}{2}\%$ of the value of each house he sells.

How much commission is earned by selling a house for £104 000?

2 Calculate the commission earned by a shop assistant who sold goods to the value of £824 if her rate of commission is 3%.

3 A car salesman is paid $2\frac{1}{2}\%$ commission on his weekly sales over £6000. In one particular week he sold two cars for £7520 and £10 640.

What was his commission for that week?

4 An insurance representative is paid a commission of 8% on all insurance sold up to a value of £4000 per week. If the value of insurance sold exceeds £4000, he is paid a commission of 18% on the excess.

Calculate his total gross pay for the 4 weeks in which his sales were £3900, £4500, £5100 and £2700.

5 Two firms place adverts for an insurance representative. Firm A offers an annual salary of £5000 plus a company car (worth £2500 per year) and 4% commission on sales over £200 000 per annum.

Firm B offers an annual salary of £7000 and 3.5% commission on sales over £150 000 per annum.

a Which is the better job if sales of £350 000 per year can be expected?

b If, in a good year, sales rose to £500 000 per year, which job would pay the higher amount and by how much?

3.4 Piecework

Some employees, particularly in the manufacturing and building industries, are paid a fixed amount for each article or piece of work they complete. This is known as **piecework**.

They may also receive a small basic wage and, in addition, some are paid a bonus if production exceeds a stipulated amount.

Example

Workers in a pottery firm who hand-paint the plates are paid a basic weekly wage of £120 and a piecework rate of 80p for every plate over 30 which they paint in a day.

Calculate the weekly wage of an employee whose daily output was as follows:

Day 1 35 plates   Day 2 38 plates   Day 3 40 plates
Day 4 45 plates   Day 5 39 plates

No. of plates over 30 painted = 5 + 8 + 10 + 15 + 9
                              = 47

Piecework bonus = 80p × 47   Weekly gross pay = £120 + £37.60
                = £37.60   = £157.60

Exercise 12

1 A firm employs casual labour to deliver advertising leaflets door to door. The rate of pay is £4.80 for every 100 leaflets delivered. How much is earned by someone who delivers 1230 leaflets?

2 Mrs Mason works a basic 40-hour week at £2.74 per hour plus a bonus of 66p for every skirt she makes over 15 per day. In five successive days she makes 18, 29, 16, 17, and 19 skirts. What is her gross wage?

3 A bricklayer receives a bonus of 54p for every 10 bricks laid in excess of 350 per day. What bonus does he receive if he lays 3070 in a six-day week?

4 A perfumery pays its packers a basic wage of £35 per week plus 2p for each bottle of perfume packaged up to a limit of 1200 per day and 4p per day for each bottle packaged over 1200.

   Calculate the gross pay for a packer whose output on five successive days of one week was 1210, 912, 1311, 1232 and 1043.

5 a A firm producing knitting patterns employs outworkers to knit sample sweaters and cardigans. The firm pays a basic wage of £20 per week plus £22 per garment. Mrs English knits 20 garments per month and Mrs Beckett knits 16 garments per month. Calculate how much they will each earn in a month.

   b In order to increase productivity among its knitters, the firm decides to abolish the basic weekly wage, but increase the piecework rate to £26.50 per garment. How much will the two ladies earn per month under the new system?

   c Comment on the consequences of the new system.

3.5 Deductions from pay

Employees do not usually receive all the money they have earned.

Certain amounts of money are **deducted** from the gross pay and the pay the employee receives is the **net pay** or **take-home pay**.

The main deductions are:
- **Income Tax**
- **National Insurance**
- **Pension** (also called **Superannuation**)

Income tax (basic rate)

Income tax is used to finance government expenditure.

It is a tax based on the amount a person earns in a tax year that begins on 6 April and ends on 5 April the following year.

Most people have income tax deducted from their pay before they receive it, by their employer, who then pays the tax to the Government. This method of paying income tax is called **PAYE** (Pay As You Earn).

Certain amounts of each person's income are not taxed. These amounts are called **tax allowances**.

The Tax Office sends the employee and the employer a PAYE Code which tells them the value of the allowances. For example, a tax code of 0342L would be given to a single person who has allowances of £342 × 10 = £3420.

$$\text{Gross income} - \text{Tax allowances} = \text{Taxable income}$$

The 1991–92 rate of income tax was 25% on taxable income up to £23 700. The allowances were:

Personal £3295
Married Couple £1720

The married couple's allowance is initially awarded to the husband, but it can be transferred to the wife if his earnings are less than £5015.

(In the year of marriage the amount awarded will be proportional to the number of months they have been married.)

Example

Mr Browne earns £19 000 per annum (per year), which is his only source of income. He is married.

Find **a** his total tax allowance, **b** his taxable income, **c** the tax paid per annum.

a Personal allowance = £3295
Married couple's allowance = £1720
Total tax allowance = £5015

b Taxable income = Gross income − Allowances
= £19 000 − £5015
= £13 985

c Tax paid = $\dfrac{25}{100} \times £13\,985$

= £3496.25

Assume that the earnings stated in the following questions are the only source of income.

1 Find (i) the total tax allowance,
  (ii) the taxable income,
  (iii) the yearly tax paid, for:

  a a single man earning £17500 per annum

  b a married man earning £12540 per annum

  c a married woman earning £64 a week.

2 Find the yearly income tax payable by Mrs Bailey, who earns £18765 per annum and has a total tax allowance of £3725.

3 Miss Davis earns £18462 per year. She is entitled to the personal allowance plus an allowance of £960 per year for expenses necessary in her work. Calculate her monthly tax bill.

4 a Jane and Denis live together. Denis earns £13185 per annum and Jane earns £10585 per annum.

  What was their total tax bill for the year 1991–92, if they claimed only the standard allowances?

  b How much tax would they have paid if they had been married?

5 If, in the next budget, the Chancellor changes the basic rate of tax from 25% to 23%, but wages remain the same, how much less tax will be paid by the employees in question 1 above?

*Income tax (higher rate)

Anyone whose taxable income is greater than the amount set by the Chancellor of the Exchequer for basic rate tax payers, has to pay a **higher rate of tax** on the **excess income**.

For the 1991–92 tax year the rates payable were:

- 25% basic rate on a taxable income up to £23700
- 40% higher rate on a taxable income over £23700

┌─ *Example* ───
Mrs Carpenter is a director of a small company and earns £35175 per year. How much income tax does she pay per month?

Her personal allowance	= £3295
Annual taxable income	= Annual gross pay − Total allowance
	= £35175 − £3295
	= £31880
Income taxable at higher rate	= Taxable income − £23700
	(maximum at basic rate)
	= £31880 − £23700
	= £8180
Income tax at basic rate	= 25% of £23700 = £5925
Income tax at higher rate	= 40% of £8180 = £3272
Total annual income tax	= £9197
Monthly income tax	= £9197 ÷ 12 = £766.42

Exercise 14

1   Calculate (i) the yearly, (ii) the monthly income tax paid in each of the following cases:

   a   Anthea Brown earns a salary of £27 000 each year.

   b   John Knight's annual salary is £30 300. He is married, but his wife is not in paid employment.

   c   Mr Woolfe's monthly salary is £3010, and his yearly tax allowance is £5060.

2   a   David has a full-time job and earns £28 000 per annum. He also has a taxable income of £2000 from shares. His wife Erica has a part-time job which pays £2500 per annum.
How much income tax do they pay per year?

   b   The income from the shares could be transferred to Erica.

How much will they save in income tax if the £2000 is added to Erica's income instead of David's?

3   Paul and his wife Rosemary have a total earned income of £56 600 per annum. Paul's salary is £29 500 per annum.
How much income tax do they pay?

4   If, in the next budget, the Government decides to have three income tax bands as follows:

25% on a taxable income up to £23 700
35% on the next £5500 of taxable income
45% on the remaining taxable income,

   a   calculate the yearly income tax which would be payable by Mr Woolfe (see question 1).

   b   Would he benefit under the new system?

*3.6   National Insurance

National Insurance (NI) is paid by both the **employee** and the **employer**. It helps to pay for:

National Health Service
Statutory sick pay
Pensions
Unemployment benefits

The amount paid in National Insurance depends on your gross pay and whether or not you are 'contracted out' of the State Pension Scheme.

Employees who are 'contracted out' pay less National Insurance. On retirement, these employees can claim only the basic State Pension, but they will receive an employment pension.
Employees who are not 'contracted out' receive the basic State Pension plus an earnings-related state pension.

The structure of employees' National Insurance contributions was reformed in the 1989 budget and from 6 April 1991 was as follows:

Weekly earnings (£)	Monthly earnings (£)	Rate of National Insurance contribution
0–51.99	0–225.99	Nil
52–390	226–1690	2% on the first £52 per week (or £226 per month) plus 9%† of remainder
Above 390	Above 1690	2% of £52 per week (or £226 per month) plus
		9%† of £338 per week (or £1464 per month)

†7% for employees who are contracted out.

Example 1

Mr Day's annual salary is £15 978. Calculate his monthly National Insurance contribution if he is contracted out.

Mr Day's monthly salary = £15 978 ÷ 12 = £1331.50

He earns over £226 per month and will pay the rate of 7% (because he is contracted out) on his earnings over £226,
i.e. on £1331.50 − £226 = £1105.50.

He pays 2% on £226   = £226 × 0.02   = £4.52
        7% on £1105.50 = £1105.50 × 0.07 = £77.39
                Total monthly contribution = £81.91

Example 2

Mrs Fielding earns a gross weekly wage of £87.90. Calculate her National Insurance contribution.

Mrs Fielding will pay 2% on £52 and 9% on her earnings over £52,
i.e. on £87.90 − £52 = £35.90.

She pays 2% on £52   = £52 × 0.02   = £1.04
        9% on £35.90 = £35.90 × 0.09 = £3.23
Total weekly contribution   = £4.27

Exercise 15

1 Using the table on p.16, calculate the National Insurance contributions for:

   a Alice, who earns £112.70 per week

   b Charlie, who earns £586.40 per month

   c David, who is contracted out and earns £971 per month

   d Anne, who is contracted out and earns £428.64 per month

   e Maria, who earns £4.30 per hour for a 28-hour week.

2 Mrs Ferry earned £52 per week for her part-time job.
   Calculate:

   a her yearly National Insurance contribution
   b her net income for the year.

3 Nadia earns £4.60 per hour for a 40-hour week. Her income tax payment for the week is £29.25. Calculate:

   a her gross weekly wage

   b her weekly National Insurance contribution

   c her net pay for the week.

4 Mr Dodds has an annual salary of £30 200 and an annual tax allowance of £5125. He is contracted out of the State Pension scheme but pays an additional pension contribution of £1680 per annum, and this additional contribution is added to his tax allowance.
   Calculate:

   a his monthly National Insurance contribution

   b his annual National Insurance contribution

   c his annual income tax

   d his total annual deductions

   e his monthly net pay.

4 Value-added Tax

Apart from income tax, the Government also raises money by taxes on goods and services. These are called **indirect taxes** as they are included in the price to the consumer. For example, the prices of alcohol, tobacco and petrol include a high rate of tax.

4.1 Prices exclusive of VAT

The main indirect tax is **value-added tax** or **VAT**. This is charged on most goods that we buy and also on services such as house improvements, garage bills, meals in a restaurant.

Certain items are **zero-rated,** i.e. no VAT is charged, and these include most food (but not luxury foods such as ice cream, confectionery and crisps), children's clothes and books.

The standard rate of VAT is 17.5% (1991–92 value)
Prices are sometimes quoted exclusive of VAT. The price you pay will include the VAT and will be 17.5% more than the price quoted.

Example

A garage bill totals £88.70. How much VAT is added to the bill?

$$\text{VAT} = 17.5\% \text{ of } £88.70$$

$$= \frac{17.5}{100} \times £88.70 \text{ or } 0.175 \times £88.70$$

$$= £15.52 \text{ (to the nearest 1p)}$$

Exercise 16

The prices of the items in questions 1–5 are quoted exclusive of VAT.
Calculate the amount of VAT payable in each case, giving your answer to the nearest 1p.

1 a typewriter priced at £152

2 a washing machine priced at £234

3 a calculator at £9.50

4 a box of chocolates at £1.44

5 a necklace at £19.20

6 A garage bill shows the cost of parts to be £30.90 and labour costs of £58.60. VAT is then added to the bill. How much is the final bill?

7 A particular calculator is offered for sale through three different catalogues at the following prices (p&p means postage and packing):

Catalogue A £12.60 exclusive of VAT, post free
Catalogue B £13.80 inclusive of VAT, plus 50p p&p
Catalogue C £12.40 exclusive of VAT, plus 40p p&p.

Calculate the price in each case and hence determine which catalogue is offering the best deal.

8 All prices at a DIY store are shown exclusive of VAT. Mrs Jennings buys eight boxes of tiles, priced at £9.90 per box, a large packet of tile cement at £2.39 and some ready mixed tile grout at £1.38.
Calculate her total bill, including VAT.

9 The cash and carry store shows prices exclusive of VAT. Mr Patel buys the following goods for his grocery shop: £45.30 worth of food which is zero-rated for tax, £27.80 worth of confectionery and £15.25 worth of toiletries. Calculate the amount of VAT payable and the total amount of his bill, including VAT.

10 The Government decides to change the rate of VAT payable on goods. Calculate the amount of VAT payable and the total cost of the following goods, including VAT at the rate given:

a a paperback at £4.95, VAT at 12%

b a television set at £340, VAT at 8%

c a personal stereo at £28.90, VAT at 9%

d a patio door at £276, VAT at 17%

e two garden gnomes at £6.87 each, VAT at 14%.

4.2 Prices inclusive of VAT

Most shop prices are quoted inclusive of VAT. An item costing £100 *exclusive* of VAT would have VAT of £17.50 added and its price *inclusive* of VAT would be £117.50. Therefore, for all prices which are inclusive of VAT,

every £117.50 of the price includes £17.50 of VAT

That is, if the price exclusive of VAT is 100%, the price inclusive of VAT is 117.5% of the price.

Example

A reclining chair costs £256.45, including VAT. Calculate: **a** the amount of VAT which was added, **b** the cost of the chair, excluding VAT.

a Price including VAT = £256.45 = 117.5% of the price

$$\therefore 1\% \text{ of the price} = \frac{£256.45}{117.5}$$

$$\text{VAT, equivalent to } 17.5\% = \frac{£256.45}{117.5} \times 17.5$$

$$= £256.45 \times \frac{17.5}{117.5}$$

$$= £38.19$$

b Price excluding VAT, equivalent to 100% $= \dfrac{£256.45}{117.5} \times 100$

$$= £256.45 \times \dfrac{100}{117.5}$$

$$= £218.26$$

Calculating another way gives:

Price excluding VAT = £256.45 − £38.19 (VAT) = £218.26

To find the VAT included in a price, divide by 117.5 and multiply by 17.5

To find the price excluding the VAT, divide by 117.5 and multiply by 100.

These calculations only apply to VAT at 17.5%
If the VAT rate has changed since 1991, can you work out how to change the figures?

Exercise 17

1 A bottle of whisky costs £8.20.
How much of this price is VAT?

2 Mr Lambert buys a motor-cruiser costing £10 710.
What is the price of the motor-cruiser excluding VAT?

3 How much VAT will a shopkeeper pay to the Government on takings of £76.80 for sweets and £43.95 for ice cream?

4 A garage sells 5000 litres of four-star petrol at 46.1p per litre and 9000 litres of unleaded petrol at 41.4p per litre.
How much VAT does the Government receive from the sale of this petrol?

5 Mrs Jali received a garage bill totalling £125.20 for parts and labour including VAT.
How much did the garage charge for parts and labour?

6 A school buys a computer which costs £899.90, including VAT. Schools are not required to pay VAT on educational equipment.

    a How much VAT can the school reclaim?

    b How much does the school pay for the computer?

7 The items listed below are on sale in a large electrical store. Assume that the Government has suddenly reduced the rate of VAT to 12%.
For each item, calculate the amount of VAT paid and the price before VAT was added (give answers to the nearest 1p):

    a steam iron, £28.99

    b hair-dryer, £17.99

    c shaver, £54.99

    d microwave, £199.99

    e washing machine, £329.99

    f vacuum cleaner, £114.99

    g television, £599.99

    h can opener, £11.99

8 The bill for a meal at a restaurant in British Columbia was $63.90, including VAT at the rate of 8%.
How much was the VAT?

5 Profit and Loss

5.1 Calculations involving one price

Everyone who is involved with buying and selling goods aims to make a **profit**.

To do so the dealer buys the goods for a certain amount, the **cost price**, and sells the same goods for a higher price, the **selling price**.

The difference between the cost price and the selling price is the profit:

$$\text{Profit} = \text{Selling price} - \text{Cost price}$$

Because it is easier to compare percentages, the **profit percent** is often required. It is usually calculated on the cost price:

$$\text{Profit \%} = \frac{\text{Profit}}{\text{Cost price}} \times 100\%$$

Sometimes the dealer makes a loss, i.e. the goods are sold for less than the cost price.
In this case, loss is the difference between selling price and cost price:

$$\text{Loss \%} = \frac{\text{Loss}}{\text{Cost price}} \times 100\%$$

Example 1

A shoe shop buys a particular style of shoe for £22 per pair and sells these shoes to the customers for £29.99 per pair.

a What profit does the shop make?

b What percentage profit does the shop make?

a Profit = Selling price − Cost price

$$= £29.99 - £22$$

$$= £7.99$$

b Profit % $= \dfrac{\text{Profit}}{\text{Cost price}} \times 100\%$

$$= \frac{£7.99}{£22} \times 100\%$$

$$= 36.3\%$$

Example 2

A car is bought for £7290 and sold three months later for £6925.
What percentage loss was made on the deal?

$$\text{Loss} = £7290 - £6925$$
$$= £365$$
$$\text{Loss \%} = \frac{£365}{£7290} \times 100\%$$
$$= 5.0\%$$

Example 3

A supermarket buys cheese for 120p per pound. When the cheese reaches
its 'sell by date' the retail price is reduced and the supermarket makes a
loss of 10%.
At what price per pound is the cheese now sold?

$$\text{Cost price (100\%)} = 120\text{p}$$
$$10\% \text{ loss} = \frac{10}{100} \times 120\text{p} = 12\text{p}$$
$$\text{Selling price} = 120\text{p} - 12\text{p}$$
$$= 108\text{p per pound}$$

Exercise 18

1 For each cost price (CP) and selling price (SP)
   given below, find: (i) the profit or loss and (ii)
   the percentage profit or loss.

   a CP = £4, SP = £5

   b CP = £20, SP = £27

   c CP = 75p, SP = 66p

   d CP = £90, SP = £108

   e CP = £525, SP = £425

   f CP = £36, SP = £39.99

2 A market trader buys kiwi fruit at £9 for a box
   of 96 and sells them at 8 for £1.
   What is his percentage profit?

3 A small greengrocer's shop buys crates
   containing 24 melons for £36 per crate. The
   melons are then sold for £1.68 each.

   a What profit does the shop make on each crate
      of melons?

   b What is this profit as a percentage of the cost
      price?

4 A car was bought for £10 565 and resold at a
   loss of 8%. What was the selling price?

5 A greengrocer buys a 10 lb box of grapes for
   £7.20. The grapes are sold to customers at £1.08
   per lb.
   If 2 lb of grapes are unsaleable, what is the
   greengrocer's percentage profit on the cost
   price?

6 A retailer buys calculators for £29.20 each plus
   17.5% VAT and resells them at £41.98 each.
   What is the retailer's profit as a percentage of his
   outlay?

7 **a** A house was bought in 1980 for £82 000 and sold 10 years later for £160 000. What percentage profit was made?

b If the purchasing power in 1990 of a 1980 £1 had eroded to 52p, what was the true profit as a percentage of £82 000?

5.2 Calculations involving more than one price

> ## Example
>
> A hardware shop buys 18 saws for £8.67 each. Thirteen are sold for £10.85, and the remaining saws are sold at the reduced price of £7.99 in the January sale.
>
> **a** Calculate the overall profit which the shop makes on the 18 saws.
>
> **b** Calculate the shop's percentage profit.
>
> **a** Purchase price for 18 saws   = £8.67 × 18   = £156.06
>
>   Amount received for 13 saws   = £10.85 × 13   = £141.05
>   Amount received for 5 saws   = £7.99 × 5   = £39.95
>
>   Total amount received   = £141.05 + £39.95   = £181.00
>   Overall profit   = £181.00 − £156.06   = £24.94
>
> **b** Percentage profit $= \dfrac{\text{Profit}}{\text{Purchase price}} \times 100 = \dfrac{£24.94}{£156.06} \times 100\%$
>
> $$= 16.0\%$$

Exercise 19

1 A chemist's shop buys 25 bottles of a new brand of perfume for £195. To attract sales, the first ten bottles are sold for £8 each. The remaining bottles are all sold at the normal price of £10 each.
Calculate:

  a the overall profit which the shop makes on this perfume

  b the percentage profit.

2 An electrical store buys fifty personal stereos for £949.50. Thirty five are sold for £24.99 each. The remaining stereos are all sold at the reduced price of £19.99 each.

Calculate:

  a the total amount received for the 50 stereos

  b the overall profit which the store makes on these 50 stereos

  c the store's profit, as a percentage of the purchase price.

3 A grocery store buys 500 tins of tomatoes for 23p per tin. The tins of tomatoes are sold to the public for 29p. Unfortunately, 36 tins have been dented and these tins are sold for 5p less than normal. Assuming all the tins are sold, calculate:

  a the shop's overall profit

  b the shop's percentage profit.

4 A garage sells bunches of fresh flowers which it buys from a local nursery.
   The garage pays £1.50 per bunch and sells them for £2.50 per bunch. Any bunches remaining after three days are sold at half price and any remaining after five days are thrown away. On a particular day, the garage buys 12 bunches of flowers. Seven bunches are sold at full price and three are sold at half price. Calculate:

   a the overall profit on these 12 bunches

   b the percentage profit.

5 The clothing department of a large store purchased 100 matching shorts and T-shirts as a special line for the summer season. The purchase prices were £7 for each pair of shorts and £5 for each T-shirt.
   The price to the general public was £9.99 for a pair of shorts and £7.99 for a T-shirt. The items could be bought separately.

Eighty pairs of shorts and ninety T-shirts were sold at these prices.
Calculate:

a the total purchase price of 100 shorts and T-shirts

b the amount the shop received for 80 pairs of shorts and 90 T-shirts.

The remaining shorts and T-shirts were all sold in the 'end of season' sale for £2 each less than the original selling price.

c Calculate the amount the shop received for the shorts and T-shirts during the sale.

d What was the total amount received by the shop for these items?

e Find the overall profit which the shop made on the shorts and T-shirts and express this as a percentage of the purchase price.

6 Savings

There are many ways people can choose to invest their money, some of which are discussed later in this unit.

Most people will invest some, if not all, of their money in a building society, post office or bank savings account.

Because they are **lending** their money they will receive **interest** on the money invested.

6.1 Simple interest

The initial sum of money invested is called the **principal.** With **simple** interest the amount of interest to be paid is always calculated on the principal, and therefore the interest remains the same every year.

In practice, this will only happen if the interest is withdrawn each year, or the interest is paid automatically, so that the amount invested is always the same.

Example 1

£3700 is invested at a simple interest rate of 6% for 5 years. Calculate the amount of interest earned.

$$\text{One year's interest} = \frac{6}{100} \times £3700$$

$$= £222$$

$$\text{Five years' interest} = £222 \times 5$$

$$= £1110$$

Simple interest can be calculated using the formula:

$$\textbf{Simple interest} = \frac{\textbf{Rate}}{\textbf{100}} \times \textbf{Principal} \times \textbf{Time}$$

$$I = \frac{R \times P \times T}{100}$$

The formula makes it easier to calculate the rate, the time or the principal.

Example 2

Calculate the length of time taken for £2000 to earn £270 if invested at 9%.

The simple interest formula is $\quad I = \dfrac{R \times P \times T}{100}$

Substituting into this gives $\quad 270 = \dfrac{9 \times 2000 \times T}{100} = 180 \times T$

$$\therefore 180 \times T = 270$$

Dividing by 180 gives $\qquad T = \dfrac{270}{180} = 1.5 \text{ years}$

▬ *Exercise 20* ▬

1 Calculate:

  a the simple interest on £1400 for 3 years at 6% per annum

  b the simple interest on £500 at 8.25% per annum for 2 years

  c the length of time for £5000 to earn £1000 if invested at 10% per annum

  d the length of time for £400 to earn £160 if invested at 8% per annum.

2 Mr Foyle invests £6750 at 8.5% per annum. How much interest has he earned and what is the amount in his account after 4 years?

3 Mr and Mrs Mahon invest £5000 at 9.25% per annum for 6 months.
How much interest will they receive?

4 Mr Allbright invested £10 800 and at the end of each year he withdrew the interest. After 4 years he had withdrawn a total of £3240 in interest. At what annual rate of interest was his money invested?

6.2   Compound interest

Calculation of compound interest

If the interest earned is not withdrawn each year, but is left in the savings account, the amount invested increases as the interest is added. This means that the interest earned the following year will also increase. In five years, with an interest rate of 10% per annum, £1000 'grows' to £1610.51:

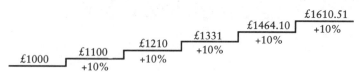

This system of paying interest is called **compound interest**. Building societies usually add compound interest to their accounts every year or every six months.

With some bank accounts, e.g. interest-paying current accounts, the interest is calculated daily and added to the account each month.

┌─ *Example 1* ──────────

£3000 is invested in an account paying 12% compound interest per year.

Find the value of the investment after 3 years.

Principal	£3000
Interest at 12% (of £3000)	360
Value of the investment after 1 year	3360
Interest at 12% (of £3360)	403.20
Value of the investment after 2 years	3763.20
Interest at 12% (of £3763.20)	451.58
Value of the investment after 3 years	£4214.78

Example 2

£2000 is invested in a high interest account for one and a half years. The interest rate is 10.5% per annum, and it is paid into the account every six months.

Calculate the value of the investment after this time and the amount of interest earned.

The rate of interest for 1 year   = 10.5%
∴ The rate of interest for 6 months =   5.25%

Principal	£2000
Interest at 5.25% (of £2000)	105
Value of the investment after 6 months	2105
Interest at 5.25% (of £2105)	110.51
Value of the investment after 1 year	2215.51
Interest at 5.25% (of £2215.51)	116.31
Value of the investment after $1\frac{1}{2}$ years	£2331.82

Compound interest earned = £2331.82 − £2000
                        = £331.82

Note that at each stage the amount of interest has been rounded to the nearest penny.

Exercise 21

1   £6000 is invested at 10% per annum compound interest which is paid annually.
How much is in the account after 3 years?

2   £2000 is invested at 10% per annum payable every 6 months.
How much is in the account at the end of $1\frac{1}{2}$ years?

3   Mrs Fletcher invests £6520 at 9.6% per annum for 18 months. The interest is added every 6 months.
Calculate the total amount of money in the account at the end of 18 months and the amount of interest accrued.

4   £12 600 is invested at 11.05% per annum for 3 years.
Calculate the interest earned over this period.
What is the average amount of interest earned per year?

5   A building society offers a rate of 6.8% per annum payable half-yearly.

   a Calculate the interest payable on £1000 at the end of 1 year.

   b What is the equivalent yearly interest rate?

6   One building society offers an interest rate of 8.5% per annum payable half-yearly. A second offers 8.75% payable annually.
Mr Dugan has £2000 to invest. Which savings account should he choose?

What other factors (apart from higher interest rate) should Mr Dugan take into account when making his choice?

Compound interest tables

Commercially, compound interest tables are used which give the growth rates for different interest rates and times, based on an investment of £1.

Rate of growth	Number of years									
	1	2	3	4	5	6	7	8	9	10
6%	1.060	1.124	1.191	1.262	1.338	1.419	1.504	1.594	1.689	1.791
7%	1.070	1.145	1.225	1.311	1.403	1.501	1.606	1.718	1.838	1.976
8%	1.080	1.166	1.260	1.360	1.469	1.587	1.714	1.851	1.999	2.159
9%	1.090	1.188	1.295	1.412	1.539	1.677	1.828	1.993	2.172	2.367
10%	1.100	1.210	1.331	1.464	1.611	1.772	1.949	2.144	2.358	2.594
11%	1.110	1.232	1.368	1.518	1.685	1.870	2.076	2.304	2.558	2.839
12%	1.120	1.254	1.405	1.574	1.762	1.974	2.211	2.476	2.773	3.106
13%	1.130	1.277	1.443	1.630	1.842	2.082	2.353	2.658	3.004	3.395
14%	1.140	1.300	1.482	1.689	1.925	2.195	2.502	2.853	3.252	3.707
15%	1.150	1.323	1.521	1.749	2.011	2.313	2.660	3.059	3.518	4.046

Example

£4620 is invested at 9% for 5 years. Calculate the approximate compound interest earned.

The growth rate per £1 at 9% for 5 years = 1.539

After 5 years, £4620 will have amounted to £4620 × 1.539

$$= £7110 \text{ (to nearest £)}$$

Compound interest = £7110 − £4620 = £2490

Exercise 22

Use the compound interest table above for the following questions.

1 Calculate the compound interest payable in Exercise 21, question 1.

2 £5000 is invested at 8% for 4 years. Calculate the compound interest earned.

3 £12 000 is invested at 12% for 7 years. Calculate the compound interest earned.

4 Mr Knowles invests £3790 at 10.5% for 8 years. How much is in his savings account at the end of this time if he has made no withdrawals?
(To find the rate of growth at 10.5% over 8 years, find the average of the rates of growth at 10% and 11% over 8 years.)

5 Mr and Mrs Butterworth buy a house for £61 000. If house prices are increasing at a rate of 14% per year, how much will the house be worth after:

a 1 year b 3 years c 10 years?

6.3 Personal savings and investments

Besides savings accounts, there are many other ways of investing money, e.g. stocks and shares, unit trusts, term shares, Premium Bonds, personal equity plans, collecting fine art or vintage wines.

You may need to do a little research in the library to help with the following exercise.

Exercise 23

1 Mr and Mrs Askew have just become grandparents. They wish to invest £2000 on behalf of their granddaughter so that she will collect a substantial sum of money on her 18th birthday.
Should they:

   a buy National Savings Certificates

   b invest the money in a children's savings account

   c buy Premium Bonds

   d buy a valuable piece of jewellery or gold coins

   e invest the money in unit trusts?

   Can you suggest a better investment for this money?

2 a What is the meaning of the term 'taxed at source'?

   b (i) Give an example of an account where the money is *not* taxed at source.

      (ii) Who would benefit by investing in this type of account?

3 Dana and her fiancé wish to save a regular amount each month towards setting up their own home.

   Should they choose:

   a a building society account

   b a unit trust

   c a personal equity plan (PEP)

   d the National Savings 'Yearly Plan'?

4 Give three reasons for investing money in a building society.

5 a For what do the letters SAYE stand?

   b How can you join this scheme?

6 a For what do the letters TESSA stand?

   b What are the conditions which must be met when investing money in this type of account?

7 Give two factors which determine the interest rate offered for your savings.

7 Banking

7.1 Cheques

All bank current accounts, and some building society accounts come with a cheque book.

A **cheque** is a written instruction to the bank to pay a sum of money from your account to yourself or to another named person, company or organisation.

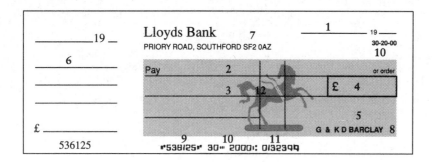

When writing a cheque you must use a pen (**never** a pencil) and fill in correctly:

(1) the date
(2) the name of the person, company or organisation to whom you are paying the cheque (the payee)
(3) the amount of money to be paid, in words (the amount of pence can be written in figures), drawing a line through any unused space
(4) the amount of money to be paid, in figures, separating the pounds from the pence with a hyphen
(5) your signature.

The cheque should be completed neatly and any alterations should be initialled.
You should also fill in:

(6) the counterfoil.

The other information on the cheque is to enable the cheque to be cleared and money transferred from and paid into the correct accounts.

The sorting is done by computers at the Clearing Bank.

The additional information is:

(7) the name and address of the branch of the bank that holds the account
(8) the name on the account
(9) the number of the cheque, which will appear on the bank statement
(10) the branch's sorting code number
(11) the account number
(12) the two lines crossing the cheque vertically – all cheques issued by the bank are 'crossed', which means the cheque must be paid into a bank account. An 'open' or 'uncrossed' cheque can be exchanged for cash. (For safety reasons, open cheques are rarely used.)

Exercise 24

1   The cheque shown below was written by R S Langdon to M F Townsend for the amount of £43, on 7 March 1992.

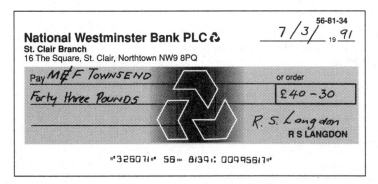

```
National Westminster Bank PLC ↻                    56-81-34
St. Clair Branch                         7 / 3 /  19 91
16 The Square, St. Clair, Northtown NW9 8PQ

   Pay M F Townsend                    or order
   Forty three Pounds                  £ 40 - 30

                                       R. S. Langdon
                                       R S LANGDON

        "326071" 58 " 8139: 00995617"
```

Unfortunately, several mistakes have been made. For each mistake you find, explain what is wrong and what should have been written.

2   Give four reasons why a bank might return a cheque unpaid.

3   What do the words 'refer to drawer' mean, when written on a cheque?

4   Explain the difference between a 'crossed' and an 'open' cheque.

5   Explain the meaning of the phrases 'A/C payee' and 'not negotiable' when written on a cheque.

6   What benefit is there in possessing a cheque guarantee card?

7   Why should a cheque book and a cheque card always be kept separate?

7.2   Current accounts

For most people, a current account is the main bank account. It is a convenient way of dealing with everyday income and expenditure because the bank offers many useful services in conjunction with a current account. Here are some of them:

- cheque book
- cheque guarantee card
- standing orders
- direct debits
- cash dispenser card
- payment debit card
- monthly statement

Standing orders and direct debits are usually made from a bank account, normally a current account, but the service is also available with some savings accounts.

Standing orders

The account holder instructs the bank to make regular payments, of a specified amount, on his/her behalf.

Standing orders can be used to pay insurance premiums, hire purchase instalments or to transfer regular amounts of money to a savings or investment account.

Because the amount to be paid can only be changed by the **payer**, standing orders are being superseded by the more flexible **direct debit**.

Direct debits

Direct debits can be used to pay regular, fixed amounts in the same way as standing orders, but, because the **payee** can change the amount to be paid, variable payments can be made at variable intervals.

The account holder (payer) is usually given 14 days notice of any change in the amount to be paid.

Direct debits are used to pay subscriptions or bills such as gas, electricity, telephone, Community Charge, where the amounts are likely to change each year.

Cash dispenser card

As an account holder at a bank or building society, you can apply for a **cash dispenser card**. You will also be given a PIN (personal identity number) which, together with the card, will enable you to obtain cash from a cash dispenser machine.

Most of these machines dispense money outside banking hours and many operate during the night.

Payment debit card

A debit card can be used to buy goods from a shop, and it works in a similar way to a cheque. The money is deducted directly from your bank account within a few days.

It can also be used to obtain money from cash dispenser machines or over the counter at a bank.

Bank statements

The bank will send an account holder a **bank statement** each month. This is a concise record of all payments into (credits) and out of (debits) his or her account during the preceding month.

The statement should be carefully checked.

Exercise 25

1   Study the bank statement opposite and then answer the following questions:

a   With which branch does the account holder bank her money?

b   What is the account number?

c   On what date was the statement produced?

d   What was the balance in the account on 13 March?

e   By how much did the account become overdrawn?

f   How much was withdrawn using Cashpoint machines?

g   What is the meaning of D/D?

h   How much was paid on standing orders?

i   What does the symbol * mean?

j   What are the numbers 412–––?

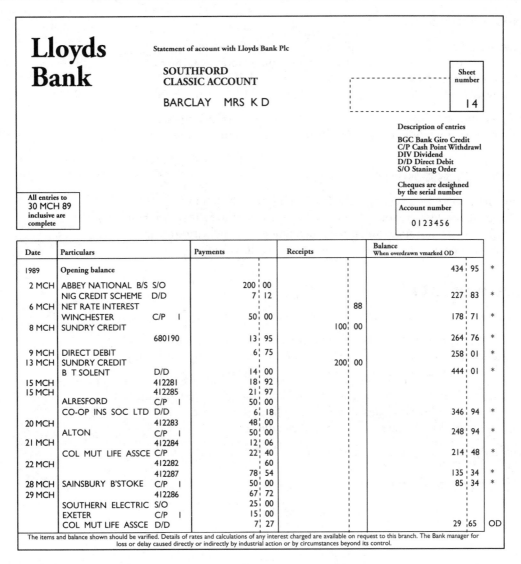

Lloyds Bank

Statement of account with Lloyds Bank Plc

**SOUTHFORD
CLASSIC ACCOUNT**

BARCLAY   MRS K D

Sheet number

| 4

Description of entries

BGC Bank Giro Credit
C/P Cash Point Withdrawl
DIV Dividend
D/D Direct Debit
S/O Staning Order

Cheques are desighned by the serial number

Account number

0 1 2 3 4 5 6

All entries to
30 MCH 89
inclusive are
complete

Date	Particulars		Payments	Receipts	Balance When overdrawn vmarked OD	
1989	Opening balance				434 95	*
2 MCH	ABBEY NATIONAL B/S	S/O	200 00			
	NIG CREDIT SCHEME	D/D	7 12		227 83	*
6 MCH	NET RATE INTEREST			88		
	WINCHESTER	C/P	50 00		178 71	*
8 MCH	SUNDRY CREDIT			100 00		
		680190	13 95		264 76	*
9 MCH	DIRECT DEBIT		6 75		258 01	*
13 MCH	SUNDRY CREDIT			200 00		
	B T SOLENT	D/D	14 00		444 01	*
15 MCH		412281	18 92			
15 MCH		412285	21 97			
	ALRESFORD	C/P	50 00			
	CO-OP INS SOC LTD	D/D	6 18		346 94	*
20 MCH		412283	48 00			
	ALTON	C/P	50 00		248 94	*
21 MCH		412284	12 06			
	COL MUT LIFE ASSCE	C/P	22 40		214 48	*
22 MCH		412282	60			
		412287	78 54		135 34	*
28 MCH	SAINSBURY B'STOKE	C/P	50 00		85 34	*
29 MCH		412286	67 72			
	SOUTHERN ELECTRIC	S/O	25 00			
	EXETER	C/P	15 00			
	COL MUT LIFE ASSCE	D/D	7 27		29 65	OD

The items and balance shown should be varified. Details of rates and calculations of any interest charged are available on request to this branch. The Bank manager for loss or delay caused directly or indirectly by industrial action or by circumstances beyond its control.

2   Explain briefly how you would check the above bank statement.

3   Name six services or facilities available when you open a bank account.

7.3   Other accounts

Deposit accounts

A deposit account is used for savings which you wish to leave untouched over a period of time. Interest, from which tax has been deducted, is paid at regular intervals, usually half-yearly.

Withdrawals of any amount can be made from the branch which keeps the account, but seven days notice is required; otherwise a penalty is charged based on the amount withdrawn. Statements are usually issued half-yearly.

Savings accounts

Savings accounts have some of the facilities of a current account, such as immediate access to your money, a cash card and regular statements. Some even offer standing orders and direct debits as a service. Interest is generally paid monthly.

Banks and building societies now offer many different types of savings account to suit different needs and these are much more widely used than deposit accounts.

Investment accounts

An investment account is for savings which are intended to be left in the account for relatively long periods of time. A higher rate of interest is paid than with other accounts, but notice is required when money is to be withdrawn, usually 90 days. Interest is lost if money is withdrawn without giving the required notice.

Budget accounts

Budget accounts enable you to plan your spending through the year.

A form is filled in detailing all regular expenses over a year – for example, standing orders, insurance premiums, subscriptions, Community Charge – and estimates are made for electricity, gas, water and telephone bills.

The total amount is averaged over 12 months and this amount is paid into the account each month. The bank pays all the listed bills as they arrive, either directly or by honouring cheques from a special cheque book.

A service charge is made for the administration of a budget account.

7.4 Banking services

In addition to those services already mentioned, banks will also:

- issue   Eurocheques, traveller's cheques, credit cards
- operate   the bank giro system, foreign exchange
- arrange   loans, overdrafts, insurance
- advise on income tax, investments, wills
- manage   family trusts, safe custody services, unit trusts.

Eurocheques

Eurocheques are used when travelling abroad. If you have a Eurocheque book and cheque card, you can use them to draw money from your account at a bank abroad or to pay for purchases from shops.

Traveller's cheques

Traveller's cheques are a safe way of taking money abroad. They can be bought in various denominations in pounds, francs, dollars, and other currencies.

They must be signed at the bank immediately by the person receiving them. They can be offered in payment for goods or services or exchanged for cash at a bank.

When they are used they must be signed by the same person who signed for them at the bank originally.

Bank giro

The bank giro system is used to transfer money from one account to another directly.

For example, a credit card company will send you a giro form with their account. The completed form plus a cheque for the required amount may be taken to the bank, and the money is then transferred directly from your account to the credit card company's account. (Alternatively you can post a cheque to the credit card company, but payment will take longer.)

▬ *Exercise 26* ▬

Copy and complete the word puzzle given below. The answer to each question is a banking term, and you may need to do a little research for Clues 4, 7, 10, and 13. Write your answer along the correct line, one letter to each square.

When completed correctly, the initial letters of each word form another banking term (two words: 8, 5).

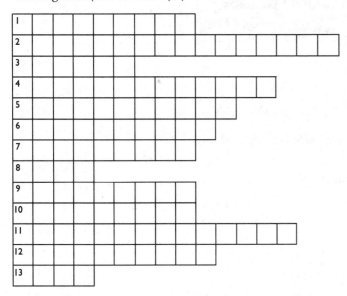

Clues

 1 A record of what has gone into and out of your bank account (9)
 2 Cheque against the bank for obtaining money abroad (10, 6)
 3 The name of a credit card (6)
 4 Two words which indicate that a cheque is paid into the payee's account (3, 10)
 5 It permits the payee to have money transferred from your account in payment of a debt (6, 5)
 6 An account for long-term savings (10)
 7 A means of depositing money with the bank after banking hours (5, 4)
 8 A system for transferring money quickly between accounts (4)
 9 Having a negative balance (9)
10 Money which is paid back (9)
11 An account which pays interest on savings (7, 7)
12 It is used to withdraw money from your account when you are abroad (10)
13 The price quoted for borrowing or lending £100 (4)

8 Borrowing and Spending

8.1   True interest rates

Someone who wants to spend money may sometimes need to borrow money or buy on credit.

The cost of borrowing can vary a great deal depending on the method of obtaining credit as well as the amount borrowed.

Usually banks, stores, and finance companies quote a flat rate of interest, but this is *not* the true price of the credit.

So that customers can accurately compare credit costs, the **APR** (**annual percentage rate**) must, by law, be quoted for all credit arrangements.

This is the *true* rate of interest, and it takes into account all the costs involved and the method of repayment. (In general, the lower the APR, the better the deal.)

If you borrow £100 for one year and the 'credit charge' is 8% you repay £108 (the loan plus interest).

But the charge of £8 would only be a true rate of 8% if you had the use of the £100 for the whole year. In fact, you have £100 at the start of the year and nothing at the end.

In practice, you begin repaying the loan almost immediately in twelve equal instalments spread over the year.

The average amount available to you during the year is therefore about half the original amount and the **true rate of interest is nearly double the quoted rate:**

$$\text{APR} \approx 2 \times \text{Flat rate of interest}$$

The APR is, in fact, between 1.8 and 2 times the flat rate of interest as the following example shows:

Suppose you borrow £100 for 1 year at a flat rate of interest of 8% per annum.

The total repayment = £100 + £8 interest = £108.
This is repaid in 12 monthly instalments of £9 (= £108 ÷ 12).

If the interest on the outstanding debt is calculated at a rate of 1.205% per month, the debt will be wiped out by the end of the year, as shown below:

	£
Initial debt	100.00
Interest in first month at 1.205% per month	1.21
Debt at end of first month	101.21
£9 repaid. Debt at beginning of second month	92.21
Interest in second month on a debt of £92.21	1.11
Debt at end of second month	93.32
£9 repaid. Debt at beginning of third month	84.32
Interest in third month	1.02
Debt at end of third month	85.34
£9 repaid. Debt at beginning of fourth month	76.34
Interest in fourth month	0.92
Debt at end of fourth month	77.26
£9 repaid. Debt at beginning of fifth month	68.26
Interest in fifth month	0.82
Debt at end of fifth month	69.08
£9 repaid. Debt at beginning of sixth month	60.08
Interest in sixth month	0.72
Debt at end of sixth month	60.80
£9 repaid. Debt at beginning of seventh month	51.80
Interest in seventh month	0.62
Debt at end of seventh month	52.42
£9 repaid. Debt at beginning of eighth month	43.42
Interest in eighth month	0.52
Debt at end of eighth month	43.94
£9 repaid. Debt at beginning of ninth month	34.94
Interest in ninth month	0.42
Debt at end of ninth month	35.36
£9 repaid. Debt at beginning of tenth month	26.36
Interest in tenth month	0.32
Debt at end of tenth month	26.68
£9 repaid. Debt at beginning of eleventh month	17.68
Interest in eleventh month	0.21
Debt at end of eleventh month	17.89
£9 repaid. Debt at beginning of twelfth month	8.89
Interest in twelfth month	0.11
Debt at end of twelfth month	9.00
£9 repaid. Debt after twelve repayments of £9	0.00

The interest rate of 1.205% per month gives a compound interest rate of 15.46% per year.

Hence a flat rate of 8% per year is equivalent to an APR of 15.46% (almost twice the percentage rate).

(How did we get that figure of 1.205% per month? That involves some tricky arithmetic which you needn't worry about!)

8.2   Buying on credit

If you do not have sufficient funds saved to buy the goods you require, or do not wish to withdraw a large sum of money from your savings, you can buy the goods on credit, i.e. borrow money which will be repaid over a period of time, usually in instalments which include interest.

There are several methods of obtaining credit:

- Hire purchase
- Credit purchase
- Loans
- Monthly accounts
- Credit card/Store card/Charge card.

Hire purchase and credit purchase

These two methods of obtaining credit are similar. A **deposit** is often required and the amount borrowed plus interest is repaid in equal instalments over a set period of time.

The main difference is that with hire purchase you do not own the goods until the last payment has been made and so cannot sell the goods while you are still paying for them. Also, the hire purchase company can repossess the goods if the repayments are not made.

Credit purchase is usually arranged with the store from which you buy the goods. You own the goods from the moment the agreement is signed. If you fail to keep up the repayments, the store will sue through the courts for the amount outstanding.

Example 1

A video recorder can be bought for a cash price of £299.95 or by credit purchase paying 12 monthly instalments of £31.50.
Calculate:

a the credit price

b the interest paid

c the flat rate of interest, as a percentage of the amount borrowed

d the approximate APR.

a Credit price = Total instalments
$$= 12 \times £31.50$$
$$= £378$$

b Interest paid = Credit price − Cash price
$$= £378.00 − £299.95$$
$$= £78.05$$

c Flat rate of interest $= \dfrac{\text{Interest}}{\text{Amount borrowed}} \times 100\%$
$$= \dfrac{£78.05}{£299.95} \times 100\%$$
$$= 26.0\%$$

d Approximate APR = 2 × Flat rate per year
$$= 52.0\%$$

Example 2

A washing machine is offered for a cash price of £329.99. It can also be bought on hire purchase for a deposit of £34.99 and 30 monthly instalments of £13.25.
Calculate:

a the total hire purchase price

b the amount of interest paid

c the amount borrowed

d the flat rate of interest

e the approximate APR.

a Total HP price = Total instalments + Deposit
    (do not forget to include the deposit)
    = 30 × £13.25 + £34.99
    = £397.50   + £34.99
    = £432.49

b Interest paid = Credit price − Cash price
    = £432.49 − £329.99
    = £102.50

c Amount borrowed = Cash price − Deposit
    = £329.99 − £34.99
    = £295.00

d Flat interest rate = $\dfrac{\text{Interest}}{\text{Amount borrowed}} \times 100\%$

    $= \dfrac{£102.50}{£295} \times 100\%$

    = 34.7%

e Flat rate per year = $\dfrac{\text{Flat rate}}{\text{Time of loan in years}}$

    $= \dfrac{34.7\%}{2.5}$

    = 13.9%

    Approximate APR = 2 × Flat rate per year
    = 2 × 13.9%
    = 27.8%

Exercise 27

1 A department store offers a hi-fi system for sale at £429.99. The customer can also buy the system on interest-free credit by paying a deposit of 10% of the cash price and 10 equal monthly instalments.

   a How much is the deposit?

   b How much is each monthly payment?

2 A rival store offers a similar hi-fi system for sale at £349.99 or 12 months free credit. The credit terms are 10% deposit and 12 equal monthly instalments.
   How much is each instalment?

3 A television set can be bought for £189.99 cash or by credit purchase.

   The credit purchase terms are: no deposit and 24 monthly payments of £9.31.
   Calculate:

   a the total credit price

   b the amount of interest paid

   c the flat rate of interest

   d the approximate APR.

4 A BMW motorbike has a cash price of £4195. It can be bought on hire purchase for a deposit of 10% and 36 monthly payments of £135.84. Calculate:

a the total hire purchase cost

b the approximate APR.

5 Mr Weston is buying a new car. The cash price of the car is £7399. He is offered a £1500 trade-in price on his old car. The hire purchase terms on the new car are a deposit of 25% of the cash price and 24 monthly payments of £290. Calculate:

a the deposit

b the amount borrowed by Mr Weston from the HP company

c the total hire purchase cost of the car

d the interest paid, as a percentage of the loan

e the approximate APR.

6 Mrs Sayeed wishes to buy a new car. The cash price of the car is £9098.

The hire purchase terms offered by the garage are: a deposit of 15% and *either* 36 monthly repayments of £271.50 *or* 60 monthly repayments of £179.12.

Find, for each deal:

a the total hire purchase price of the car

b the approximate APR.

7 The same camera is available in two different shops at the same price of £98.60, but with different credit terms.

Shop A requires a deposit of 10% and twelve monthly repayments of £8.43.

Shop B requires a deposit of 15% and ten monthly repayments of £9.60.

By calculating the true rates of interest (APR), decide which shop is offering the better deal.

Loans

If a large amount of money is required, for example, for a car or holiday or home improvements, a common form of credit which is applied for is a **loan**.

A **bank loan** or **personal loan** can be arranged with the bank manager or with a finance company. The interest rates can be lower than for other forms of credit.

It is important to choose a reputable company and to compare interest rates by looking at the quoted APR.

A bank loan is applied for through the bank manager. The amount borrowed is added to the borrower's account as a single payment and is then repaid over a fixed period of time in instalments. One advantage of this type of loan is that the interest is fixed in advance so that the amount to be repaid is known exactly.

An **overdraft** is a temporary, and sometimes expensive, method of obtaining credit used for a short-term loan.

The bank will charge interest on the amount of the overdraft plus 'transaction charges' on all cheques, standing orders, direct debits and cashpoint withdrawals during this period.

Exercise 28

1 The table below is published by a finance company. It shows the monthly repayments required for a given personal loan and the APR charged.

Months	12 APR 23.9%	24 APR 23.9%	36 APR 23.9%
Amount of loan	Monthly repayments	Monthly repayments	Monthly repayments
£300	28.02	15.50	11.39
£400	37.36	20.67	15.19
£500	46.70	25.84	18.99
£600	56.04	31.01	22.79
£700	65.38	36.18	26.59
£800	74.73	41.35	30.39
£900	84.07	46.52	34.19
Amount of loan	APR 21.9%	APR 21.9%	APR 21.9%
£1000	92.61	50.87	37.14
£1100	101.88	55.96	40.86
£1200	111.14	61.05	44.57
£1300	120.40	66.14	48.29
£1400	129.66	71.23	52.00
£1500	138.92	76.31	55.72
£1600	148.19	81.40	59.73
£1700	157.45	86.49	63.14
£1800	166.71	91.58	66.86
£1900	175.97	96.67	70.57
£2000	185.23	101.75	74.29
£2100	194.49	106.84	78.00
£2200	203.76	111.93	81.72
£2300	213.02	117.02	85.43
£2400	222.28	122.11	89.15
£2500	231.54	127.19	92.86

For each of the following loans, calculate:

(i) the total amount to be repaid
(ii) the total interest
(iii) the flat rate of interest.

Compare the flat rate of interest with the given APR:

a £1000 for 12 months   d £300 for 24 months

b £1000 for 24 months   e £2400 for 36 months

c £1000 for 36 months   f £1700 for 24 months.

2 Explain what is meant by the phrases:

a 'security for a loan'

b 'an uninsured loan'.

3 Mr and Mrs D'Angelo wish to borrow £5000 to be repaid over a period of 5 years. After visiting several banks and building societies, they find that the best terms are:

a monthly repayments of £123.83 for a 'secured' loan

b monthly repayments of £147.85 for an 'insured' loan

c monthly repayments of £132.56 for a straightforward repayment loan.

  (i) Calculate the total cost of each of the above loans.

  (ii) What are the advantages and disadvantages of each type of loan?

4 Jamal's bank manager agrees to an overdraft of up to £100 on his account each month. There is an arrangement fee of 1% of the amount allowed for this service.

Because Jamal is a student, the interest charged is only 1.2% (APR 15.3%) and there are no transaction charges. At the end of January his account is overdrawn by £80 and at the end of February by £74.20.

a How much was the arrangement fee?

b How much does he pay each month for his overdraft?

5 Mr and Mrs Madden buy their clothing through a mail order catalogue using their 12 month credit scheme.
Their costs are:

- the catalogue price for the goods, *plus*
- handling and delivery charges:
  £1.99 for 1 item, £2.80 for 2 items, £3.30 for 3 or more items.

A service charge of 2.4p in the pound is added each month to the outstanding balance.

a At the end of June their account is clear. During July they buy a dress and a pair of jeans priced at £29.99 and £34.99 respectively.

  (i) What is the total price charged?

  (ii) They make a first repayment of £6.
How much interest is added to the account they receive one month later?

b The bill is repaid by means of a further ten repayments of £6 (eleven altogether) and one of £13.74.

  (i) What was the total cost of the two items?

  (ii) What was the credit charge?

  (iii) Find an approximate value for the APR.

8.3 Credit cards

Credit cards, store cards and charge cards can be used to pay for a wide range of goods and services.

Credit cards

Access cards and *some* Visa cards are credit cards. (Other Visa cards are debit cards or Automatic Telling Machine cards – they are not considered here.) With credit cards you receive a statement every month which lists the purchases made in the previous month.

The bill can be settled in full, within 25 days, in which case there are no interest charges to be paid.

Alternatively a minimum payment of £5 or 5% of the total bill (whichever is the greater) can be made, or any larger repayment. Interest will then be charged each month on the outstanding amount.

Typical interest charges are 2.2% per month (APR 29.8%) or 1.9% per month (APR 25.3%), plus an annual charge of £12.

Many cards offer different rates and terms, and some offer free insurance for goods bought using them.

Store cards

A store card operates in a similar way to a credit card, but it is only accepted in the store or group of stores in which it was issued.

The interest rates are usually higher than for credit cards. An APR above 30% is not uncommon.

To make matters more complicated, some store cards (e.g. Marks and Spencer's cards) are credit cards, some are charge cards and some are budget accounts.

Charge cards

An annual fee is usually paid for a charge card (e.g. American Express and Diner's) and the bill must be settled in full each month.

Example

Mr Buckley borrows £450, on average, each month on his credit card and interest is charged at 2.2% per month.

If he changed credit card companies, he would pay an annual fee of £12 and interest charges of 1.9% per month.

Would he benefit by changing?

Comparing the monthly charges for the two companies:

2.2% interest on £450 = £9.90

1.9% interest on £450 = £8.55
average monthly fee   = £1.00
                         _____
Total   = £9.55

Conclusion:
Mr Buckley would benefit by changing his credit card.

Exercise 29

1 What advantages does the possession of a credit card offer the card holder?

2 What additional 'perks' are available to credit card holders?

3 What is meant by the term 'credit limit'.

4 What should a credit card holder do in the event of a card being lost or stolen?

5 What is the minimum age for a credit card holder?

6 What disadvantages are there in possessing a credit card?

7 Mr Johnstone receives a bill for £250 from his credit card company. He makes the minimum repayment and is charged interest of 2.2% per month on the balance.

   a What repayment does he make?

   b How much is the balance?

   c What interest charges will be added to his next statement?

8 Mrs Magee receives her credit card statement which shows a total bill of £262, including her annual £12 fee. She pays £24.50 by the due date and is charged interest of 1.9% per month on the balance.

   a What is the balance?

   b What interest charges are added to the next statement?

   c Should either Mrs Magee or Mr Johnstone (see question 7) change credit card companies?

9 **a** Mr and Mrs Collier see the following advertisement in their paper:

CD COMPACT MIDI
£399.99
FREE HEADPHONES

   They make some inquiries and discover that they will need to make 35 repayments of £17 and a final payment of £8.49 in order to pay for the CD system.
Calculate:

   (i) the credit charge
   (ii) the approximate APR.

   b The Colliers then find the same equipment for sale through a mail order catalogue with weekly payments of £4.85 over 2 years (i.e. 104 repayments).
Calculate:

   (i) the credit charge
   (ii) the approximate APR.

   c Compare the two deals.

8.4   Discrimination in spending

It pays to be discriminating when spending money. Taking time to make comparisons means getting the best value for your money.

You can save money by looking for special offers, buying during the sales (if you are sure you are getting a bargain) or buying in bulk.

Example

Mrs Newman is shopping for soap powder and finds a new brand on the supermarket shelves.

The powder is sold in three packet sizes:

a starter pack: weight 145 g, price 39p
a medium size: weight 800 g, price £1.55
a large size: weight 2 kg, price £3.79

Which size should Mrs Newman buy to give her the best value for her money?

There are two methods which can be used to compare value for money on the basis of cost.

Method 1
Prices are compared by calculating the cost of 1 g of powder for each size (given here to 3 significant figures).

1 g costs:
- (i) Starter pack $\dfrac{39p}{145} = 0.269p$
- (ii) Medium size $\dfrac{£1.55}{800} = 0.194p$
- (iii) Large size $\dfrac{£3.79}{2000} = 0.190p$

Packet type:	*Starter*	*Medium*	*Large*
Price per g:	0.269p	0.194p	0.190p

This shows that the large size gives the best value for price.

Of course, price is not the only thing to be taken into account when making a purchase.

It is also important to consider, where appropriate:

- the quality of the product
- the length of time for which non-durables can be stored
- personal preferences of taste, colour, style etc.
- the manufacturer's reputation for reliability
- the shop's reputation for service.

Method 2
Value is compared by calculating the amount of powder which can be bought for the same amount of money (e.g. £1) in each case (again to 3 significant figures).

Amount of powder for £1:
- (i) Starter pack $= \dfrac{145}{0.39} = 372g$
- (ii) Medium size $= \dfrac{800}{1.55} = 516g$
- (iii) Large size $= \dfrac{2000}{3.79} = 528g$

Packet type:	*Starter*	*Medium*	*Large*
Powder per £1:	372g	516g	528g

This also shows that the large size will give Mrs Newman more for her money.

For example, Mrs Newman might consider that the starter pack will be the best value at the moment. If she finds that she prefers her old powder, she has not spent much money on a product she does not want.

The medium size packet costs very little more per kg than the large size and is lighter to carry. It also takes up less storage space.
For some people, this size may be the best buy.

Exercise 30

1 **a** A well-known brand of margarine sells at 38p for 250 g and 67p for 500 g.

> (i) For how much per kg is each size sold?
> (ii) Which size is the better value for money?

  b The price of the smaller size of margarine is reduced by 5p on 'special offer'. Which size is the best buy?

2 Mayonnaise costs 99p for a 400 g jar and £1.39 for a 600 g jar. Mrs Kaplan has a voucher for '10p OFF' a jar of mayonnaise.

  a If she uses the voucher, how many grams of mayonnaise will she be able to buy for 1p:

> (i) with the 400 g jar
> (ii) with the 600 g jar?

  b Which jar gives the best value for money?

  c What other factor(s) should Mrs Kaplan take into consideration before making her purchase?

3 The village shop stocks soap powder in three sizes:

  E3   contains 1.05 kg of powder and costs £1.22
  E10 contains 3.50 kg of powder and costs £3.79
  E15 contains 5.25 kg of powder and costs £5.49.

  a Calculate comparative costs for each size, based on the E15 size.

  b Which size would you recommend as the best buy for the following people, giving at least one reason for each choice?

> (i) Mr and Mrs Donovan and their three children
>
> (ii) Mrs Kyle, a 66-year-old widow
>
> (iii) Mr and Mrs Groves, a newly married couple with no children, but a large mortgage
>
> (iv) Mr Hambledon, a bachelor living in a small flat.

4 Compare the following advertisements and decide which is the best value for money, giving clear reasons for your choice.

£339.99
• Free 5 year guarantee
• 24 wash programmes
• 11 lb wash load
• 1300 rpm spin speed

£329.95
• 5 year guarantee only £90
• 21 programmes
• 10 lb wash load
• 300/1000 rpm spin speed

5 Mr and Mrs Palfrey decided to find out which nearby supermarket gives the best value for money. They made a shopping list of items which they buy regularly and then visited the four nearest supermarkets in turn making a note of all the prices.

The results of their survey are shown below:

Super-market	A	B	C	D
Apples (2 lb)	58p	70p	72p	64p
Bread (large)	46p	47p	41p	45p
Butter (250 g)	59p	62p	62p	63p
Cereal (pkt)	72p	75p	72p	71p
Chicken (3 lb)	£2.37	£2.94	£2.85	£2.79
Coffee (100 g)	£1.92	£1.91	£1.93	£1.89
Fish (1 lb)	£2.60	£2.19	£2.82	£2.39
Meat (1 lb steak)	£3.40	£3.46	£3.69	£3.66
Potatoes (3 lb)	60p	66p	66p	75p
Tea bags (80)	£1.27	£1.36	£1.29	£1.30

a (i) If they had bought all the groceries listed above in the same supermarket, at which supermarket would they have paid the least amount?

 (ii) Assuming each grocery item on the list above was bought at the most economical price, calculate the lowest price they could have paid for the above items, buying from these supermarkets.

b The distance from the Palfreys' home to each of the supermarkets is:

A 20 miles
B 3 miles
C 1 mile
D 10 miles.
Supermarket D also sells petrol which is, on average, 5p per gallon cheaper than any local petrol station.

Which supermarket should Mr and Mrs Palfrey use:
 (i) if they shop weekly
 (ii) if they shop monthly
 (iii) if they need a few extra items?

9 Travel

9.1 Foreign currency

Currency exchange

If you plan a trip abroad for business or holiday, you must work out the money you will need and the form in which you will take it.

Most people take a limited amount of cash in the currency of each country to be visited. They take the remainder in the more secure form of traveller's cheques. Alternatively, they use Eurocheques or credit cards to obtain cash.

In the UK British currency (pounds sterling) is usually exchanged for foreign currency at a bank or travel agency. Exchange rates between one currency and other currencies change frequently and are published daily in newspapers and displayed where money is exchanged.

By consulting the **selling price** you can calculate the amount of foreign currency you will be sold for your pounds and from the **buying price** you can calculate the amount of pounds you will receive in return for your foreign currency.

Example

Mr Oztürk is to travel on a business trip to Turkey. He changes £350 into Turkish lira on a day when the bank selling price is 3750.

On returning home he changes his remaining 262 500 lira into sterling. The bank buying price is 4150 to £1.

Calculate:

a the amount of lira he receives
b the amount of pounds he receives on his return.

a The bank pays 3750 lira for every £1 it buys. For £350 he will receive 3750 × 350 lira = 1 312 500 lira

b The bank charges 4150 lira for every £1 it sells. For 262 500 lira he will receive $\frac{£262\,500}{4150}$ = £63.25

Exercise 31

1 Using the bank selling price, change:

  a £12 to Austrian schillings
  b £140 to Spanish pesetas
  c £96.50 to American dollars.

2 Using the bank buying price, calculate the sterling equivalent of:

  a 200 French francs
  b 11 100 Italian lira
  c 650 Japanese yen.

EXCHANGE RATES

	Bank Buys	Bank Sells
Australia $	2.33	2.18
Austria Sch	20.85	19.65
Belgium Fr	62.40	58.50
Canada $	2.08	1.98
Denmark Kr	11.38	10.78
Finland Mkk	7.00	6.60
France Fr	10.05	9.45
Germany Dm	2.978	2.798
Greece Dr	292	266
Italy Lira	2205	2075
Japan Yen	266	250
Netherlands Gld	3.33	3.15
Norway Kr	11.48	10.82
Portugal Esc	262.5	246.50
Spain Pta	190	178
Swedan Kr	10.86	10.20
Switzerland Fr	2.618	2.458
USA $	1.77	1.67

3 Geraldine and Peter ate a meal in a restaurant while on holiday in Rhodes.
The meal for two costs 4256 drachma. Use the bank selling price to calculate the cost of the meal in pounds.

4 The Andersons spent a holiday touring in Yugoslavia. While travelling they used 200 litres of petrol which cost 1236 dinar per litre.
The exchange rate was 4330 dinar to £1.

a How much did the petrol cost them in pounds?

b What was the price per litre of the petrol in pounds?

5 Before going on holiday to Germany and Austria, the Williams family changed £600 into German marks. While in Germany they spent 824 DM and then changed their remaining marks into Austrian schillings as they crossed the border. The exchange rate was 7.047 Sch to 1 DM. Calculate:

a the number of marks they received

b the number of schillings they bought.

6 If Mr Oztürk (in the example on p.48) had postponed his trip to Turkey until the following week the bank selling price would have been 3915 lira to the pound.
On his return, the buying price would have been 4515 lira to the pound.

a How many lira would he have received?

b How many lira would he have had left on his return to Britain (assuming that he would have spent the same amount)?

c How many pounds would he have received on his return?

d How much money would he have saved by travelling the following week?

7 a Mr Elton changed £100 into French francs for a day trip to France. How many francs did he receive?

b Unfortunately the excursion was cancelled and so he changed all his francs back to pounds. How many pounds did he receive?

c How much money did he lose because of the cancellation?

Commission

In practice, the banks also charge commission for each currency exchange.

The commission is £1 on currency exchanges up to the value of £200 and 0.5% of the value above £200.

On traveller's cheques the commission is £2 for up to £200 in value and 1% of the value above £200.

Questions 1 and 2 refer to Exercise 31.

1 **a** How much commission did Mr Elton pay when changing pounds to francs?

   b How much commission did he pay when changing the francs back to pounds?

   c How much did his cancelled trip cost him, including commission charges?

2 **a** How much commission did the Williams family pay for their Deutschmarks?

   b How much commission would they have paid if they had taken the £600 in traveller's cheques?

3 **a** For a three-week holiday, touring Germany, Switzerland and Austria, a group of four friends decided to take the equivalent of £900 abroad.

At the bank they changed £110 into Deutschmarks, £100 into Swiss francs and £120 into Austrian schillings. How much of each currency did they receive (after commission was deducted)?

   b After buying the foreign currencies, they changed as much as possible of the remaining money into traveller's cheques.

   The smallest value of traveller's cheque which can be bought is £10.

   (i) How much did they exchange for traveller's cheques?
   (ii) How much commission did they pay for the cheques?

   c What was the total cost per person of the foreign currencies and cheques?

9.2 Time

There are two methods in general use for showing the time:

 (i) the 12-hour clock   (ii) the 24-hour clock.

With the 12-hour clock the day is divided into two periods, from midnight to noon (am) and noon to midnight (pm).

The 24-hour clock uses a single period of 24 hours, starting at midnight. The time is written as a four-digit number, without a decimal point, and there is no need to specify morning (am) or afternoon (pm).

This method is always used for timetables and is in common use on video recorders and digital clocks.

Example

Write the times
a ten to nine in the evening,
b twenty past seven in the morning, using both the 12-hour and 24-hour clocks.

a The 12-hour clock time is 8.50 pm
    The 24-hour clock time is 2050 (i.e. 8.50 + 12 hours)

b The 12-hour clock time is 7.20 am
    The 24-hour clock time is 0720 (note the zero at the beginning to make a four-digit number)

Exercise 33

In the following exercise the time is written using words, the 12-hour clock or the 24-hour clock.

For each question, give the time using the other two methods.

1   Six fifteen in the morning

2   11.10 pm

3   0930

4   1.40 am

5   1350

6   2220

7   Ten to ten in the evening

8   10.45 am

9   Twenty-five past midnight

10   1656

Example

The 0745 train from Newcastle is scheduled to arrive in Southampton at 1516.

How long is the journey?

Method 1

From 0745 to 0800	= 15 mins
From 0800 to 1500	= 7 hours
From 1500 to 1516	= 16 mins
Total journey time	= 7 hours + 15 mins + 16 mins
	= 7 hours 31 mins

Method 2

	Hours	Mins		Hours	Mins	
Train arrives	15	16	=	14	76	(76 = 16 + 60)
Train departs	7	45		−7	45	
Time taken			=	7	31	

The journey time is 7 hours 31 minutes.

Exercise 34

1   Mr Little catches the 0654 train to London.
If he arrives at the station at twenty to seven, how long does he wait?

2   Miss Brothers travels on the 0704 train. She arrives at the station nine minutes before the train is due.
At what time does she arrive at the station?

3   Mr Ghosh arrives at the airport at 8.35 am and his plane takes off 55 minutes later.
At what time does the plane take off?

4   At what time does the 0929 train at Coventry arrive if it is 13 minutes late?

5   a   Mrs Richardson's afternoon train is scheduled to arrive in Birmingham at eight minutes past four, but it is 15 minutes late.
At what time, on the 24-hour clock, does she arrive?

   b   Her connecting train leaves at 1651.
How long does she have to wait?

6 Chris has an appointment in London at 2.15 pm. The train journey from his local station takes 1 hour and 10 minutes. He allows a further 20 minutes to travel on the underground to his place of appointment. In case there are any delays on the journey, he allows an extra 15 minutes travelling time.
What length of time should Chris allow from his local station?

7 **a** The boat from Dover to Calais departs at 1715. The crossing takes $1\frac{1}{2}$ hours and French time is 1 hour ahead.
At what time does the boat dock in France, French time?

   b The same boat leaves Calais for Dover at 2000. At what time does it arrive in England, British time?

9.3 Timetables

All timetables use the 24-hour clock.

Exercise 35

Mondays to Saturdays

BASINGSTOKE (Bus Station)							0730			0840	0935		1035	1135		1235	1335		1435	1535
Basingstoke (Winton Square)							0735			0845	0940		1040	1140		1240	1340		1440	1540
Worting Road (South Ham)							0739			0849	0944		1044	1144		1244	1344		1444	1544
Worting (White Hart)							0743			0853	0948		1048	1148		1248	1348		1448	1548
Newfound (Fox Inn)							0747			0857	0952		1052	1152		1252	1352		1452	1552
Deane Gate							0753			0903	0958		1058	1158		1258	1358		1458	1558
Overton (Post Office)							0758	0756		0908	1003		1103	1203		1303	1403		1503	1603
Laverstoke (Mill)							0802	0802		0912	1007		1107	1207		1307	1407		1507	1607
Whitchurch (Square)							0809	0809		0919	1014		1114	1214		1314	1414		1514	1614
Whitchurch (Bere Hill Estate)					0650		0813			0923			1118			1318			1518	
Whitchurch (Square)					0654	0809	0817			0927	1014		1122	1214		1322	1414		1522	1614
Hurstbourne Priors (Portsmouth Arms)					0659	0814	0822			0932	1019		1127	1219		1327	1419		1527	1619
The Middleway							0821				1026			1226			1426			1626
Longparish (Plough Inn)							0708		0831		0941			1136			1336			1536	
Longparish (Station Hill)							0712		0835		0945			1140			1340			1540	
London Road (Admirals Way)							0723*	0825	0843		0953	1030		1148	1230		1348	1430		1548	1630
ANDOVER (Bridge Street) arr.							0727	0829	0847		0957	1034		1152	1234		1352	1434		1552	1634
ANDOVER (West Street) arr.							0729C				0959C			1154C			1354C	

ANDOVER (West Street) dep.	0624		0734			0911		1111			1311			1511			
ANDOVER (Bridge Street) dep.	0626		0736		0831	0913	1031	1113		1231	1313		1431	1513		1631	1735
London Road (Admirals Way)	0630		0740		0835	0917	1035	1117		1235	1317		1435	1517		1635	1739
Longparish (Station Hill)						0925		1125			1325			1525			1747
Longparish (Plough Inn)						0929		1129			1329			1529			1751
The Middleway	0634		0744		0839		1039			1239			1439			1639	
Hurstborne Priors (Portsmouth Arms)	0641		0751		0846	0938	1046	1138		1246	1338		1446	1538		1646	1800
Whitchurch (Square)	0646		0756		0851	0943	1051	1143		1251	1343		1451	1543		1651	1805
Whitchurch (Bere Hill Estate)						0947		1147			1347			1547			1809
Whitchurch (Square)	0646		0756		0851	0951	1051	1151		1251	1351		1451	1551		1651	1813
Laverstoke (Mill)	0653		0803		0858	0858	0958	1058	1158	1258	1358		1458	1558		1658	1820
Overton (Post Office)	0657		0807		0902	0902	1002	1102	1202	1302	1402		1502	1602		1702	1824
Deane Gate	0702		0812		0907	0907	1007	1107	1207	1307	1407		1507	1607		1707	1829
Newfound (Fox Inn)	0708		0818		0913	0913	1013	1113	1213	1313	1413		1513	1613		1713	1835
Worting (White Hart)	0712		0822		0917	0917	1017	1117	1217	1317	1417		1517	1617		1717	1839
Worting Road (South Ham)	0716		0826		0921	0921	1021	1121	1221	0321	1421		0521	1621		1721	1843
Basingstoke (Winton Square)	0720		0830		0925	0925	1025	1125	1225	0325	1425		0525	1625		1725	1847
BASINGSTOKE (Bus Station)	0725		0835		0930	0930	1030	1130	1230	0330	1430		0530	1630		1730	1852

1 Use the bus timetable to answer the following questions.

   a Mr Tully arrives at Basingstoke bus station at 10.15 am. How long does he have to wait for a bus?

   b How many buses from Basingstoke stop at Longparish?

   c Miss Dawes catches the 0908 bus at Overton. To get to Salisbury she must change buses at Andover. The Salisbury bus leaves Andover at 1038.

   How long does she have to wait at Andover?

   d Mrs Goff lives in South Ham and visits her mother in Whitchurch for at least 3 hours every Wednesday. If she catches the 1044 bus from Worting Road, which buses can she catch from The Square in order to be home before 5 o'clock?

2   Assume that the time in France and Belgium is 1 hour ahead of British time.

Dover/Boulogne [1¾ hours]	0030	0330	0630	0930	1230	1530	1830	2130		
Boulogne/Dover	0130	0430	0730	1030	1330	1630	1930	2230	2359	

Dover/Calais [1¼ hours]	0200	0400	0600	0730	0900	1030	1200	1330	1500	1630	1800	1930	2100	2230
Calais/Dover	0015 0200	0400	0600	0730	0915	1045	1215	1345	1515	1645	1815	1945	2115	2245

Dover/Zeebrugge [4 hours]	0530	0830	1130	2030	2340
Zeebrugge/Dover	0100	0400	0700	1300	1600	1900

a How many ferries leave Dover in the evening between 7 o'clock and 10 o'clock?

b At what time does the 8.30 pm boat arrive in Zeebrugge?

c The 1330 ferry from Dover arrives in Calais at 1610.
How many minutes late is it?

d The Carmichael family have hired a chalet in Boulogne, but it is not available until after 12 noon.
Assuming it takes 45 minutes to pass through customs and drive to the chalet, what is the earliest ferry they should catch from Dover?

e Mr and Mrs Davenport plan to return to England on the 0130 ferry from Boulogne but arrive just as the boat is leaving.
How much time will they save if they catch the next available ferry from Calais?
(Assume that it will take more than half an hour, but less than 2 hours to drive from Boulogne to Calais.)

f The 1215 ferry from Calais to Dover leaves 20 minutes late.
Because of heavy seas, the crossing takes 45 minutes longer than usual.
At what time does the ferry arrive in Dover?

10 Household Costs

10.1 Mortgages

Loans based on value of property

A **mortgage** is a loan from a building society, bank or other financial institution to buy a house.

The size of the mortgage obtained depends on:

- the value of the property to be purchased,
- the earnings of the prospective purchaser(s).

The loan is normally up to 80% of the value of the property (which is usually less than the purchase price), but can be as much as 100%.

The remainder of the purchase price must be paid before the buyer can occupy the house. It is called the **deposit**.

Example

This semi-detached, four-bedroom house is sold for £160 000. The building society surveyor values the property at £150 000. The building society agrees to a mortgage of 90%.

Calculate the amount of the loan and the deposit.

Loan $= 90\%$ of the valuation

$$= \frac{90}{100} \times £150\,000$$

$$= £135\,000$$

Deposit $=$ Purchase price $-$ Loan

$$= £160\,000 - £135\,000$$

$$= £25\,000$$

Exercise 36

A bank offers mortgages of 95% to buyers of properties valued at £70 000 or less, 90% for properties not exceeding £120 000 in value and 80% if the value is more than £120 000.
For each of the following houses, calculate the maximum loan available from this bank and the deposit required:

1   a semi-detached house sold for £78 500, valued at £75 000

2   a four-bedroom family house sold for £110 000, valued at £106 000

3   a three-bedroom bungalow sold for £61 000, valued at £62 500

4   a five-bedroom house sold for £147 000, valued at £142 000

5   a town house sold for £260 000, valued at £200 000.

Loans based on income

The same bank will give a loan of up to three times the borrower's gross basic annual income. For joint borrowers, the maximum loan is *either* three times the larger income plus the smaller income *or* 2.25 times the joint income.

Example

A couple require a mortgage to buy a house. Their basic incomes are £12 500 and £14 000. What is the maximum loan they can obtain from this bank?

Arrangement 1. Loan = 3 × £14 000 + £12 500

$\qquad\qquad\qquad$ = £54 500

Arrangement 2. Loan = 2.25 × (£14 000 + £12 500)

$\qquad\qquad\qquad$ = £59 625

The maximum loan is therefore £59 625.

Exercise 37

1 Find the maximum loan obtained by someone whose basic income is £17 000 per annum.

2 A man earns £14 000 per annum and his wife earns £21 000 per annum.
   What is the maximum loan they can obtain?

3 Two friends decide to become joint owners of a property. Their basic annual incomes are £15 000 and £10 000.
   What is the maximum amount of money they can borrow?

4 A couple require a mortgage from the bank. If the smaller basic income is £10 000, determine (by drawing a graph, or otherwise) the values of the larger income for which it is better to take 2.25 times the joint incomes.

5 A couple's basic annual incomes are £22 000 and £15 000. They wish to purchase a £97 000 house which has been valued at £92 000.
   By considering the three different arrangements they could make with the bank, find the largest possible loan they could obtain.

Mortgage repayments

Once the mortgage is arranged and the property has been bought the mortgage, plus interest, must then be repaid over a number of years. The payments are made monthly and will depend on:

● the size of the loan
● the number of years over which the loan is to be repaid
● the current rate of interest.

There are many different types of mortgage available, the two most common being a **repayment** and an **endowment** mortgage.

A repayment mortgage   is the original type of mortgage whereby the repayments are made over a set number of years. A single payment is made each month, part of which is interest on the loan and part repayment of the capital.

This is the most flexible type of mortgage, but as time passes the repayment of interest becomes smaller and the repayment of capital is then a larger part of the monthly premium. As tax relief is given on the interest, this means that the monthly premiums increase over the period of the loan.

It is also necessary to take out a mortgage protection policy.

With an endowment mortgage   two payments are made each month. One is to the lender to repay the interest on the loan and the other is to an insurance company. When the insurance policy matures the loan is repaid. The insurance policy also includes full life cover.

The monthly repayments vary during the period of the loan according to the current interest rate. Initially the premiums for an endowment mortgage are higher than for a repayment mortgage.

Tax relief is allowed on the interest paid on the loan. For mortgages of £30 000 or less this tax relief is deducted at source i.e. it is deducted from the monthly premium. This scheme is called MIRAS (Mortgage Interest Relief At Source).

Which mortgage you choose depends on long-term inflation and your personal circumstances. It is difficult to obtain impartial advice from specialists because many of them are paid by commission or tend to deal with one particular insurance company.

The following table gives the monthly repayments, after tax relief, for a loan of £1000.

No. of years	10	15	20	25	30
Monthly repayment	£12.07	£10.14	£9.26	£8.84	£8.64

(In real life the monthly repayments may vary depending on the current interest rate.)

Example

A loan of £25 200 is made on a house over 25 years. Find:
a the monthly repayment,
b the total amount paid.

a $\qquad$ Monthly repayment $= \dfrac{25\,200}{1000} \times £8.84$

$\qquad\qquad\qquad\qquad = 25.2 \times £8.84$

$\qquad\qquad\qquad\qquad = £222.77$

b Total repayments in 1 year $= 12 \times £222.77 = £2673.24$

$\quad$ Total repayments in 25 years $= 25 \times £2673.24 = £66\,831$

1 For each of the following loans find:

  (i) the monthly repayment
  (ii) the total amount repaid.

  a £15 000 over 20 years

  b £12 000 over 30 years

  c £29 500 over 25 years

  d £30 000 over 15 years

2 Mr and Mrs Yardley pay a deposit of 20% on a house costing £34 000. They obtain a 25-year mortgage on the remainder.

  a What is their monthly premium?

  b How much will they have paid for the house at the end of 25 years?

3 A young couple purchase a flat for £31 000 which was valued at £29 500. They obtain a 95% mortgage over 30 years. Calculate:

  a the amount of the loan

  b the monthly repayments

  c the total repayments

  d the amount of the deposit

  e the total cost of the flat.

4 The following table shows the monthly repayments for a loan of £1000 before tax relief has been deducted:

No. of years	15	20	25	30
Monthly repayment	£12.91	£12.04	£11.66	£11.45

For the following loans, calculate:

  (i) the monthly premium
  (ii) the total amount repaid.

  a £45 000 over 25 years

  b £36 000 over 20 years

  c £32 500 over 15 years

  d £51 000 over 30 years

10.2 The cost of renting

Many people rent accommodation rather than buy their own property. For some this is a necessity because they cannot afford the high cost of a mortgage; others choose to rent because there are fewer responsibilities, and sometimes because it is more economical if a job requires frequent moves to new areas. The cost of renting often includes a charge for gas, electricity and water rates. Repairs and redecoration are usually the responsibility of the landlord.

1 Mr and Mrs Evans rent a furnished flat at a cost of £120 per week.
How much is this per year?

2 Daniel rents furnished accommodation at £370 per month.
What is the annual cost of renting?

3 Four students share a flat. The monthly rent is £576.
How much per week should each student contribute to cover the cost of the rent?

4 Anwen is a student looking for accommodation for two years. She sees the following advertisements in a local newspaper:

Which arrangement will cost her the least amount?

5 a The Jacksons live in a council house and pay £470 per month in rent. How much do they pay per year?

b They have the opportunity to buy their council house for £61 000. They can obtain a 90% mortgage and the repayment rate is £8.80 per month per £1000.
  (i) How much would they need for the deposit?
  (ii) How much would they pay per year in mortgage repayments?

c Assuming they have sufficient money saved for the deposit, should the Jacksons buy their council house? Give reasons for your decision.

6 a What advantages are there in renting accommodation?

b What advantages are there in buying a house?

10.3 The community charge

In 1990 the old system of charging each householder domestic rates was replaced by a **community charge**, more commonly known as the **poll tax**.

Everyone over the age of 18 years, with a few exceptions, now contributes towards the cost of local government.

Local government is responsible for certain services:

- Education
- Planning and transportation
- Police
- Public protection (fire, waste disposal, etc.)
- Recreation (libraries, country parks, museums, etc.)
- Social services.

Charge payers in the same district pay about the same – there are slight variations due to different charges levied by parish councils.

Charges between different areas vary a great deal.

In addition to the personal charge, paid in the area of the principal residence, a **standard charge** is also payable if more than one home is owned.

The **standard charge** is a multiple of the personal community charge and is set by the local council.

Full-time students and student nurses who study under Project 2000 pay 20% of the community charge, and each counts as 0.20 of a charge payer. Some people on low incomes can claim **benefit** and are entitled to a reduction in their poll tax of up to 80%.

Poll tax payments can be made as a lump sum, payable by 25 April, or in instalments over 10 months.

> **Example**
>
> Mr and Mrs Collier live in Winchester where the personal community charge is £364.07 per year. They have three children aged 19, 16 and 13 years. The eldest daughter is studying at Southampton University, but lives at home.
> Calculate:
>
> **a** the family's total community charge,
>
> **b** the amount of each monthly instalment.
>
> **a** Charge for two adults   = 2 × £364.07 = £728.14
>   Charge for one student = $\frac{1}{5}$ × £364.07 = £72.81
>   Total community charge   = £800.95
>
> **b** Monthly instalment = £800.95 ÷ 10 = £80.10

Note. In 1991 the Government announced its intention to replace the community charge with a council tax.

Exercise 40

1 Mr and Mrs Wolski live in an area where the community charge is £288.62 per person.

  a How much is the annual charge?

  b How much is each monthly instalment?

2 Rhys is a student living in Swansea. The personal community charge (tal cymunedol personal) is £264.
How much does he pay?

3 Mr and Mrs Yaqub have two children over 18 years of age: Selma who works, and Tariq who is a student living at home. The poll tax is set at £314.20 per year.
How much does the family pay?

4 The parish of Micheldever has 773.40 registered community charge payers. The contribution from poll tax to the parish is £4.31 per head.
How much does the parish of Micheldever receive from the community charge?

5 **a** The parish of Otterbourne needs to raise £14 045. There are 1114 registered community charge payers.
How much will need to be levied per head for the parish?

  b If the total poll tax is £361.08 per person, what percentage of the tax will the parish receive?

6 Miss Miles qualifies for a 30% reduction on her poll tax bill of £320.
How much does she pay?

7 Mr Chalmers is on income support and is entitled to the maximum benefit of 80% on his community charge of £360.
How much community charge does he pay each month if he pays in monthly instalments?

10.4 Household bills

Apart from the community charge, substantial bills which a householder can expect to receive each year are:

- Water rates
- Electricity bill
- Gas bill
- Telephone bill

The electricity, gas and telephone bills arrive quarterly; the water rates bill arrives annually.

Water rates

Water rates are still levied by some local councils at so much in the pound of rateable value plus a standing charge.

The rateable value is an amount assigned by the local district valuer to all properties in the council's area.

The charges for Winchester in 1990–91 are shown below:

Service	Annual standing charge	Charge per £ of rateable value	Minimum charge
Water supply	£13.00	28.1p	£23.00
Wastewater	£20.00	29.4p	£23.00

Example

The average customer in this area has a rateable value of £210 and receives all water services.

Calculate:

a the average annual water rates bill

b the approximate daily cost for water.

a Total standing charge   = £33.00
   Water supply   = 28.1p × 210   = £59.01
   Wastewater (sewerage)   = 29.4p × 210   = £61.74

   Total annual charge   = £153.75

b Approximate daily cost $= \dfrac{£153.75}{365}$ = 42.1p

Exercise 41

1 Calculate the yearly water rates for a bungalow with rateable value £425.

2 The occupier of a house with rateable value £234 pays the water rates in two half-yearly instalments.
How much is each payment?

3 A cottage has a septic tank for sewerage, which is emptied by private contract.
Calculate the occupier's yearly water rates if the rateable value is £194.
How much will the owner pay monthly if she opts to pay in 10 instalments?

4 In the previous year, the standing charge for water services was £11. The owner of the cottage (see question 3) paid £49.61 for water. How much (to the nearest 0.1p) was the charge per pound of rateable value for water supply?

The following information refers to questions 5, 6 and 7.

Many people now pay for water according to the amount they actually use. The amount of water used is measured by a water meter. Typical charges are shown below:

Service	Annual standing charge	Pence per cubic metre
Water supply	£28.00	35.6p
Wastewater	£30.00	51.9p

5 Calculate the total water bill, including all services, for the following amounts of water used:

a 50 cubic metres

b 10.9 cubic metres

c 31.6 cubic metres

d 9000 gallons

(Assume 1000 gallons = 4.5 m³.)

6 The owner of the bungalow (see question 1), estimates that the quarterly water consumption is about 38 cubic metres.
How much would the water bill have been if the water supply had been metered?

7 The owner of the cottage (see question 3), estimates the annual water consumption to be 52 cubic metres.
Should she have a water meter installed?

Electricity bills

The amount of electricity or gas which a household uses is recorded by a meter.

There are two types of meter in use.

The digital meter is a line of figures:

The clock meter has dials which are read from left to right. Each dial is numbered from 0 to 9 and the hands move clockwise and anticlockwise alternately.

When a hand is between numbers the lower number is recorded. (When a hand is between 9 and 0, record 9.)

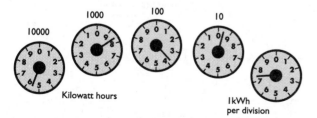

To calculate an electricity bill the price per unit is multiplied by the number of units used and a standing charge is added.

Example 1

What is the reading on the set of dials on the left?

The reading is 58397.

Example 2

In February, the reading on Mr Spencer's meter is 28279 and the previous reading, in November was 26814.

The cost of electricity is 5.44p per unit and the quarterly standing charge is £6.27.

Calculate:

a the number of units of electricity used

b the cost of electricity used

c the total electricity bill.

a	Number of units used	= 28279 − 26814	= 1465
b	Cost of electricity	= 1465 × 5.44p	= £79.70
c	Total bill	= £79.70 + £6.27	= £85.97

Exercise 42

1 What is the reading on the following sets of dials?

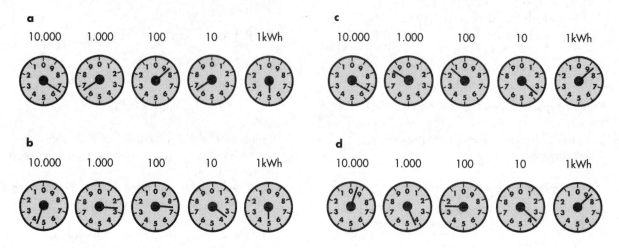

a

| 10.000 | 1.000 | 100 | 10 | 1kWh |

b

| 10.000 | 1.000 | 100 | 10 | 1kWh |

c

| 10.000 | 1.000 | 100 | 10 | 1kWh |

d

| 10.000 | 1.000 | 100 | 10 | 1kWh |

2 Calculate the total electricity bills for the following amounts:

	Present reading	Previous reading	Cost per unit	Standing charge
a	37162	35841	5.90p	£9.70
b	86719	85017	6.83p	£9.90
c	26341	24863	5.70p	£7.67
d	70905	69438	6.04p	£8.45
e	42186	41902	7.63p	£9.20

3 Mr Sinclair receives the following electricity bill based on an estimated reading:

Previous reading	Present reading	Tariff	Units	Price per unit	Amount £	p
77945	78616E	DOMESTIC	671	6.04p	40	53
QUARTERLY CHARGE					8	45
		TOTAL THIS ACCOUNT			48	98

The true reading was 78262.

a By how much has Mr Sinclair been overcharged?

b Draw a set of dials and put in arrows to show the true meter readings.

4 Mrs Yamoto receives an electricity bill for £59.08.
The standing charge is £7.67 and the cost per unit is 5.70p.

a How many units of electricity were used?

b If the present reading is 66112, what was the previous reading?

Gas bills

Gas bills are calculated in a similar way to electricity bills, but the number of units used is the number of cubic feet of gas used.

This number has then to be converted into therms and multiplied by the price per therm.

A standing charge is added.

Typically 105 therms = 100 cubic feet of gas.

Example

In one quarter Mr and Mrs Blake used 260 hundred cubic feet of gas. The cost of the gas was 36p per therm and the standing charge was £8.98 per quarter.

Calculate the amount paid for gas.

Amount of gas used = 260 hundred cubic feet of gas

$$= \frac{260}{100} \times 105 \text{ therms}$$

$$= 273 \text{ therms}$$

Cost of gas $= 36p \times 273$

$= £98.28$

Standing charge $= £8.98$

Amount paid $= £107.26$

Exercise 43

1   In the following questions, the cost of gas is 39.8p per therm and the standing charge is £8.70 per quarter.

Calculate the quarterly gas bills for the following people:

a Mr Hudson, who used 401 therms

b Mrs Snell, who used 275 therms

c Miss Dawe, who used 392 therms

d Mr Ghosh, who used 456 therms.

2   Mrs Kirk used 370 hundred cubic feet of gas in one quarter. How many therms did she use?

3   Mr McBain used 95 therms of gas at a cost of 34.2p per therm. The standing charge was £9.00 per quarter.
How much was his gas bill?

4 Miss Hennigan's gas meter reads 3469 on 10 December and 3617 on
   10 March. The charge per therm is 43p and the quarterly standing charge is
   £9.40.
   Calculate:

   a the number of units of gas consumed

   b the number of therms of gas consumed

   c the cost of the gas

   d the total gas bill for the quarter.

5 a The reading on Stephen Bond's gas meter at the beginning of the quarter
      was 5628.

      At the end of the quarter the reading is as shown below:

      (i)   What reading is shown on the meter?
      (ii) How many units (i.e. hundreds of cubic feet of gas) have been used
            during the quarter?

   b The cost of gas is 39.8p per therm and there is a standing charge of
      £8.70 per quarter.

      (i)   How many therms have been consumed?
      (ii)   What is the cost of the gas consumed?
      (iii) What is the total gas bill?

Telephone bills

Telephone bills are sent three or four times per year. You are charged for the
use of the system, the hire of the telephone and the calls which have been made.

Some people have now bought their own telephone, and their bills will not
include the charge for the hire of the telephone.

VAT is added to telephone bills as a service charge.

Exercise 44

1 Mrs Kingsley has her own telephone. She pays the
   system charge of £13.95 and the cost of 415 units
   at 4.40p per unit. Calculate her total bill
   (excluding VAT).

2 A meter reading on 1 March was 27258, and on
   31 May it was 27623. The cost per unit was
   4.40p and the standing charges totalled £15.35.
   Calculate the total bill (excluding VAT).

3 a Mr Stears' telephone bill is made up of the
      following charges:

      ● quarterly rental, £18.20
      ● 196 metered units @ 4.80p per unit
      ● operator calls, £1.68

      Calculate the total bill (excluding VAT).

   b VAT at 17.5% is added to the bill.
      Calculate the final bill which Mr Stears
      receives.

4 Complete the following telephone bill:

	Quarterly rate			£
○	SYSTEM			13.95
	APPARATUS			3.10
	TOTAL			**a**

	Date	Meter reading	Units used	
○	13 MAR	024367		
	12 JUN	024536	**b**	

○	UNITS AT 4.40p	**c**
	TOTAL (EXCLUSIVE OF VAT)	**d**
	VAT AT 17.5%	**e**
○	TOTAL PAYABLE	**f**

*10.5 Redecoration

Part of the upkeep of your property involves redecoration. Painting and wallpapering need to be done at regular intervals. Recarpeting is undertaken at much longer intervals, whereas tiling is likely to be a one-off job.

Painting

On each tin of paint is stated the 'coverage', i.e. the area of wall which that amount of paint is expected to cover. By estimating the area of the walls to be painted, the amount of paint required can be calculated.

Example 1

Melanie decides to repaint the walls of her bedroom. The size of the room is 4.0 m × 3.5 m and the room is 2.7 m high. It will require two coats of paint and 1 litre of paint covers 12 square metres.

Calculate the amount of paint Melanie must buy.

$$\text{Area of walls} = (4.0 \times 2.7) + (3.5 \times 2.7) + (4.0 \times 2.7) + (3.5 \times 2.7)\,\text{m}^2$$
$$= (4.0 + 3.5 + 4.0 + 3.5) \times 2.7\,\text{m}^2$$
$$= 40.5\,\text{m}^2$$

$$\text{Area to be painted (two coats)} = 2 \times 40.5\,\text{m}^2$$
$$= 81\,\text{m}^2$$

$$\text{No. of litres of paint} = \frac{81}{12}$$

$$= 6.75\,\text{litres}$$

Example 2

Melanie realises that the door and window in the room do not need to be painted. The door is $2\,\text{m} \times 0.9\,\text{m}$ and the window is $1.4\,\text{m} \times 1.2\,\text{m}$. How much paint does she require if she takes this into account?

Area of door and window $= (2 \times 0.9) + (1.4 \times 1.2)\,\text{m}^2 = 3.48\,\text{m}^2$

Area to be painted $= 2 \times (40.5 - 3.48)\,\text{m}^2$

$= 74.04\,\text{m}^2$

Paint required $= \dfrac{74.04}{12}$

$= 6.17\ \text{litres}$

Exercise 45

1 Gordon paints one wall of his study with two coats of emulsion. The wall is 4.3 m long × 2.8 m high.
If 1 litre of paint covers 12 square metres, how much paint will he require?

2 A particular type of paint has a coverage of $17\,\text{m}^2$ per litre and is available in the following sizes at the prices given.

Size	Price
1 litre	£2.99
2.5 litres	£5.99
5 litres	£10.69
10 litres	£19.75

Find the best combination of cans of paint to cover the following areas at the most economical price:

a 9 sq metres   **d** 66 sq metres

b 36.5 sq metres   **e** 27.8 sq metres

c 115 sq metres   **f** 180 sq metres.

3 A rival brand of paint is available in two sizes. 1 litre costs £3.99 and $2\frac{1}{2}$ litres cost £7.99.

1 litre covers 150 square feet.

Find the most economical cost of each of the following painting jobs:

a a single wall 16′ × 9′ with two coats of paint

b two walls 18′ long and 15′ long by 8′6″ high, with two coats of paint

c the walls of a room 20′ × 16′ × 9′ high containing a door 7′ × 3′ and a patio door 14′ × 8′ with one coat.

4 Mr and Mrs Freeman decide to repaint all the bedrooms with one coat of silk emulsion.

The rooms are all 2.4 m high and the sizes are:

Bedroom 1   4.2 m × 4.0 m
Bedroom 2   3.8 m × 3.0 m
Bedroom 3   3.3 m × 2.2 m

To make calculation easier they decide to ignore the doors and the windows, which are small.

a If 1 litre of paint covers $12\,\text{m}^2$, calculate the total amount of paint required for each room.

b What is the total amount of paint required?

The cost of the paint is shown below:

Size	Price
$\frac{1}{2}$ litre	£1.99
1 litre	£2.99
2.5 litres	£6.99
5 litres	£10.99

c What is the lowest cost of painting each room a different colour?

d What is the lowest cost of painting all the rooms the same colour?

Wallpapering

A standard roll of wallpaper is 0.53 m wide and 10.05 m long. The number of rolls of wallpaper required can be calculated once the size of the wall to be papered is known.

No allowance is made for doors and windows as extra paper is needed for matching patterns.

Example

A room 3.8 m × 2.8 m and 2.9 m high is to be papered with wallpaper costing £5.25 per roll.
How much will the job cost?

$$\text{Perimeter of room} = 2 \times (3.8 + 2.8)$$

$$= 13.2\,\text{m}$$

$$\text{No. of strips of paper required} = \frac{\text{Width of wall}}{\text{Width of roll}} \text{ and round } up$$

$$= \frac{13.2}{0.53} = 24.9 = 25 \text{ whole strips}$$

$$\text{No. of strips per roll} = \frac{\text{Length of roll}}{\text{Height of room}} \text{ and round } down$$

$$= \frac{10.05}{2.9} = 3.46 = 3 \text{ whole strips}$$

$$\text{No. of rolls required} = \frac{25}{3} = 8.3 = 9 \text{ whole rolls}$$

$$\text{Cost of wallpaper} = 9 \times £5.25 = £47.25$$

Exercise 46

1   Calculate the number of rolls of wallpaper required to paper walls with the following dimensions:

	Height	Length
a	2.8 m	5.6 m
b	2.7 m	4.9 m
c	2.9 m	2.4 m
d	3.2 m	6.2 m
e	2.4 m	4.5 m

2   Find the cost of papering a wall 3.2 m long × 3.0 m high with wallpaper costing £4.50 per roll.

3   Mrs Heron's lounge is a square room with walls 5.2 m long and 2.8 m high. She intends papering two adjacent walls with wallpaper costing £6.20 per roll.

a   How many rolls of paper are required?

b   How much will the paper cost?

4   a   Mr Kerr plans to wallpaper two adjacent walls in his bedroom. The room measures 4.3 m × 3.2 m × 2.6 m high.
How many rolls of paper are required?

b   The paper costs £7.25 per roll.
In addition he requires wallpaper paste.
A packet which will hang 4 rolls costs £1.15
A packet which will hang 6 rolls costs £1.60

How much does it cost to wallpaper the two walls?

Wallpaper charts. In any DIY book you will find a wallpaper chart which makes calculating the number of rolls of wallpaper required much simpler. However, the charts assume the existence of doors and windows which are taken into account. When papering complete walls it is safer to calculate the number of rolls required by the previous method or make an additional allowance on the amount given by the chart.

	WALLPAPER CHART										
					Perimeter (m)						
Height (m)	10	11	12	13	14	15	16	17	18	19	20
2.0	4	4	5	5	5	6	6	6	7	7	8
2.2	4	5	5	5	6	6	7	7	7	8	8
2.4	5	5	5	6	6	7	7	8	8	9	9
2.6	5	5	6	6	7	7	8	8	9	9	10
2.8	5	6	6	7	7	8	8	9	10	10	11
3.0	6	6	7	7	8	8	9	10	10	11	11
3.2	6	7	7	8	8	9	10	10	11	11	12
3.4	6	7	8	8	9	10	10	11	12	12	13

Example

Using the wallpaper chart, calculate the number of rolls of paper required for the room in example 1, which is 3.8 m × 2.8 m and 2.9 m high.

Height   = 2.9 m   (for height 2.9 m read 3.0 m on the chart)

Perimeter = 13.2 m = 14 whole metres

From the chart:
a room of height 3.0 m and perimeter 14 m requires 8 rolls

No. of rolls required = 8

This result, which includes allowances for doors and windows, is (as expected) less than the result calculated previously.

Exercise 47

1   Use the wallpaper chart to calculate the number of rolls of wallpaper needed for the following rooms:

	Length	Width	Height
a	4.0 m	3.0 m	2.4 m
b	4.3 m	3.6 m	2.7 m
c	3.2 m	2.7 m	2.8 m
d	2.9 m	2.5 m	2.5 m

2 Julia has just bought a small, one-bedroom flat which requires decorating.
   The ceilings are 2.6 m high.

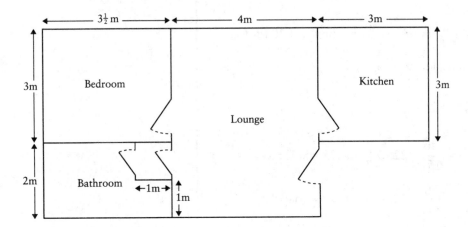

How many rolls of wallpaper will be needed for:

a the lounge

b the bedroom

c the bathroom?

3 In order to save money, Julia decides to paper all three rooms with the
   same inexpensive, plain paper which can then be painted.
   How many rolls of wallpaper are required?

4 The more expensive wallpaper would cost Julia £7.30 per roll. The
   cheaper, plain paper costs £2.99 per roll. The emulsion paint covers
   12 square metres and costs:

   £2.99 for 1   litre
   £5.95 for 2½ litres
   £9.95 for 5   litres.

   a How much would it cost to paper the three rooms differently using the
      more expensive paper?

   b How much would it cost to paper the three rooms using the inexpensive
      paper?

   c How much would it cost to paint each room a different colour?

   d Which is the most economical price for redecoration?

Tiling

Unlike wallpaper, which can be cut and matched, for example, under windows
or at corners, the pieces of tile which are cut off are usually wasted. Each wall,
therefore, needs to be treated separately.

Mistakes and breakages occur when cutting tiles so the estimate of the number
of tiles required needs to be generous.

Example

a A wall 3.4 m × 2.8 m is to be covered with tiles measuring 15 cm × 15 cm.
How many tiles are required?

b The tiles can be bought singly for 95p each or in boxes of 20 for £17 per box.
Calculate the most economical price for the job.

a No. of tiles per row $= \dfrac{340}{15} = 22.67$ $= 23$ whole tiles

No. of rows of tiles $= \dfrac{280}{15} = 18.67$ $= 19$ whole tiles

Total no. of tiles $= 23 \times 19$ $= 437$

b No. of boxes of tiles $= \dfrac{437}{20}$ $= 21$ boxes $+ 17$ tiles

Cost of 21 boxes + 17 tiles $= (£17 \times 21) + (£0.95 \times 17)$

$= £373.15$

Cost of 22 boxes $= £17 \times 22$

$= £374$

The least price for tiles is £373.15, but in reality you would probably buy 22 boxes as the three extra tiles cost less than one tile.

Exercise 48

1 Julia decides to make a splash-back above her washbasin, using nine rows of small tiles. The area is 750 mm wide and the tiles she has chosen are 6 cm square.
How many tiles should she buy?

2 Nusrat tiles the wall above his gas hob with rectangular tiles measuring 15 cm × 11 cm. The tiles cover an area 1.75 m long × 1 m high.
Estimate the number of tiles needed if:

a the shorter side is the height of the tile

b the longer side is the height of the tile.

3 a Floor tiles measuring 15 cm square are to be laid on a kitchen floor which is 4 m × 2.6 m. How many tiles are required?

b When the tiles are laid spaces are left between them for grouting. If the width of the grouting is 0.4 cm, what width does each tile plus grout occupy?

c Using the dimensions of a tile plus grout, calculate a more accurate estimate of the number of tiles required.

4 Tiles are sold in boxes of 18 for £12.60 or singly for 80p per tile.
Calculate the most economical price for the following numbers of tiles:

a 40 **b** 160 **c** 100 **d** 13 **e** 285 **f** 196.

5 **a** Mrs Pinner has just had a new shower installed. The dimensions of the shower are 85 cm × 75 cm × 210 cm high.

The front of the shower, which is 85 cm wide, has a shower door fitted. The other three walls are to be tiled.

How many tiles, 15 cm square, are required for the job?

b The tiles are sold in boxes of 20 at a cost of £16.80 per box.
  (i) How many boxes are required?
  (ii) What is the cost of the tiles?

Carpeting

Floor coverings such as carpet, underlay and lino are usually sold in standard widths.

The required *length* is bought, but the cost of carpeting depends on the *area* of carpet bought.

The direction in which a carpet is laid in a room can be chosen for reasons of economy (which way causes the least amount of wastage?) or aesthetics (which way looks the best?).

A carpet should never be joined if there is a method of laying which will avoid a join.

If a join is unavoidable, the lengths of carpet must be laid in the same direction.

Example

A rectangular room is 3.8 m long and 2.7 m wide. The carpet chosen for this room is sold in rolls 3 m wide and 4 m wide.

The cost is £10.99 per square metre.

Calculate the least cost of carpeting the floor if there are to be no joins in the carpet.

There are two ways of carpeting the floor:

  (i) buy the 4 m width and lay it parallel to the shortest side
  (ii) buy the 3 m width and lay it parallel to the longest side.

Area of carpet required for (i) = 4 m × 2.7 m = 10.8 m^2

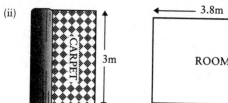
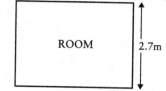
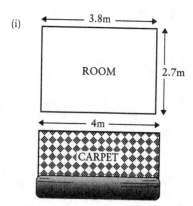

Area of carpet required for (ii) = 3.8 m × 3 m = 11.4 m^2

Method (i) requires the least area of carpet.

Therefore least cost of carpet = £10.99 × 10.8 = £118.69

Exercise 49

1 Carpet can be bought in 3 m or 4 m widths. Find, for each floor with the following dimensions, the length(s) of carpet required so that the minimum amount of carpet and the minimum number of joins would be used.

a 5.4 m × 3.2 m   d 3.8 m × 2.5 m

b 4.9 m × 3.7 m   e 6.6 m × 7.4 m

c 3.4 m × 2.9 m   f 7.9 m × 6.2 m

2 The carpets in question 1 are priced per square metre as follows:

a £13.99   c £7.59   e £17.69

b £5.99   d £2.99   f £12.50

Calculate the cost of each carpet.

3 a Mr and Mrs Donkin's living room is 15 ft × 12 ft. The new carpet they have chosen for this room is 13 ft wide and costs £15.99 per square yard.
Calculate the cost of the carpet.
(9 sq ft = 1 sq yard.)

b The Donkins also require underlay which is sold in 4 ft 6 in widths and costs £2.99 per square yard.
Calculate the minimum cost of the underlay.

4 The Kowalczyk's house has a living/dining room as shown below:

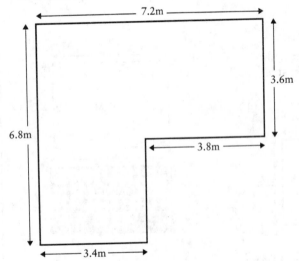

Carpet is 4 m wide and costs £14.59 per square metre.
Find the least cost of the carpet if it has one join.

5 a A rectangular room, $5\frac{1}{2}$ m × $3\frac{1}{2}$ m, is to have the floor covered with underlay and then have carpet laid on top. The carpet is 4 m wide and costs £11.99 per square metre.
The underlay costs £1.99 per square metre and is costed on the floor to be covered.
Calculate:
(i) the cost of the underlay
(ii) the cost of the carpet.

b Double-sided tape is laid round the perimeter of the room to hold the carpet in place. The price of the tape is £1.30 for a roll 4.6 m long. Calculate the cost of the tape required.

c The carpet-layers find they need two metal door strips costing £3.35 each.
What is the total cost for materials?

d The carpet-layers charge 95 pence per square, based on the floor area, to fit the carpet. How much is the their bill in total? (All prices include VAT.)

e If the bill is settled within one month of presentation, a $7\frac{1}{2}$% discount is allowed. What is the reduction in cost for early settlement?

*10.6   Repairs and improvements

Exercise 50

1 **a** Mr and Mrs Flood are advised to have their loft insulated.

   The job will require:
   12 packs of fibre glass @ £12.20 per pack
   4 packs of pipe insulation @ £7.92 per pack.
   VAT at 17.5% must be added to all costs.

   Calculate the cost of insulating the loft.

   b As the loft has not been insulated previously, the Floods are entitled to a grant of two-thirds of the cost of insulation.
   How much do they pay for loft insulation?

2 Mrs Redmond needs a footpath 900 mm wide from her back door to the end of the garden. The distance is 15 m.
   Paving slabs can be bought in the following sizes:

   Size 1: 600 mm × 600 mm costing £2.99 each
   Size 2: 600 mm × 300 mm costing £1.89 each
   Size 3: 300 mm × 300 mm costing £1.39 each

   Mrs Redmond sketches the following designs for her path:

   A   B   C

   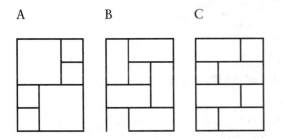

   a For each design, calculate:

   (i) the number of each size of paving slab required to make the path, without cutting any slabs
   (ii) the cost for each design of path.

   b Sketch a design of your own for the path and find the cost of the paving slabs required.

3 Mr Barlow wishes to lay a new lawn in his garden. The size of the lawn will be 8 m × 6 m.

   a Grass seed can be bought as follows:

   500 g packets costing £3.25
   2 kg packets costing £7.09
   3 kg bags costing   £9.62

   1 kg covers approximately 20 m².

   (i) What quantity of seed will Mr Barlow require?
   (ii) What is the minimum cost of the grass seed?

   b Alternatively Mr Barlow can buy turves. Each turf measures 1 m × 0.5 m and they cost £44 per 100 or 92 pence per square metre.

   (i) How many turves will Mr Barlow require?
   (ii) How much will it cost to turf the lawn?

4 **a** The Costain's fence is badly damaged in a gale. The fence posts are still in place, but four fence panels need to be replaced.

   To hold the panels in place they are nailed to the post using six nails at each end of the panel.

   The panels are then treated with wood preservative. It requires 0.75 litres of preservative to treat one panel.

   (i) How many nails are required?
   (ii) What quantity of wood preservative is required?

   b Use the following information to calculate the cost of replacing the four panels:

   Fence panels   £9.75 each
   Nails   75p   for 100
   Wood preservative £1.20 for 2 litres

11 Insurance

Is insurance really necessary?

In the UK:

- every 10 minutes a house catches fire
- every 2 minutes a road accident occurs
- every 40 seconds a car is stolen
- every 30 seconds a house is burgled

so you may begin to understand why insurance *is* necessary.

These misfortunes don't always happen to someone else, and so some form of insurance cover is necessary for practically everyone.

Most people will need insurance for:

- *Their home.* House buildings insurance will cover the house against possible damage from fire, flood, storm or subsidence
- *Their possessions.* House contents insurance will cover such items as furniture, carpets, clothes, books, jewellery against loss or damage
- *Their life.* Life insurance will protect dependents from financial hardship in the event of the death of the insured
- *Their car or motorbike.* Car or motorbike insurance is required by law
- *Their holiday.* Holiday insurance is strongly recommended when booking a package holiday ·

11.1 House buildings and contents insurance

A buildings policy covers the following against damage: the fabric of your home (i.e. the foundations, roof, walls, floors, doors and windows), the fixtures (kitchen units, fitted wardrobes, light fittings, etc.), and any outbuildings.

The contents can be insured against loss or damage on either a new-for-old basis or at their true value, i.e. the new price less a deduction for wear and tear.

A new-for-old policy is a sensible choice for most people as the full costs of repairs are met by the insurance company. For example, if you burn a hole in a two-year-old carpet which might have been expected to last for ten years, a new-for-old policy will give you the money to buy a new carpet. A basic policy will give you only eight-tenths of the new price.

The cost of contents insurance depends on the type of policy (new-for-old costs more) and the area in which the house is situated.

Many insurance companies put each area of the country into a low, high or top risk category depending on the likelihood of a housebreaking or burglary occurring. The premium payable is usually decided by the post code and will be quoted as a fixed rate per £1000 (or £100) of cover.

The policy might cost extra if:

- there are lots of easily stolen, valuable items
- there have been claims made in the past.

There are optional extras which will increase the cover provided by your policy and increase the premium you pay. For example:

- An 'all-risks' extension will cover loss or damage to jewellery and such items as spectacles, contact lenses, sports equipment, cameras, which may be lost or damaged outside the home.
- Freezer contents cover will insure you in case of a power cut or breakdown.
- Bicycles can be covered for loss or damage anywhere in the UK, provided that, if a bike is stolen, it was securely locked at the time.

Policies should be index-linked so that the amount of cover increases with the rate of inflation. There is then less danger of being underinsured.

Insurance based on risk

Below are examples of the prices offered by several companies. Buildings and contents can be insured with the same company or with two separate companies.

Company	Buildings (Cost per £1000 of cover)	House contents					
		Basic			New-for-old		
		Top risk	High risk	Low risk	Top risk	High risk	Low risk
Company A	2.10	10.00	4.45	2.50	14.00	6.00	3.00
Company B	1.90	7.00	4.00	3.00	8.50	4.50	3.50
Company C	2.10	7.50	4.40	2.40	10.00	5.20	3.00
Company D	1.80	8.20	2.50	2.50	10.00	3.00	3.00
Company E	2.04	7.40	3.33	2.70	9.96	4.44	3.48

Example

Mr and Mrs Perry own a house in a top-risk area. They wish to insure the house for £75 000 and the contents for £14 000 (on a new-for-old basis) with Company B. Calculate the annual premium payable.

Premium for buildings $= \dfrac{75\,000}{1000} \times £1.90 \qquad = £142.50$

Premium for contents $= \dfrac{14\,000}{1000} \times £8.50 \qquad = £119.00$

Total annual premium $= £145.50 + £119.00 \qquad = £261.50$

Exercise 51

1 What is the yearly premium for £14 000 of basic cover for house contents for someone living in a high-risk area and insuring with Company B?

2 A couple insure their house with Company C for £40 000.
What is their yearly premium?

3 a How much does it cost per year to insure contents (basic cover) for £15 000 with Company C, in a low-risk area?

   b How much extra would a couple living in a top-risk area pay if the contents were insured for £15 000 and were covered by a new-for-old policy?

4 Miss Morgan lives in a top-risk area. She decides to insure the house with Company D and the contents with Company B. The house is to be covered for £95 000 and the contents (on a new-for-old basis) for £24 000.
How much will she pay for insurance:

   a yearly   b monthly?

5 Mr and Mrs Samra live in a top-risk area. They take out new-for-old contents insurance with Company E for £18 000.

   a Calculate their annual premium.

   b They decide to pay the premium monthly. What is their monthly premium?

Insurance zones

Many insurance companies divide the country into zones, which also depend on the post code. The premium payable is then decided by the zone number.

Below is a table used by one insurance company.
(All rates are per £100 sum insured.)

BUILDINGS			ALL ZONES 13.5p				
CONTENTS	ZONE 1	ZONE 2	ZONE 3	ZONE 4	ZONE 5	ZONE 6	ZONE 7
	25.5p	28.5p	32.5p	40.5p	52.5p	65.0p	81.0p

Rates can be reduced by the following discounts:
  $7\frac{1}{2}\%$ if you agree to pay the first £50 of any claim, or
$17\frac{1}{2}\%$ if you agree to pay the first £100 of any claim.

Example

Mr and Mrs Perkins live in Sunderland (Zone 4) and wish to insure their house for £45 000 and its contents for £11 500. They agree to pay the first £50 of any claim.
Calculate their annual premium.

Premium for buildings	$=\dfrac{45\,000}{100} \times 13.5\text{p}$	$= £60.75$
Premium for contents	$=\dfrac{11\,500}{100} \times 40.5\text{p}$	$= £46.58$
Total premium	$= £60.75 + £46.58$	$= £107.33$
$7\frac{1}{2}\%$ discount	$= £107.33 \times \dfrac{7\frac{1}{2}}{100}$	$= £8.05$
Total premium payable	$= £107.33 - £8.05$	$= £99.28$

Exercise 52

Exercise 52

1 Ms Robinson rents a house in Oxford (Zone 4) and she wishes to insure the contents for £16 700.

Calculate the annual premium payable.

2 Miss Lewis owns a house in Penzance (Zone 1). She decides to insure the house for £75 500 and the contents for £18 500. In addition she agrees to pay the first £50 of any claim.

Calculate the annual premium payable.

3 Mr Jackson insures his house in Kingston (Zone 3) for £94 800 and its contents for £18 400. He agrees to pay the first £100 of any claim on the contents insurance.

Calculate his monthly premium to the nearest penny.

4 Mrs Burgess moves from Walsall (Zone 3) to Glasgow (Zone 5). She insured her old house for £72 000 and its contents for £18 000. She intends to insure her new house and its contents for the same amounts.

Calculate the increase in her annual premium and express this as a percentage of the original premium.

5 Mr and Mrs Carr live near London (Zone 6). Their house is insured for £95 000 and the contents for £25 000. What insurance premium do they pay?

The following year they increase the contents insurance by £1000, as they have just bought a word processor, and the insurance company increases its rates by 8%.
By how much will their premium increase?

11.2 Life insurance

Four out of five people have life insurance, which is usually referred to as **life assurance**.
The majority take it out on their own lives in order to protect their dependents when they die.

Before insuring your life you need to decide:

- if you require insurance at all
- how much insurance you require
- what type of insurance you require.

There are several types of insurance to choose from.

Term assurance. With this type of policy you pay a fixed amount in premiums for a fixed length of time and the policy pays out to your dependents if you die during this term. It is the cheapest form of insurance as the insurance company will not have to pay out if you live longer than the agreed term.
This type is often taken out as a 'mortgage protection' policy.

Whole-life assurance. With this type of policy you pay premiums either for a fixed period or for the rest of your life. The insurance money is paid to your family, or next of kin, when you die, however far into the future this occurs.
This type of insurance is cheaper than the endowment type.

Endowment assurance. This type of policy is taken out as an investment. You pay a fixed premium for a fixed length of time and, with a 'non-profit' policy, at the end of this time you receive the amount for which you are insured. If you die before the end of the fixed period, the money is paid to your next of kin.

The usual type of investment policy is a **with-profits** policy. This guarantees you not only a fixed amount of life assurance but also a share of the company's profits which is added each year to the sum insured in the form of a bonus. When the policy matures you receive more than the amount for which you were insured. You do however pay a higher premium for a with-profits policy.

This is a long-term investment. Commission paid to the life insurance salesperson is high, and if you stop your payments after one or two years you are unlikely to receive any money back.

The three tables below show the monthly premiums per £1000 for an endowment policy, a whole-life policy and a term assurance policy.

ENDOWMENT ASSURANCE WITH PROFITS					WHOLE-LIFE		TERM ASSURANCE			
Age next birthday	10 yrs	15 yrs	20 yrs	25 yrs	Age n.b.d.	Rate	Age n.b.d	5 yrs	10 yrs	15 yrs
20	9.15	6.04	4.47	3.53	20	1.10	25	0.168	0.204	0.216
21	9.15	6.04	4.47	3.53	21	1.11	26	0.171	0.213	0.225
22	9.15	6.05	4.48	3.54	22	1.13	27	0.174	0.221	0.235
23	9.15	6.05	4.48	3.54	23	1.15	28	0.178	0.229	0.245
24	9.15	6.05	4.49	3.55	24	1.17	29	0.182	0.237	0.253
25	9.16	6.06	4.50	3.55	25	1.20	30	0.186	0.243	0.261
26	9.16	6.06	4.50	3.55	26	1.22	31	0.191	0.249	0.268
27	9.16	6.07	4.51	3.56	27	1.25	32	0.194	0.255	0.274
28	9.16	6.07	4.52	3.56	28	1.27	33	0.199	0.261	0.282
29	9.17	6.07	4.52	3.57	29	1.30	34	0.202	0.266	0.290
30	9.17	6.08	4.53	3.57	30	1.33	35	0.207	0.274	0.299
31	9.17	6.08	4.53	3.57	31	1.35	36	0.213	0.282	0.309
32	9.17	6.09	4.54	3.58	32	1.38	37	0.219	0.291	0.321
33	9.17	6.09	4.55	3.58	33	1.42	38	0.226	0.302	0.336
34	9.18	6.09	4.55	3.59	34	1.45	39	0.234	0.316	0.352
35	9.18	6.10	4.56	3.60	35	1.49	40	0.243	0.329	0.370
36	9.18	6.11	4.57	3.61	36	1.52	41	0.253	0.346	0.394
37	9.19	6.11	4.58	3.62	37	1.56	42	0.266	0.364	0.421
38	9.20	6.12	4.60	3.64	38	1.61	43	0.280	0.386	0.451
39	9.20	6.13	4.61	3.66	39	1.65	44	0.296	0.409	0.485
40	9.21	6.14	4.63	3.67	40	1.70	45	0.312	0.436	0.523
41	9.22	6.16	4.64	3.69	41	1.75	46	0.330	0.465	0.565
42	9.23	6.17	4.66	3.72	42	1.80	47	0.351	0.498	0.611
43	9.24	6.18	4.69	3.75	43	1.86	48	0.374	0.535	0.663
44	9.25	6.20	4.71	3.78	44	1.92	49	0.399	0.577	0.719
45	9.26	6.22	4.74	3.80	45	1.99	50	0.426	0.622	0.782
46	9.28	6.24	4.77	3.83	46	2.06	51	0.462	0.679	0.849
47	9.29	6.27	4.81	3.87	47	2.13	52	0.498	0.740	0.922
48	9.31	6.30	4.85	3.92	48	2.21	53	0.538	0.807	1.003
49	9.33	6.33	4.90	3.97	49	2.29	54	0.581	0.879	1.092
50	9.35	6.36	4.95	4.03	50	2.38	55	0.629	0.959	1.190
51	9.38	6.40	5.01	4.09	51	2.47	56	0.681	1.045	1.297
52	9.40	6.44	5.07	4.16	52	2.57	57	0.736	1.139	1.415
53	9.43	6.49	5.13	4.24	53	2.67	58	0.797	1.243	1.544
54	9.46	6.54	5.21	4.31	54	2.78	59	0.860	1.358	1.685
55	9.50	6.60	5.29	4.43	55	2.90	60	0.929	1.486	1.840
56	9.54	6.66	5.39	4.53	56	3.03	61	0.998	1.619	2.009
57	9.59	6.74	5.49	4.64	57	3.16	62	1.074	1.768	2.193
58	9.64	6.82	5.61	4.76	58	3.31	63	1.157	1.929	2.396
59	9.70	6.91	5.73	4.90	59	3.46	64	1.250	2.103	2.617

The rates in the table opposite apply to **male non-smokers**. Other rates are obtained as follows:

male-smokers by **adding** 6 years to age
female non-smokers by **deducting** 5 years from age
female smokers by **deducting** 1 year from age

Example

Mrs Singh is 40 years old and a non-smoker. She wishes to insure her life for the sum of £15 000. She decides to take out endowment insurance, with profits, for 20 years as the policy will then mature just before she retires.

Calculate her monthly premium.

For female non-smokers deduct 5 years: $40 - 5 = 35$

Premium per £1000 over 20 years, for a 35-year-old male = £4.56

Mrs Singh's monthly premium = $£4.56 \times \dfrac{15\,000}{1000}$

$= £68.40$

Use the tables to calculate the premiums required in the exercise which follows.

Exercise 53

1 Calculate the monthly premium for each of the following:

	Type of policy	Age next birthday	Term	Sum assured	Sex	Smoker
a	Whole-life	24	–	£15 000	M	✓
b	Endowment	30	25 yr	£8500	M	✗
c	Term	42	10 yr	£12 000	F	✗
d	Endowment	21	15 yr	£20 000	M	✗
e	Whole-life	51	–	£17 000	F	✓

2 Ayshe is 33 and decides to take out an endowment policy which will mature in 25 years time. She is a non-smoker and can afford to pay about £18 per month.
For how much can she insure herself?

3 Mr and Mrs Jarman decide to insure their lives for £15 000 each. Mr Jarman is 42 and smokes. Mrs Jarman is two years younger and does not smoke. Assume that each is taking out a whole-life policy for £15 000.
What is the difference in the monthly premiums they will each pay?

4 Miss Menzies is a 39-year-old non-smoker. She wishes to insure her life for £17 000 by taking out an endowment policy over 20 years.
What will her monthly premium be?
How much will she pay over the 20-year period?

5 What is the difference between assurance and insurance?

6 What happens to the monthly cost of assurance as you get older?
Explain the reason for this.

11.3 Car insurance

The law requires that every car driver is insured against damage to another person's property or vehicle or injury to any person.

Your basic premium for car insurance will depend, not only on the **insurance company** you choose, but also on:

- the type of cover you take out
- the make and model of car you own
  Insurance companies classify cars into groups which depend on the original cost, the performance, the cost of repairs and spare parts. The lower the group number the cheaper the insurance which reflects the insurers' belief that these cars are less likely to be involved in accidents and/or are cheaper to repair.
- the area in which you live
  Accidents are more likely to occur in areas where traffic is heavy. Some areas have a higher rate of theft or vandalism.
- your age, sex and driving experience
  There is usually an extra premium to pay for drivers under 25 years of age as this age group has a higher accident rate.
  Women drivers are considered a better insurance risk than men drivers and some insurance companies will give a discount to women drivers or will use a separate table of premiums.
  Convicted drink-drivers will have to pay a much larger premium.
- whether you use the car for business or pleasure
- the type of cover:

  Third party is the minimum insurance allowed by law. It covers injury or damage to other people and their property caused by your car. It is the cheapest form of insurance.

  Third party, fire and theft also covers damage to your own car from fire, explosion, lightning, theft or attempted theft and loss of your car by theft. It is more expensive than third party insurance.

  Comprehensive insurance is the most expensive type of car insurance. This also covers you for damage to your own car in an accident, whether or not it was your fault, and often provides additional cover for loss of contents.

Table I shows, for a sample of cars, the groups to which cars are assigned:

Table I

Group 1	Citroen 2CV, Fiat Uno 45, Ford Fiesta 950, Renault 4, Metro 1.0L, Mini City
Group 2	Fiat Uno 60S, Ford Fiesta 1100L, Lada 1200L, Lancia Y10, Peugeot 205XL, Renault 5TR
Group 3	Daihatsu Charade, Fiat Strada, Nissan Micra, Peugeot 309, Maestro 1.3, Metro MG
Group 4	Ford Cortina 1600, Fiat Tipo, Mazda 1.3, Nissan Cherry, Renault 18, Rover 216S, Ambassador
Group 5	Ford Cortina 2000GL, Honda Accord 1600, Nissan Prairie, Toyota Carina, Vauxhall Cavalier 20GLi
Group 6	Alfa 33, Ford Granada, Honda Civic 1.3DX, Renault Espace, Rover 820Si, VW Scirocco
Group 7	Audi Coupe, BMW 520i, Mercedes-Benz 190E, Volvo 240DL
Group 8	Jaguar 5.3, Saab 900 Turbo

Table II shows the areas, for insurance purposes, to which some of the counties
and cities of mainland Britain are assigned.

Table II

Area 1	Berkshire, Central Scotland, Cornwall, Dorset, Gwent, Lincolnshire, Norfolk, Powys
Area 2	Avon, Cumbria, Dyfed, Hampshire, Gloucestershire, Northumberland, Shropshire, Wiltshire, Yorkshire
Area 3	Durham, Lancashire, Strathclyde, Tyne and Wear
Area 4	Buckinghamshire, Merseyside, Oxfordshire, Tayside
Area 5	Birmingham, Outer Glasgow, Manchester
Area 6	Inner Glasgow, Liverpool, Greater London
Area 7	Inner London

Table III shows a typical basic comprehensive premium payable, depending on
age, make of car and place of residence. The same table is used for men and
women drivers.

Table III

Car group	Age or above	Area 1	Area 2	Area 3	Area 4	Area 5	Area 6	Area 7
1	20	536	574	614	657	702	752	804
	25	367	395	427	461	498	537	580
	30	337	364	393	424	458	495	535
	35	306	330	357	385	416	450	486
2	20	611	654	700	749	800	857	917
	25	419	457	487	525	568	612	661
	30	384	415	448	483	522	564	610
	35	349	376	407	439	474	513	554
3	20	730	785	840	899	960	1028	1100
	25	502	556	584	630	682	734	793
	30	461	498	538	580	626	677	732
	35	419	451	488	527	569	616	665
4	20	869	931	996	1066	1138	1219	1305
	25	595	659	693	747	809	870	940
	30	547	591	638	688	742	803	868
	35	497	535	579	625	672	731	789
5	20	1030	1104	1181	1264	1350	1446	1548
	25	706	782	821	886	959	1032	1115
	30	649	701	757	816	881	952	1030
	35	589	635	686	741	798	866	935
6	20	1222	1310	1410	1500	1601	1715	1836
	25	837	928	974	1051	1138	1224	1323
	30	769	831	898	968	1044	1129	1221
	35	649	755	814	879	946	1028	1109
7	20	1450	1553	1662	1779	1899	2033	2177
	25	993	1099	1155	1246	1349	1452	1569
	30	912	986	1064	1147	1239	1339	1448
	35	829	892	966	1043	1122	1219	1316
8	20	1719	1842	1971	2110	2253	2412	2582
	25	1177	1304	1370	1478	1600	1722	1861
	30	1082	1169	1262	1361	1469	1589	1718
	35	983	1059	1145	1237	1381	1495	1560

Discounts

You can reduce the cost of your insurance by obtaining discounts on the basic premium.

Most insurance companies will give discounts for the following:

- **limiting the number of drivers** to yourself or self and spouse.

- **a voluntary excess,** that is agreeing to pay the first part of any claim.

- **no-claims bonus.** If you have not made a claim on your policy over a period of time the insurers will award you a 'no claims' discount. The longer the claim-free period, the larger the discount, usually up to a maximum of 60%.

Example

Megan Jones is 27 and the proud owner of a Renault 5TR. She lives in Gwent and has been driving for 5 years without having an accident.
She agrees to limit the driving to herself and for this she is allowed a 15% discount on the basic premium.
Calculate her comprehensive car insurance premium.

From Table I: a Renault 5TR is in Group 2
From Table II: Gwent is in Area 1
From Table III: a 27-year-old, with a car in Group 2, living in
                       Area 1 pays a basic premium of £419

Discount for limiting to one driver = 15% of £419   = $0.15 \times £419$
                                                       = £62.85
            Premium now = £419 − £62.85   = £356.15

As Miss Jones has made no claim on her insurance for more than 4 years, she is entitled to the maximum no claims bonus of 60%. (The no claims bonus is nearly always deducted last.)

No claims bonus = 60% of £356.15 = 0.6×356.15 = £213.69
Comprehensive premium = £356.15 − £213.69   = £142.46

(**Note. Discounts are made consecutively. Do *not* add the percentages.**)

Exercise 54

1 Use Tables I, II and III to find the basic comprehensive premium for:

   a the 25-year-old owner of a Ford Fiesta 1100L living in Hampshire

   b the 22-year-old owner of a Mini City living in Birmingham

   c the 40-year-old owner of a Volvo 240DL living in Wiltshire

   d the 34-year-old owner of a Jaguar 5.3 living in Greater London.

2 a Because the owner of the Mini (see question 1b) limits driving to himself and his wife, he is allowed a 10% discount on the basic premium. What is his premium now?

b He agrees to pay the first £50 of any claim for damage and is allowed a further 10% discount.
Calculate the new premium.

c Because he has driven for three years without making a claim he receives a no claims bonus according to the scale:

- 30% after 1 claim-free year
- 40% after 2 claim-free years
- 50% after 3 claim-free years
- 60% after 4 claim-free years.

How much does he now pay?

3 The owner of the Ford Fiesta (see question 1**a**) decides to limit the driving to herself, for which she is allowed a 15% discount. She has six claim-free years of driving.
Calculate her insurance premium.

4 a Mr Flynn is 21 and has only been driving for one year. He lives in Manchester and drives a Peugeot 309.
What is his basic premium?

b He agrees to a voluntary excess of £100, for which he receives a discount of 10%.
What is his premium now?

c He has made no claims since he began driving. Find his actual car insurance premium.

In questions 5 and 6, calculate the net insurance premium.

5

Policy holder's name and address	Car details:
S W REID	
27 HILL ROAD	Registration: **D174 BEF**
GATESHEAD	Type of Body: **HATCHBACK**
TYNE AND WEAR	Make and Model: **V. CAVALIER 20GLi**
Date of Birth: **16/4/51**	Cover: **Comprehensive**

Discounts:
Named Driver: **10%**
Voluntary excess: **15%**
No-claims Bonus: **50%**

6

Policy holder's name and address	Car details:
M A ROBBINS	
BRIAR COTTAGE	Registration: **G541 LJH**
OVERTON	Type of Body: **4 DOOR SALOON**
HANTS	Make and Model: **FORD CORTINA 1600**
Date of Birth: **28/7/46**	Cover: **Comprehensive**

Discounts:
Named Driver: **–**
Voluntary excess: **10%**
No-claims Bonus: **60%**

12 Further Percentages

*12.1 Inverse percentages

Inverse percentages were used in Chapter 5 to find the original price of an item when the price inclusive of VAT was known. In this unit the same method is used in a more general context.

Example

Andy bought a guitar which he later sold to a friend for £57.40, making a loss of 18% on the amount he paid.
How much did the guitar originally cost?

The **original price** is always represented by **100%**

Andy's selling price is represented by $\quad 100\% - 18\% = 82\%$

So 82% represents the selling price of $\quad$ £57.40

$$1\% \text{ represents} \quad \frac{£57.40}{82}$$

$$100\% \text{ represents the cost price of} \quad \frac{£57.40}{82} \times 100 = £70$$

Therefore the guitar originally cost £70.

If preferred, the information can be represented on a table instead:

	%	£
Cost price	100	?
Loss	18	?
Selling price	82	57.40

$$\text{Cost price} = \frac{£57.40}{82} \times 100$$

$$= £70$$

If required, the amount of money lost on the deal could be found by a similar method:

$$\text{Loss} = \frac{£57.40}{82} \times 18$$

$$= £12.60$$

---

*Exercise 55

1   In the following questions you are given the percentage profit and the selling price. For each case find the original cost price.

   a 20%, £30   **d** 110%, £466.20

   b 28%, £69.12   **e** 28.7%, £57.92

   c $12\frac{1}{2}$%, £1662

2   In the following questions you are given the percentage loss and the selling price. For each case, find the original cost price.

   a 10%, £54   **d** 28%, £864

   b $7\frac{1}{2}$%, £31.45 **e** 8.25%, £77.99.

   c 5%, 96p

3   Doris is a pensioner and does not pay income tax. Her bank account earns interest of £540 from which tax at a rate of 25% has been deducted. How much tax can she reclaim?

4   Mario's wage is increased by 9.2%. He now earns £235 per week.
How much did he earn previously?

5   After one year Mrs Brennan's car is valued for insurance at £5600, a depreciation on the price when new of 22%.
How much, to the nearest £10, did Mrs Brennan pay for the car?

6   A garden centre buys plants and resells at a profit of 28%.
How much was the original price of a rose bush which is sold for £3.40?

*12.2   Compound percentages

> ### Example
>
> A wholesaler adds 20% profit to the price of goods when he sells to a retailer. The retailer then adds 15% profit to the same goods before selling to the customer. What percentage above the original price does the customer pay?
>
> The answer is NOT 35%!
>
> The original price is equivalent to 100%
> The wholesale price is equivalent to 100 + 20% of 100 = 120%
> The retail price is equivalent to 120 + 15% of 120 = 138%
>
> The customer pays 38% above the original price.

Exercise 56

1   The wholesale price of a certain item is 25% more than the manufacturing cost. The retail price is 20% more than the wholesale price.
What is the percentage difference between the retail price and the manufacturing cost?

2   Mr Miles invests a sum of money for 2 years at an interest rate of 10% per annum.
By what percentage has his original sum of money increased after two years?

3   The menswear department is having a sale. The normal selling price of a particular style of shirt includes a profit of 30%. During the sale the normal price is reduced by 15%.
What percentage profit does the shop make on these shirts during the sale?

4   Last year when Mrs Berry had her car serviced, 80% of the cost of servicing was for labour and 20% was for parts. In one year the labour costs increased by 9% and the cost of parts by 12%.
What is the percentage increase in Mrs Berry's bill compared to the previous year?

5   The manufacturing cost of a child's toy is made up of 50% labour, 30% materials and 20% overheads.
The factory improves the machinery which increases the overheads by 80%, but decreases the labour costs by 25%. At the same time the cost of materials increases by 10%.
What is the overall percentage increase in the manufacturing cost?

12.3   Depreciation

Most possessions, such as cars, caravans, electrical equipment, depreciate in value as time passes.

For example, at the end of each year the value of a car will be less than its value at the beginning of the year.

Depreciation is calculated by a similar method to compound interest.

Example

A car was bought for £6500 in 1990. During the first year of ownership its value depreciated by 20% and during each subsequent year by 15%.
Calculate the value of the car three years later.

	£
Cost of car	6500
Depreciation of 20% (of £6500)	1300
Value after 1 year	5200
Depreciation of 15% (of £5200)	780
Value after 2 years	4420
Depreciation of 15% (of £4420)	663
Value after 3 years	3757

The value of the car after 3 years is £3757.

Exercise 57

1   A small business buys a computer costing £4500. The rate of depreciation is 20% per annum.
What is its value after three years?

2   A motorbike is bought second hand for £795. Its price depreciates by 11% per year.
For how much could it be sold two years later? (Give your answer to the nearest pound.)

3   Mr and Mrs Parsons' carpet is accidentally damaged by fire. It was bought only three years ago for £450. The insurance investigator (loss-adjuster) decides that it will have depreciated in value by 10% each year.

What was the value of the carpet just before the accident?

4   Mr Carey invests £3000 in unit trusts in the hope that they will appreciate in value. Unfortunately, although they appreciate in value by 8% during the first year, they depreciate in value during the second year by 8% and during the following two years depreciate by 10% of the value at the beginning of each year.
How much are the unit trusts worth after four years?

13 Statistical Terms

In 1834, the Royal Statistical Society was founded, and defined statistics as 'using figures and tabular exhibitions to illustrate the conditions and prospects of society'. Statistics is now used to deal with the collection, classification, tabulation and analysis of information and opinions.

Data. Data is the information which you have obtained (or have been given). The word 'data' is a plural, really, and the singular is 'datum' (a single piece of information). These days, though, nearly everyone says 'The data *is* . . .' rather than 'The data *are* . . .', so we shall do the same.

Variables. Something which can change from one item to the next is a *variable*. A variable can either be **quantitative** (i.e. numerical like the number of people on a bus), or **qualitative** (i.e. non-numerical, like colour).

There are two types of quantitative variables:

(i) *Continuous*. A continuous variable is a variable which could take all possible values within a given range e.g. the height of a tree.

(ii) *Discrete*. A discrete variable is a variable which increases in steps (often whole numbers), e.g. the number of rooms in a building.

A discrete variable does not have to consist only of whole numbers. For example, the size of shoes is also a discrete variable, and the sizes go up in steps of halves ($5, 5\frac{1}{2}, 6, 6\frac{1}{2}$, etc.)

In real life, you can measure continuous variables to a certain degree of accuracy. For example, the height of a tower can be measured to the nearest metre. Although the measurements recorded produce discrete data, the height is still a continuous variable.

Population. The term '*population*' means everything (or everybody) in the category you are considering. For example, if you were studying the length of life of light bulbs from a factory, the population would be all such light bulbs.

Exercise 58

For each population below state whether the variable given is qualitative or quantitative. If it is quantitative, state whether it is discrete or continuous.

1 the number of people in a room

2 the height of a person

3 the age of a person

4 the number of people on a train

5 the colour of the shoes of people on a train

6 the speed of a car

7 the number of dresses sold in a shop

8 the number of cars in a car park

9 the number of children in a family

10 the newspaper which a person reads on a Sunday

11 the weight of a person

12 types of shrub sold in a garden centre.

STATISTICS

14 Sampling, Surveys, Questionnaires

One of the problems with statistical surveys involving people is, that whatever your opinion, there are likely to be many other people with the same opinion. If you ask only these people, your opinion will be seen to be that of the whole population. If you ask only people with the opposite opinion, you will be seen to be in a minority. Therefore, you must ask a variety of people, so that you have a true picture of the population.

Remember, however, that in statistics, the term **population** does not necessarily refer to people. If you wished to survey the ages of cars on the road, your population might be all the cars in Britain.

14.1 Surveys

When you record any information – for example, about other people's opinions or numbers of surviving African elephants or types of road accidents – you are carrying out a **survey**. The survey results may be obtained by asking questions, by observation or by research.

To obtain completely accurate information, you would have to ask *everybody* (in your town or country or whatever), and receive answers from everybody, or observe *all* the elephants in Africa.

14.2 Censuses

When information is gathered about all the members of a population, the survey is called a **census**.

A national census is carried out every ten years. The last one was in 1991. Every adult in Britain is asked a large number of questions on mainly factual matters, for example the number of rooms in their house, their age, and the number of cars they possess.

A national census is a very large undertaking, and the results, though accurate, take a substantial length of time to be produced. Apart from the vast number of people to be asked, and the placing of their answers in computers, it is very difficult to ensure that every adult has in fact replied. It costs the country a great deal of money to complete a national census.

14.3 Samples

It is usually impossible for firms, newspapers, biologists, medical researchers, etc. to obtain

information about the whole population, because the survey:

- may be expensive
- may take a long time
- may involve testing to destruction – e.g. if you wish to find out how long batteries last, you test them until they run out
- may be impossible to carry out for every member of the population – e.g. a survey to find the weights of trout in Scottish rivers.

A small part of the population is chosen for the survey and this is called a **sample.**

The statistician then assumes that the results for the sample are representative of the population as a whole. The larger the number of people asked, the more likely their response is to be a valid result for the whole population.

Clearly it is vital that for the survey to be accurate the sample you choose must be representative of the whole population.

To achieve this, every member of the population must have an equal chance of being chosen.

14.4 Sampling methods

Random sampling

A random sample is one in which every member has an equal chance of being selected.

Campaign groups for or against a particular issue (such as the possible siting of a new supermarket near a park) can often obtain a large majority for their point of view simply by selecting which passers-by to question (perhaps the people living near the park who will be worried about the possibility of noise). By careful selection, majorities as high as 70% can easily be obtained both for and against the same issue! (Some people may well want a supermarket behind their back garden.)

The simplest way to obtain a random sample is to give every member a number, and to select numbers from tickets in a box (as in a raffle), or (if there are too many for this method), select numbers by computer. (Random numbers can also be obtained by using the RAN button on some calculators.)

It is common to use the electoral roll of a suitably sized area (on which every adult is listed) to obtain a numbered list from which to select a sample.

Periodic sampling

With periodic or systematic sampling, a regular pattern is used to pick the sample, for example, every hundredth firework on a production line. This can give a unrepresentative sample if there is a pattern to the list which is echoed by the sample.

Stratified random sampling

A stratified sample is more accurate than a random sample, and is used in opinion polls, when 1 or 2% accuracy is important. A stratified sample (or **strata sample**) is one in which the population is divided into categories. The sample should then be constructed to have the same categories in the same proportions.

Random sampling is then used to select the required numbers in each category.

For example, if you wished to find out about the earnings of students in a sixth-form college, it would be sensible to have both lower sixth and upper sixth students represented. You may also wish to make sure that one-year students, males and females are fairly represented. Suppose there are 1000 students in college, of whom 220 are lower sixth one-year students, 420 are lower sixth two-year students and 360 are upper sixth students.

A sample of 50 would contain the following numbers:

LVI one-year students $= \dfrac{220}{1000} \times 50 = 11$

LVI two-year students $= \dfrac{420}{1000} \times 50 = 21$

UVI students $\qquad = \dfrac{360}{1000} \times 50 = 18$

The eleven LVI one-year students would be randomly chosen from the 220 students in college. The other two strata would be chosen in the same way.

Quota sampling

For a quota sample, a manufacturer may determine the proportions of each group to interview.

For example, if a manufacturer wishes to launch a new chocolate bar on the market, it may be more important to canvass the opinions of children and those who do the shopping than any other sector of the market.

A market researcher paid to survey a sample of 100 people could be instructed to ask, say, 20 people under the age of 18, 30 in the age range 19 – 40 who do the family shopping, 10 in the same age range who don't, 30 in the age range over 40 who do the family shopping, and 10 in this age range who don't. The researcher will probably use convenience sampling (see below) to choose who to ask, but once one of the quotas is filled, no more people in that category may be asked. The researcher will continue to ask people in the other categories until the sample of 100 has been surveyed.

This is a common method used for market research, but inexperienced (or lazy!) researchers may choose an unrepresentative sample.

Convenience sampling

The most convenient sample is chosen, which, for a sample of size fifty, usually means the first fifty people you meet. There is obviously no guarantee that this sample will be representative. In fact it is highly likely that it won't be.

14.5 Bias

The results of a survey are biased if the sample is not representative of the whole population. Bias can be introduced if:

- the sample is unrepresentative. Even when using random sampling an unusual sample may be chosen, and this is just bad luck.
- an incorrect sampling method is used. Sampling methods, other than random, or stratified random sampling, are very likely to produce biased samples.

If you wanted to know people's views on drinking, a survey held outside a public house at closing time would clearly produce a different response from one held outside the office of the 'Teetotallers' League'! Neither would be representative of the complete population. Both of these samples would be biased.

- the questions asked in the survey are not clear or are leading questions (see section 14.6).

Exercise 59

In questions 1 to 8:

a identify the population

b criticise the method of obtaining the sample

c recommend an alternative way of obtaining a sample.

1 A journalist at a local newspaper wants to canvass popular opinion about plans for a new shopping centre in town. He goes into the High Street, and asks people, until he has asked 50.

2 Stephanie wishes to find out the earnings of college students. She goes into a college common room, and asks 40 girls.

3 The police wish to ascertain how many cars have a valid tax disc. One day, they set up a survey point on a road out of a town, between 5pm and 6pm. They stop a car, and check its tax disc. As soon as it has left, they stop the next car.

4 A geography student needs to collect five soil samples from his garden for a project. He stands in the middle, and throws a coin in the air. Where it lands, he takes a sample.

5 For a survey into the smoking habits of teenagers, Carol went to a tobacconist's near a school at 3.30pm, which was when the school day ended. She asked everyone entering the shop how much they spent on cigarettes in a week.

6 To find out how many homes in a telephone area have central heating, a salesgirl telephones 100 people, picked at random from a telephone directory.

7 To find out the make of car that people in an area of town use, Peter went out after lunch and knocked on doors until he had one hundred responses. He was pleased with his efficiency, as he had finished by 4pm.

8 To investigate what influenced people in their decision on mode of transport to work, John went to the station just before the 8.15 train departed, and asked as many people as he could.

14.6 Questionnaires

If your questions are written down and given to people to complete, the list of questions is called a **questionnaire**. The questions you ask must be chosen with care. They must:

(i) **not give offence**. Some people do not wish to give their precise age, or social class, so you *either*: (*a*) find an alternative question, (e.g. 'Which of these age ranges applies to you?'), *or* (*b*) fill in the information by using your own judgement.

(ii) **not be leading**. 'What do you think of the superb new facilities at . . .' will *lead* most people to agree they are better than the old facilities. People do not usually want to contradict the questioner. However, the point of the survey should not be to obtain agreement with your view, but to obtain other people's opinions.

(iii) **be able to be answered quickly**. The person answering the questions will often have only a small amount of time to spare and will not want you to write long sentences on their point of view. To obtain information easily from the survey it is helpful to have Yes/No answers or 'boxes' for the answers which are ticked. Here is an example:

How many different television sets does your household possess?						
0	1	2	3	4	5	More
☐	☐	☐	☐	☐	☐	☐

A questionnaire must also be easy for anyone to understand. The questions themselves must also be designed carefully. The question 'How much do you watch TV?' could result in the following types of response:

'A lot', 'Not much',
'Every night', 'twice a week'
'For two hours a night'
'Whenever there's sport, a film, . . .'

A better question is 'How many hours do you spend watching TV?', but this may encourage wild guesses because of poor memory.

An even better question to ask is 'How many hours did you watch TV *last* night?'. You can then offer a range of possible answers such as:

'Not at all'
'Up to $\frac{1}{2}$ hour'
'$\frac{1}{2}$ to 1 hour'
'1 to 2 hours'

and so on.

If you suspect different times are spent on different days, it is up to you, as a statistician, to ask a few people each day over a period of a week.

All surveys are open to error. The larger the sample, the more accurate the result.

14.7　Pilot surveys

It is common for companies to carry out an initial survey on a small area of the country in order to identify potential problems with the questions and to identify typical responses. This limits the errors in expensive large-scale surveys.

Exercise 60

Criticise the questions asked in this exercise and suggest questions which should be asked to find the information required.

1　What do you think of the improved checkout facilities?

2　Do you agree that BBC2 programmes are the best on TV?

3　What is your date of birth?

4　Sheepskin coats are made from sheep. Do you wear a sheepskin coat?

5　Dolphins are wild animals. Do you enjoy watching dolphins perform?

6　Sunbathing causes skin cancer. Do you sunbathe?

7　Vitamin D is obtained from sunlight. Do you sunbathe?

8　Is the new decor a major improvement on the old?

9　Would you rather use your local shops than a major supermarket miles away?

15 Classification and Tabulation of Data

Data can be classified in many different ways; the most common are
quantitative and qualitative (as defined in Chapter 13).

15.1 Tabulation

The purpose of tabulation is to arrange information, after collection and
classification, into a compact space so that it can be read easily and quickly. It
then may be represented pictorially to enable relevant facts to be seen readily,
as explained in the next chapter.

Tabulation consists of entering the data found in columns or rows.

Example

The numbers of students living in certain villages were:

Village	Number of students
Ashurst	31
Botleigh	15
Crow	28
Downton	24
Eaglecliffe	19
Fillingdales	33
Total	150

The data can be made more detailed by subdividing the rows and/or
columns to give more precise information.

An example would be:

Village	Number of students	
	Male	Female
Ashurst	15	16
Botleigh	10	5
		etc.

It is important that the tables produced are neat, all rows and columns
are clearly identified, and that units (where appropriate) are given.

15.2 Tally charts

It is common to record the data by means of a **tally chart**. Suppose you were
noting the speeds of cars at a particular point. You would draw up a list of
possible speeds and as each car went by, you would record a *I* . To enable
you to total your results quickly, you would mark every fifth *I* in a row
horizontally to produce a block of five ⊬⊢T .

A section of your results for this exercise would look like this:

Speed of car (mph)	Tally	Total
61	卌 卌 |||	13
62	卌 卌 |	11
63	||||	4
64	卌 卌 卌 |||	18

15.3 Frequency tables

The above table (with or without the tally) shows the raw data obtained. It is called a **frequency table.**

The grouping of the data to identify how many are in each category produces a **frequency distribution.**

It is often found useful to combine this data into a more compact form by grouping the particular values, as shown on the right:

Speed of cars (mph)	Frequency
0–39	0
40–49	7
50–59	54
60–69	75
70–79	25
80–89	2

In order to complete such a table, it is necessary to collect the raw data, fix the magnitude of each class interval, and group the data accordingly.

Exercise 61

1 In a board game a die was thrown several times. Here is a sequence of the scores:

3, 4, 1, 5, 6, 1, 2, 3, 2, 4,
5, 4, 3, 1, 2, 5, 6, 3, 1, 4,
2, 5, 6, 4, 5, 1, 6, 5.

Use a tally chart to obtain a frequency table for this data.

2 The number of people on the 87 bus was counted each time it passed the City Centre. Here is the data:

11, 25, 60, 16, 23, 2, 44, 26, 49, 58,
29, 8, 14, 24, 7, 16, 47, 5, 30, 34,
9, 12, 33, 10, 55, 21, 56, 32, 19, 6,
1, 21, 21, 42, 9, 35, 25, 55, 37, 46,
32, 14, 59.

With intervals 0–10, 11–20, 21–30, 31–40, 41–50, 51–60, use a tally chart to obtain the frequency distribution.

3 The numbers of seats *not* occupied on 80 transatlantic flights in one day were:

34, 8, 9, 6, 12, 30, 9, 11, 5, 39,
6, 25, 26, 42, 33, 16, 13, 30, 5, 29,
43, 34, 11, 26, 2, 39, 35, 19, 20, 40,
15, 11, 20, 34, 31, 17, 23, 2, 17, 15,
32, 3, 44, 6, 1, 7, 26, 35, 18, 25,
37, 4, 39, 37, 34, 26, 33, 7, 21, 16,
18, 15, 29, 35, 21, 6, 40, 39, 13, 12,
4, 4, 38, 39, 12, 0, 4, 33, 34, 18.

Summarise the information into class intervals 0–4, 5–9, 10–14, 15–19, 20–24, 25–29, 30–34, 35–39, 40–44.

4 A specialist in 'vowel research' counted the number of times each vowel was used on the page of a book. This is what she found:

A 卌 卌 卌 | O 卌 卌 卌 ||
E 卌 卌 卌 卌 || U |||
I 卌 卌 ||

Then she realised that she'd missed the last paragraph on the page. Here it is:

'Look out, Danny! There's some broken floorboards here! Come back!'

Continue the tally, and hence obtain the frequencies of the use of each vowel.

16 Pictorial Representation of Data

The presentation of data in the form of tables has been considered in Chapter 15. However, most people find that the presentation of data is more effective, and easier to understand, if the data is presented in a pictorial or diagrammatic form.

The pictorial presentation used must enable the data to be more effectively displayed and more easily understood. The diagrams must be fully labelled, clear and should not be capable of visual misrepresentation (see section 17.3). Types of pictorial representation in common use are the pictogram, bar chart, frequency polygon, and histogram.

16.1   Pictograms

In a **pictogram** data is represented by the repeated use of a pictorial symbol. The example below shows how a pictogram works.

Example

A survey of 1000 people living in Freeton was taken, to see what colour of cars they owned.

Represent this data in the form of a pictogram. The results of the survey were:

Colour of cars	Number of cars
Red	60
White	100
Blue	200
Grey	50
Gold	80
Black	30

Here is one possibility. A full car symbol represents 20 cars; half a car represents 10 cars. It is not possible to show small fractions of a symbol accurately, and the detail required should not normally be to more than half of a symbol (but certain symbols may allow for a quarter).

Colour of car

Red

White

Blue

Grey

Gold

Black

Key: = 20 cars

For each question illustrate the data given by means of a pictogram.

1 The numbers of bottles of champagne sold in five villages in 1991 were:

Abbotshurst	60	Tobbenham	80
East Lynne	120	Westering	100
Marlinsby	50		

2 The makes of a number of cars passing a junction were:

Ford	35	Citroën	20
Rover	30	Renault	10
BMW	5	Vauxhall	25

3 The number of flights for each airline out of Gatwick in a one-hour period was:

British Airways	8	Aer Lingus	2
Swissair	1	Britannia	6
Virgin Atlantic	1	Monarch	3

4 The workforce of a factory was asked by which mode of transport they came to work. The results were:

Car	35	Motorcycle	15
Train	25	Walk	20
Bus	10		

5 The contents of a fruit bowl were:

Apples	7	Bananas	3
Pears	5	Peaches	7
Kiwi fruit	6	Oranges	2

6 Students in a department of a college were asked about the type of accommodation in which they lived. The data was:

Flat	25	Semi-detached house	40
Maisonette	5	Detached house	30

7 Forty children were asked their favourite type of bread.
The answers were:

White (sliced)	16	Brown	14
White (unsliced)	6	French stick	4

16.2 Bar charts

A **bar chart** is a diagram consisting of columns (i.e. bars). Usually the bar chart shows the bars arranged vertically, with the heights of the bars indicating the frequencies.

A bar chart should always have a heading describing exactly what information it is illustrating. Each column or bar must be labelled, and the scale must be stated. It is common to leave 'gaps' between the bars.

Example

Fifty households were surveyed, and the number of children in each family was recorded as follows:

Children in family	Frequency
0	8
1	11
2	17
3	8
4	5
5	1

Represent this data by means of a bar chart.

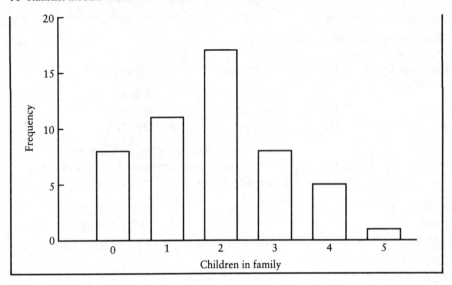

Dual bar charts

Dual bar charts are used when two different sets of
information are given on connected topics.

Example

The number of people over 17
years old, and the number of
people holding driving licences in
a particular street were found
over a period of years.
These are as shown on the right.

Year	1986	1987	1988	1989	1990	1991
No. of people over 17	32	27	29	31	33	39
No. of people with driving licence	12	17	19	11	24	28

Represent this data by means of a dual bar chart.

☐ No. of people over 17

■ No. of people over 17 with driving licence

Sectional bar charts

Sectional bar charts, or **component bar charts**, are used when two, or more,
different sets of information are given on the same topics. They are particularly
useful when the *total* of the two or more bars is also of interest.

Example

The numbers of saloons and hatchbacks sold by a garage were recorded. They were:

Month	Jan	Feb	Mar	Apr	May	Jun
Saloons	18	7	8	12	10	13
Hatchbacks	16	12	9	7	9	8

Represent this data by means of a sectional bar chart.

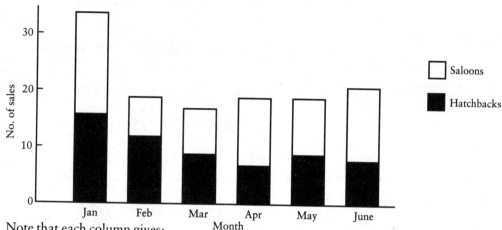

Note that each column gives:

(i) the number of saloons sold,

(ii) the number of hatchbacks sold, and

(iii) the total number of cars sold during that month.

All three sets of information can rapidly be compared by using the same diagram.

Exercise 63

1 The goals scored in 38 football league matches on Saturday 24 February 1990 were:

Number of goals in match	0	1	2	3	4	5
Number of matches	3	6	10	12	5	2

Illustrate this information by means of a bar chart.

2 200 people were asked their favourite television channel.
The results were:

BBC1	78
BBC2	24
ITV	75
Channel 4	23

Illustrate this information by means of a bar chart.

3 On a walk through a forest the following types of tree were seen:

Oak	80	Beech	35
Elm	6	Conifer	70
Chestnut	18	Cedar	21

Illustrate this data by means of a bar chart.

4 The holiday destinations of 100 people entering an airport were:

	France	Spain	Greece	Italy	Morocco	USA
Male passengers	3	18	9	8	5	9
Female passengers	4	11	15	2	6	10

Draw a sectional bar chart to illustrate this data.

5 Twenty people noted the television channel they were watching at 8.15pm on two successive nights. The results were:

	First night	Second night
BBC 1	6	7
BBC 2	4	1
ITV	7	6
Channel 4	3	4
Satellite	1	2

Draw a suitable bar chart to illustrate this data.

6 The papers sold in one day were:

	Male buyer	Female buyer
The Sun	8	4
Daily Mail	3	7
Daily Mirror	4	2
Daily Telegraph	3	1
The Times	2	3

Draw a sectional bar chart to illustrate this data.

7 The days with more than one hour of sun, and the number of days with rain, were recorded as:

	Sunny days	Rainy days
January	1	27
February	11	21
March	18	12
April	14	14
May	20	7
June	27	4
July	28	8

Draw a dual bar chart to illustrate this data.

16.3 Pie charts

A **pie chart** is another type of diagram for displaying information. It is particularly suitable if you want to illustrate how a population is divided up into different parts and what proportion of the whole each part represents. The bigger the proportion, the bigger the slice (or 'sector').

Example

Represent by a pie chart the following data.

The mode of transport of 90 students into college was found to be:

Walking	12
Cycling	8
Bus	26
Train	33
Car	11
Total	90

Represent this data by means of a pie chart.

A circle has 360°. Divide this by 90 to give 4°. This is then the angle of the pie chart that represents each individual person.

Since 12 people walk to college, they will be represented by $12 \times 4° = 48°$.

Similarly for the others:

	Angle in pie chart
Walking	12 × 4 = 48°
Cycling	8 × 4 = 32°
Bus	26 × 4 = 104°
Train	33 × 4 = 132°
Car	11 × 4 = 44°
	Total = 360°

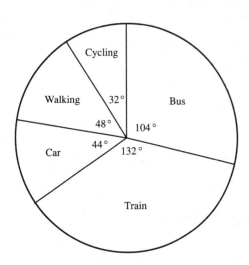

Method for calculating the angles on a pie chart

Here is a summary of how to work out the size of each bit of 'pie'.

(i) Add up the frequencies. This will give you the total population (call it p) to be represented by the pie.

(ii) Divide this number into 360.

(iii) Multiply each individual frequency by this result. This will give you the angle for each section of the pie chart.

Exercise 64

By means of pie charts, illustrate the data given below:

1 The types of central heating used by households in a village were:

Solid fuel	14
Gas	105
Electricity	41
None	20

2 The holiday destinations of 60 people were:

France	Spain	Greece	Tunisia	USA	Caribbean	Portugal
21	15	6	3	8	5	2

3 240 students were asked what they were intending to do during the next year. The results were:

80 going to university
86 staying at college
64 going into employment
10 no firm intention.

4 The numbers of bedrooms in 720 houses were recorded as:

1 bedroom	80
2 bedrooms	235
3 bedrooms	364
4 bedrooms	39
5 bedrooms	2

5 At a sports centre, the ages of 100 people were recorded as follows:

Under 20 years	30
20 – 29 years	15
30 – 39 years	12
40 – 59 years	14
60 years and over	29

6 Each pound spent at the Winchester Theatre Royal box-office is used to meet the theatre's expenses as follows:

Performance fees	60p
Salaries	17p
Premises and depreciation	10p
Administration	5p
Publicity	5p
Equipment	3p

16.4   Line graphs

A bar chart can be replaced by a line graph. In this case the data is plotted as a series of points which are joined by straight lines.

Line graphs are usually associated with time when they are called **time-series graphs**.

They are used, for example, by geographers to illustrate monthly rainfall or yearly crop yield etc., and by businesses to display information about profits or production over a period of time.

They show trends and have the advantage that they can be easily extended.

Example

The numbers of cars sold by a garage during the first nine months of 1990 were:

Month	Jan	Feb	Mar	Apr	May	Jun	Jul	Aug	Sep
Number sold	32	25	17	10	14	5	4	48	27

Represent this data by means of a line graph.

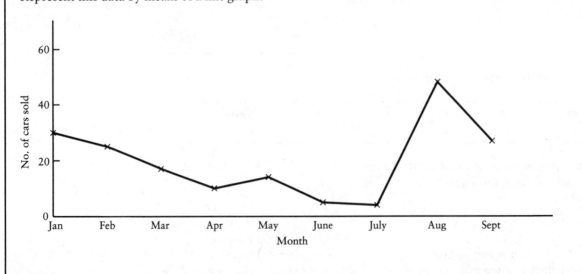

Illustrate the data given using a line graph.

1   The rainfall during a period of six months was:

Month	Jan	Feb	Mar	Apr	May	Jun
Rainfall (mm)	75	192	86	89	25	19

2   The numbers of cars sold per month at a garage were:

Month	Jul	Aug	Sep	Oct	Nov	Dec
Number of cars	15	92	27	18	21	11

Comment on any trends you notice.

3   The numbers of passengers carried by an airline (in thousands) were as follows (Sp = spring, Su = summer, etc.):

1988				1989				1990			
Sp	Su	Au	Wi	Sp	Su	Au	Wi	Sp	Su	Au	Wi
21	48	31	17	22	49	29	18	23	41	24	25

Comment on any trends you notice.

4   The hours of sunshine per month on Bliss Island were:

Month	Mar	Apr	May	Jun	Jul	Aug	Sep	Oct
Hours of sunshine	145	195	241	304	310	307	261	175

5   The numbers of ferries per day sailing from Dover by a certain shipping company were:

Month	Jan	Feb	Mar	Apr	May	Jun
Number of ferries	22	23	26	27	31	35

6   The maximum temperatures for six successive months at Sunbourne were:

Month	Apr	May	Jun	Jul	Aug	Sep
Temperature (°F)	61	74	72	91	85	56

16.5   Histograms

A **histogram** looks similar to a bar chart. However, a bar chart is often used when the variables have distinct, precise values (**discrete data**) for non-numerical (**qualitative**) data.

When the data given has a spread of values a histogram should be used. In a histogram the *area* of the bar represents the frequency, whereas in a bar chart the *height* of the bar represents frequency.

The scale on the horizontal axis must be continuous and the bars of a histogram must be drawn without gaps between them.

Example 1

The diameters of 140 apples in a box were measured in millimetres.
The diameters were recorded as follows:

Diameter of apple (mm)	No. of apples
30–31	5
31–32	7
32–33	8
33–34	12
34–35	25
35–36	31
36–37	22
37–38	18
38–39	8
39–40	4

Draw a histogram to represent this data.

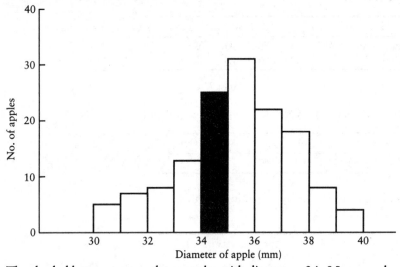

The shaded bar represents those apples with diameters 34–35 mm and
has a height of 25, this being the frequency of these apples.

The data could be grouped by combining two or more of the individual bars as
illustrated in Example 2. If we do this, the area will still represent the
frequency, so it will be necessary to divide the frequency by the width of the
bar to obtain the height of the bar. The vertical axis must now be labelled
'frequency per unit class interval', or frequency density, to specify precisely the
data plotted.

Example 2

The diameters of 140 apples in a box were measured in millimetres.
The diameters were recorded as follows:

Diameter of apple (mm)	No. of apples
30–32	12
32–34	20
34–36	56
36–38	40
38–40	12

Draw a histogram to represent this data.

The width of each bar is 2, and hence the frequencies must be divided by 2 before drawing the histogram.
Note that each bar of the second histogram represents the combination of two bars in the first histogram.

Diameter of apple (mm)	No. of apples	Frequency / Class width
30–32	12	6
32–34	20	10
34–36	56	28
36–38	40	20
38–40	12	6

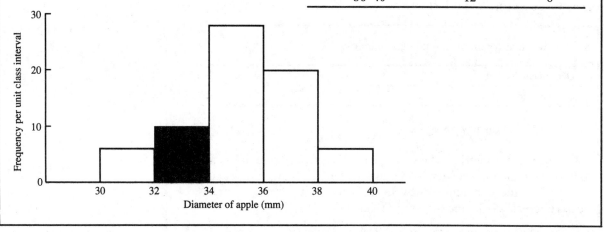

Although the two histograms in Examples 1 and 2 appear, at first glance, to be different, they both represent the same data.

The *areas* of the bars tell us how many apples are in that size range.

For example, the shaded bar represents 10 apples per unit class interval and its area, 20, is the number of apples in the interval from 32 mm to 34 mm.

The **class interval** is the interval chosen for classification.

The **class frequency** is the number of times the item appears in the class interval.

If you are given a histogram, the frequency is obtained by multiplying the frequency density by the width of the class interval.

Class boundaries

When a histogram is drawn, there must not be a gap between the bars, and it is necessary to obtain the exact class boundaries carefully.

For example:

  (i) The heights of trees are measured to the nearest metre. The class 4–7 m contains trees of height 3.5–7.5 m, and the class interval is 3.5–7.5 m.
 (ii) The ages of students are recorded. The class 17–19 years has class boundaries of 17 and 20 years (students are 19 until their 20th birthdays).
(iii) Heights are measured to the nearest 0.1 metre. The class 4–7 m has class boundaries 3.95 m and 7.05 m.

To find the heights of bars in a histogram, *divide* the frequency by the width of the class interval.

Frequency polygons

A frequency polygon is usually superimposed on a histogram by joining the mid-points of the tops of the bars with straight lines. Alternatively, the frequency polygon can be drawn without first drawing the histogram.

For ungrouped data, the frequencies are plotted as points. For grouped data, which is more usual, the frequencies are plotted against the mid-point of the class interval. In both cases the points are joined with straight lines.

Frequency polygons are used to compare frequency distributions, i.e. to compare the 'shapes' of the histograms.

Example 3

a Draw a histogram of the data given below:

Height of conifer (m)	1–3	4–6	7–9	10–12	13–15
No. of conifers	12	24	27	30	15

b Draw the frequency polygon.

The interval 4–6 metres includes trees of heights from 3.5 to 6.5 m, and hence it has an interval width of 3 units. To find the frequency per class interval, divide each frequency by 3. For example, the group 1–3 m has a class interval 0.5–3.5 metres, with a frequency of 12.

The frequency per unit class interval (frequency density) is $\dfrac{12}{3} = 4$

The table now becomes:

Height of conifer (m)	1–3	4–6	7–9	10–12	13–15
Frequency per unit class interval	4	8	9	10	5

The bar relating to height 1–3 m covers the true heights of 0.5 m (the lower class boundary) to 3.5 m (the upper class boundary).

The bar is drawn between these two limits, as shown:

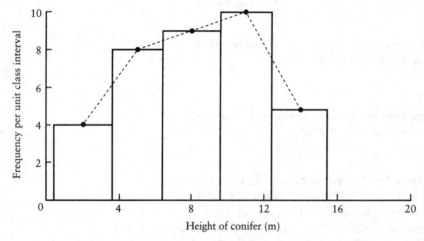

b The frequencies 4, 8, 9, 10, and 5 are plotted against the mid-points of the class intervals, which are 2 m, 5 m, 11 m, and 14 m. This is shown in the graph above by a series of dashed lines.

*Histograms with bars of unequal widths

The histogram does not have to have equal class intervals of equal width.

The data in the example opposite could be grouped as follows:

Height of conifer (m)	1–2	3–6	7–9	10–11	12–14	15
No. of conifers	6	30	30	28	12	5

Group   1–2   now has class interval 0.5–2.5, class width = 2
       3–6   now has class interval 2.5– 6.5, class width = 4
       7–9   now has class interval 6.5– 9.5, class width = 3
    10–11 now has class interval 9.5–11.5, class width = 2
    12–14 now has class interval 11.5–14.5, class width = 3
    15   now has class interval 14.5–15.5, class width = 1

The data to be plotted therefore is:

Height of conifer (m)	1–2	3–6	7–9	10–11	12–14	15
Frequency density	3	7.5	10	14	4	5

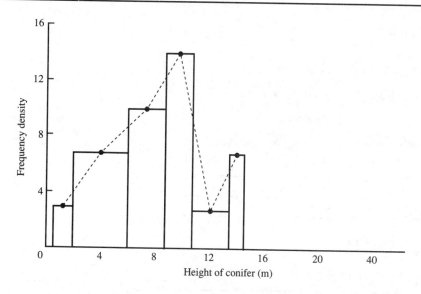

Height of conifer (m)

Exercise 66

1   The speeds of 100 cars on a motorway were recorded. The data found was:

Speed (mph)	30–40	40–50	50–60	60–70	70–80	80–90
No. of cars	2	11	35	42	9	1

Represent this data by means of a histogram.

2   The heights of 80 students were recorded. The data was:

Height (cm)	150–160	160–170	170–180	180–190	190–200	200–210
No. students	4	7	15	47	6	1

Represent this data by means of a histogram and draw a frequency polygon.

3 Below is a table showing the number of seats not booked on an airline's daily flight between London and Miami over 10 weeks.

19	1	8	11	15	19	21	17	1	23	19	11	12	15
21	11	8	4	15	27	21	20	14	18	7	11	23	21
8	17	1	19	12	16	21	25	28	29	17	15	11	8
16	8	2	8	6	10	11	15	9	8	6	2	3	8
21	18	27	32	37	4	11	19	21	34	21	15	12	11

By means of a tally chart, find the frequencies in the class intervals 0–4, 5–9, 10–14, . . ., 30–34, 35–39. Represent this grouped data by means of a histogram.

4 A vending machine in a college sells chocolate bars. The catering staff restocked the machine every day, and noted the number of packets sold. Over a period of time, these were:

18	31	27	38	31	26	39	37	36	38	37	14	28	34	12
30	27	30	18	25	27	35	34	26	23	36	30	32	2	29
24	34	13	21	32	30	18	38	21	14	23	8	29	25	17
13	35	5	24	12	4	11	28	1	2	33	8	1	20	7

By means of a tally chart, find the frequencies in the following class intervals:

0–4, 5–9, 10–14, 15–19, 20–24, 25–29, 30–34, 35–39.

Represent this data by means of a histogram and draw a frequency polygon.

*5 The number of accidents in High Town was recorded over two years, and the information was grouped in weekly periods.

Number of accidents	0–1	2–3	4–5	6–9	10–14	15–17	18–20
Number of weeks	1	7	12	36	28	19	1

Represent this data by means of a histogram.

*6 The IQs of 100 students were measured and the results were as follows:

IQ	100–109	110–119	120–129	130–139	140–159
No. of students	12	34	38	13	3

Represent this data by means of a histogram and draw a frequency polygon.

*7 The wages of 50 workers were as below. Represent this data using a histogram.

Weekly wage (£)	70–80	80–100	100–150	150–175	175–200	200–300
No. of workers	2	8	18	14	5	3

17 Interpretation of Statistical Diagrams

17.1 Reading and interpreting diagrams

We have seen in Chapter 16 that there are many ways of presenting data in pictorial form. It is clearly necessary to be able to interpret correctly any diagrams given. Examples of pie chart and histogram interpretation are given below.

Interpreting pie charts

The initial interpretation is the fact that the largest portion of a pie chart relates to the largest group, and the smallest portion to the smallest group. However, if any of the data is known, the rest of the data can be calculated.

Example

The pie chart below shows the number of students in different sections of a college. 220 students are in the Construction department.

a How many students are there in the college, and

b how many students are there in Catering?

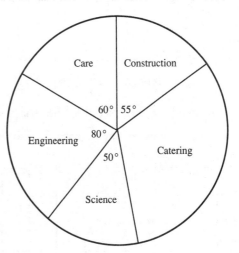

a 55° represents 220 students.

$\therefore$ 1° represents $\dfrac{220}{55} = 4$ students.

The complete circle (360°) represents $4 \times 360 = 1440$ students.

$\therefore$ There are 1440 students in the College.

b The angle representing Catering is
$360 - (80 + 55 + 60 + 50) = 115°$.

$\therefore$ The number of students in Catering is $4 \times 115 = 460$.

To interpret histograms

Example 2

The histogram shows the heights (in metres) of trees in a plantation.
There are 40 trees with heights between 10 m and 12 m.

a How many trees are there with heights between 4 and 6 m?

b How many trees are there in the plantation?

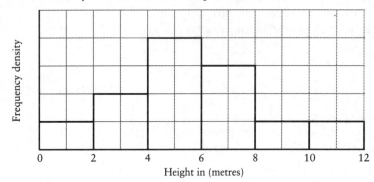

The frequency in a histogram relates to the area of each bar. The 10 to 12 interval has an area of 2 large squares, i.e. each large square represents 20 trees.

a The bar for the 4–6 metre interval has an area of 8 large squares.
There are 8 × 20 = 160 trees with heights between 4 and 6 metres.

b The bars have a total area of 24 large squares.
There are 24 × 20 = 480 trees in the plantation.

Example 3

The histogram represents the speeds of 200 cars along a motorway.

a How many cars are exceeding the 70 mph speed limit?

b What percentage of cars is travelling within 10 mph of the speed limit?

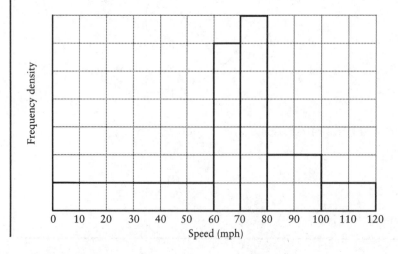

When the histogram has bars of unequal widths, the method described in Example 1 is still used.

The total number of large squares is 25.

This represents 200 cars.

∴ 1 large square represents $\dfrac{200}{25} = 8$ cars.

a The cars exceeding 70 mph are represented by 13 large squares.

∴ $13 \times 8 = 104$ cars are exceeding the 70 mph speed limit.

b Cars travelling between 60 and 80 mph are within 10 mph of the speed limit.

These cars are represented by 13 large squares.

∴ There are $13 \times 8 = 104$ cars within 10 mph of the speed limit.

∴ The percentage of cars within 10 mph of the speed limit $= \dfrac{104}{200} \times 100 = 52\%$

Exercise 67

1   The pie chart shows the different drinks sold at lunchtime in college. 720 drinks were sold in total.

Find the number of

a Coke

b orange

c coffee

d chocolate

drinks sold during lunchtime.

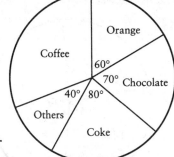

2   The pie chart shows the number of passengers flying from London to Miami on one afternoon. 1800 passengers in total flew this route on that afternoon.

Find the number flying:

a Virgin

b American Airlines

c British Airways.

The plane used by Virgin is a Boeing 747 seating 370. What percentage of the Virgin seats were occupied?

3   A bookshop sold 1080 books, and noted the types of book sold.

Find the number of books
sold which were classed as:

a Thriller

b Hobby

c Travel

d Science.

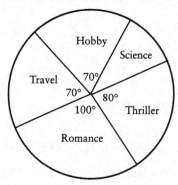

4   The pie chart shows the results of an election in a constituency. There are
72 000 voters of whom 80% voted. Which party won, and what was the
winning margin?

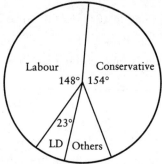

5   The pie chart shows the different petrols sold in one week.

The garage sold 25 000 gallons of diesel.

a How much unleaded
petrol was sold?

b How much 4-star petrol
was sold?

c What were the total sales?

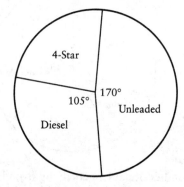

6   The pie chart shows the types of dwelling in which people in a village live.
There are 720 dwellings in the village. By measuring the angles, find how
many are:

a detached houses

b bungalows

c semi-detached houses.

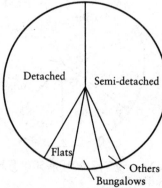

7 The pie chart shows the use of agricultural land in South Australia.

   a Find the percentage of land used for wheat.

   b The total acreage is 3 000 000 acres. What area is used for hay?

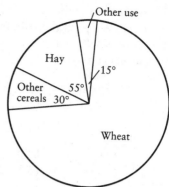

8 The pie chart shows details of the 200 pets kept by people living in a village.

   How many

   a dogs

   b cats

   c rabbits

   were there?

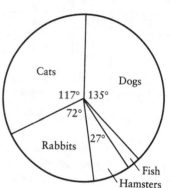

9 The histogram shows the number of lunches of various prices sold in a restaurant.
   Reconstruct the frequency table.

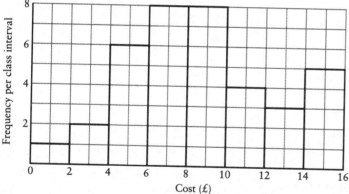

10 The histogram shows the distance travelled by 200 lobsters in one week.
    Reconstruct the frequency table.

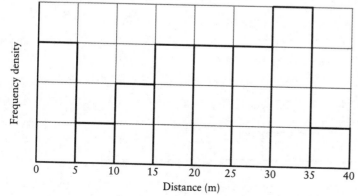

11 The histogram shows the number of cars arriving at a ferry terminal in the last hour before departure. 250 cars arrived in this hour. The latest official check-in time is half an hour before departure.
How many cars arrived after the official check-in time?
What percentage of the cars arrived within 10 minutes of the check-in time?

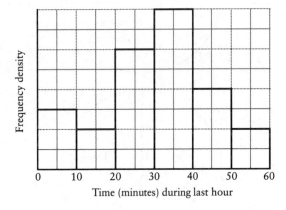

Time (minutes) during last hour

12 At a fête, the number of peas in a jar was guessed. The histogram represents the guesses made.
Reconstruct the frequency table.

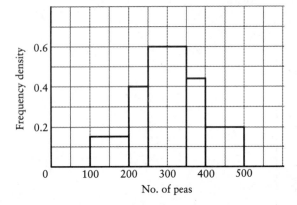

No. of peas

13 The histogram represents the number of pupils in a school.
Calculate the number of pupils in each class interval, and hence find the total number of pupils in the school.

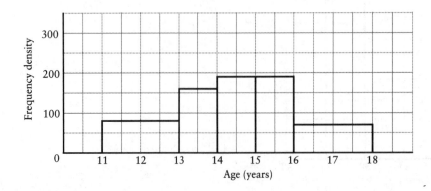

Age (years)

*14 The histogram below represents the distribution of ages in a small village.
Find the number of people who live in the village.

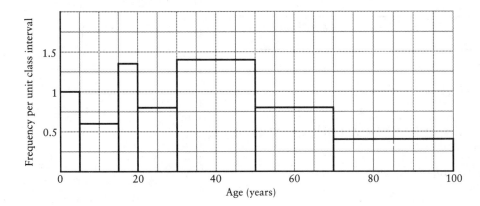

*15 The following histogram shows the number of metal rods, of different lengths, made in a factory in one week.
Find the total number of rods.

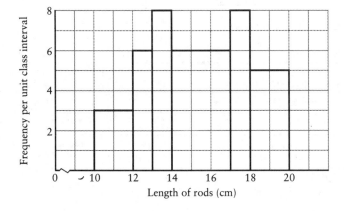

17.2 Drawing inferences from diagrams

When you can readily transfer data to pictorial form, and can convert pictorial representation back into numerical data, you can start to draw inferences from the data given in either form.

Drawing inferences from statistical data is not an exact science. There is rarely a correct, precise answer. If some of the data does not fit the overall pattern, this should be noted. Then, the reasons for the apparent contradiction should be considered.

We can illustrate this by an example.

Example

The number of people each day visiting an open-air swimming pool in August were:

341, 352, 347, 355, 361, 341, 344, 352, 344, 360, 371, 347, 329, 351, 621, 357, 348, 359, 354, 372, ...

What can be inferred from this data?

The 15 August figure of 621 is clearly exceptional and should be easily identified.

The reasons why this figure is exceptional would be unlikely to be found without further questioning – for example:

(i)   was an alternative swimming facility closed for the day?
(ii) was the weather on the 15th substantially hotter than on the other days?
(iii) does the data refer to a country with a Bank Holiday on 15 August (e.g. France)?

These are all possible reasons and you may be able to suggest others.

Some data gives results from which suitable inferences can be fairly quickly drawn. Here is an example.

Example

A men's clothes shop has two branches: one in a city centre, and one at an out-of-town shopping complex. The weekly sales of the two shops over a period of three months are given below:

Date (week commencing)		December				January				February			March	
		7	14	21	28	5	12	19	26	2	9	16	23	2	9
Sales in	City shop	18	21	23	14	29	31	28	7	5	8	6	4	5	8
thousands	Out-of-town	22	27	52	15	4	5	6	8	9	8	10	9	7	8
of pounds	shop														

What can be inferred from this data?

From the above data, the following inferences could be made:

(i)   Sales immediately prior to Christmas are higher than at other times in the three-month period.

(ii) Sales in the out-of-town complex are generally higher than those in the city shop.

(iii) The city shop had a 'sale' during the first three weeks of January.

(iv) During the 'sale' the trade at the out-of-town shop was reduced.

In reality, the firm could well investigate the types of garments sold over the year to decide whether or not to promote the same articles in both shops at the same or different times. Computerisation of sales enables shops to keep far better checks on stock sold. This enables them to react more quickly to consumer demand and to supply each individual shop with the goods which its specific customers require.

17.3 Dangers of visual misrepresentation

Statistics can very easily be presented in a form which, although correct, is misleading. The simplest way in which this occurs is by the use of the 'false origin'.

For example, when a rail line is electrified, the number of cars per day on the parallel road reduces from 38 500 per day to 38 200.

A correct bar chart would show:

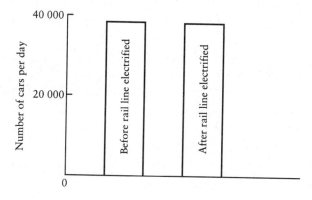

From this, readers would draw the conclusion that there has been little change.

Carefully deleting most of the vertical scale, and starting at 38 000 produces the bar chart below:

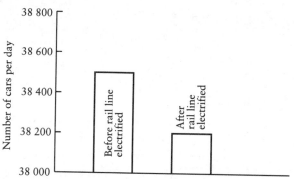

To a casual reader, the conclusion would be quite clear: there has been a significant reduction in the number of cars using the road.

Although this would rarely be spotted, the scale can be more subtly altered as below:

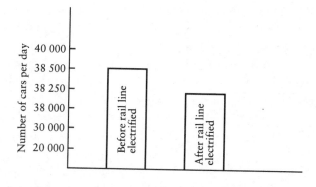

This over-expansion of the relevant part of the vertical scale can be justified as giving prominence to that part of the graph in which we are interested. However, its effect is to mislead readers.

If a section of a scale is to be deleted, it should be clearly identified – usually with a squiggle on the axis as shown on the right:

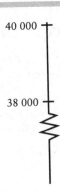

Bars of different widths can also mislead:

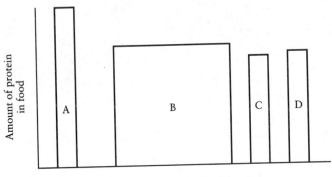

Types of food

Although B does not have the highest amount, a casual reader's eye is drawn to it as its 'area' is much greater than any other bar.

The phrase 'amount of protein in food' is also open to misrepresentation, as it does not specify how much of food A is compared with how much of food B. Are they proportions, or amount per penny, or amount per gram, or something else?

False impressions can also be given by careful selection of which figures to show. In a time-series, you have to have a start and end date. In the presentation of information demanding an increase in wages, salaries or subsidies, it is to be expected that the data will start at a peak year. In other cases, the data will stop at a favourable moment.

Example

The diagram shows the Japanese share index for 1981–90. The diagram on the right shows the index stopping in Summer 1987, when it would show an unbroken climb.

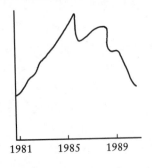
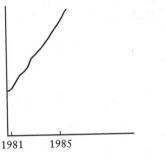

All diagrams should:

(i) be clearly labelled and titled

(ii) have the scales clearly identified

(iii) have the units given

(iv) should be drawn as outlined in Chapter 16.

Any diagram which is not so drawn may, intentionally or not, mislead.

Exercise 68

Criticise the diagrammatical representations shown in the following questions:

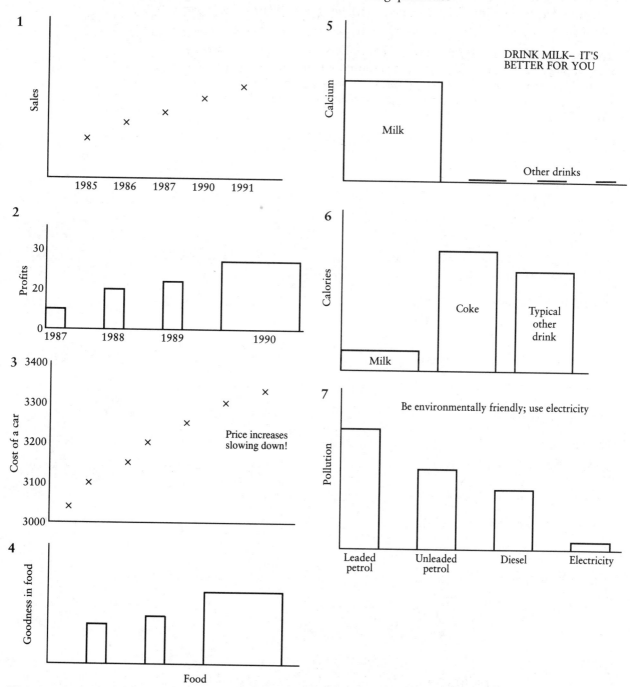

17.4 Interpretation of statistical information

All statistical information needs to be considered carefully before any inference can be made. It is rare for a simple set of statistics to prove anything.

An advertisement states '60% of dog owners, who expressed a preference, liked Dogamix'. It could be that out of 1000 owners, only 5 showed a preference; i.e. 3 out of 1000 liked it and 2 out of 1000 disliked it.

If a packet states average net weight of 250 grams, about half of the packets bought will have weight less than 250 grams, and half will have weight over 250 grams. Hence it is difficult for shoppers to tell whether they are being sold too little.

Professional statisticians are often involved in work of this nature. They are often employed to obtain, and present, the information in as favourable a way as possible for their employer. Detailed work in this area is beyond the scope of this course, but readers should always approach statistical information with care.

18 Averages

Sets of data can be compared, as mentioned previously, by comparing their frequency distributions or frequency polygons. It is also very useful to be able to compare a single, 'typical', statistic from one set of data with a single, 'typical', statistic from another set of data.

This statistic must be representative of the distribution. For this reason it is usually located at or near the centre of the distribution and is called a **measure of central location** or **average**.

The most commonly used averages are the **mean, mode** and **median**.

18.1 The arithmetic mean

The **arithmetic mean**, which is usually just referred to as the mean, is the most widely used average.

To calculate the mean, the total of the values is found and this is 'shared out' equally by dividing by the total number of values.

Example

Felicity sat six test papers and her marks, out of 50, were 17, 23, 27, 29, 30, 36.

Find her mean mark.

$$\text{Total marks} = 17 + 23 + 27 + 29 + 30 + 36$$
$$= 162$$

Number of marks $= 6$

$$\therefore \text{Mean mark} = \frac{162}{6} = 27$$

This means that if Felicity's performance had been the same in all the tests, she would have gained 27 marks for each one.

The mean is often given in formula form:

$$\text{Mean} = \frac{\Sigma x}{n}$$

Σ is a capital Greek letter (sigma). Σx means the sum of all the terms and n is the number of terms.

1 The weekly wages of ten workers were:

   £110, £115, £135, £141, £119, £152, £144, £128, £117, £139.

   Find the mean wage.

2 The speeds (in mph) of sixteen cars were recorded as:

   83, 75, 61, 72, 64, 51, 41, 89, 92, 71, 68, 66, 67, 64, 69, 72.

   Find the mean speed.

3 The wind speed (in mph) at 8 am on a particular day, was recorded at a number of measuring stations as:

   88, 74, 61, 92, 48, 59, 71, 80, 70, 51, 48, 45, 75, 80, 82.

   Find the mean wind speed.

4 A rugby team scores 37, 21, 64, 0, 18, 7, 35, 49, 28, 51, 82, 71 points in 12 successive matches. What is its mean score?

5 Eight people were asked their ages, and the replies were 37, 41, 29, 17, 15, 21, 32, 38. John claims that their average age is over 29. Why is he correct?

6 Nine workers have a mean wage of £190 per week.

   a What is their total wage?

   A new worker starts at the firm, and is given a wage of £110 per week.

   b What is the total wage for the ten workers?

   c What is the mean wage for the ten workers?

7 Five men have a mean height of 1.95 m.

   Four women have a mean height of 1.72 m.

   What is the mean height of the nine people?

8 A train of 12 carriages separates for the last part of its journey. There was a mean of 52 people in each carriage before it separated, and the four carriages going to Weymouth contained a mean of 63 people per carriage.

   What was the mean number in the carriages *not* going to Weymouth?

18.2 The mode

The **mode** is the number which occurs most frequently. Suppose seven students scored as follows in a test:

$$2, 3, 6, 7, 7, 8, 9$$

The mean score here is $\dfrac{42}{7} = 6$.

The **mode**, the number which occurs most frequently, is 7.

Some distributions can have more than one mode. For example, the numbers of people in twenty minibuses were recorded as:

3, 3, 4, 5, 7, 8, 8, 8, 10, 11, 13, 13, 14, 15, 15, 15, 17, 17, 18, 18.

Both 8 and 15 are modes.

This distribution is said to be **bimodal**.

┌─ *Example* ─────────────────────────────────

Find the mode of the numbers:

5, 7, 12, 12, 12, 17, 17, 19, 21.

12 occurs three times, which is more often than any other number. Hence the mode is 12.

Modal class

When the values are grouped the mode is replaced by a **modal class**, which is the group of values occurring most frequently.

Example

What is the modal class in the following table?

Height of trees (in m)	Frequency
0–2	4
2–4	8
4–6	11
6–8	4
8–10	2

There are 11 trees in the 4–6 category.

This is the greatest frequency.

∴ The class 4–6 is the modal class.

Exercise 70

1 A die was thrown 12 times, and the scores were

2, 4, 1, 3, 4, 1, 5, 6, 6, 4, 2, 5.

What is the modal score?

2 The price (in pence) of a loaf of bread at 10 shops was found to be 44, 49, 51, 68, 62, 44, 69, 51, 44, 47. What is the mode?

3 The heights of 40 trees were measured, and the data was:

Height (m)	Frequency
2–4	2
4–6	7
6–8	11
8–10	13
10–12	7

What is the modal class?

4 The IQs of 70 students were recorded as:

IQ	Frequency
95–99	5
100–104	7
105–109	18
110–114	21
115–119	7
120–124	4
125–129	5
130–134	1
135–140	2

What is the modal class?

5 Fifteen cars have the following colours: blue, black, red, white, red, white, gold, blue, red, grey, red, grey, black, blue, purple.

What is the modal colour?

18.3 The median

The **median** is the value of the 'middle' item when the items are placed in numerical order.

To calculate the median, you put all the quantities given in order (usually ascending), and the median is the value of the middle one. Let's look at an example.

Example 1

Find the median of the numbers 17, 18, 24, 27, 28.

The 'middle' number is 24, so the median is 24.

There is a complication with the definition of a median when there is an even number of quantities, as the following example shows.

Example 2

Find the median of the numbers 13, 15, 18, 24, 32, 35.
There are 6 numbers, so there is no 'middle' number. The third number is 18, the fourth is 24, and the convention is to average these two 'middle values'.

The average (mean) of 18 and 24 is $\dfrac{18 + 24}{2} = 21$.

$\therefore$ The median is 21.

A quick rule for finding the middle quantity

To find the middle quantity, add one to the number of quantities given, then divide by 2, and take the quantity corresponding to this position.

In the first example above, there were 5 items. Add 1 to get 6. A half of 6 is 3. The 3rd number is 24, which is the median.

In the second example above there were 6 items, add 1 to get 7. A half of 7 is $3\frac{1}{2}$, which is half way between 3 and 4. Therefore the median lies half way between the 3rd and 4th number.

18.4   The use of mean, mode, and median

The three averages are useful in different contexts.

- If you were an employer considering the production capacity of your works, it would be helpful to use the *mean* of past production, as this would give you a good idea of the number of goods you can produce.

- If you were a shopkeeper wanting to keep a minimum stock of shirts to sell, the *mode* would be the best to use, as this will tell you which shirts you are most likely to sell.

- If you were a union wage negotiator, the *median* salary would be appropriate to use, because the few high wage earners would not then affect your 'average' of the wages paid.

Exercise 71

1   The numbers of people on nine buses were recorded as:

17, 31, 11, 3, 51, 49, 52, 47, 34.

Find the median number of people.

2   The numbers of cars per hour on a country road during the hours of daylight were recorded as:

11, 13, 15, 9, 17, 12, 18, 14, 7, 9, 14, 16, 7, 11.

Find the median number of cars.

3   A die was thrown 12 times, and the scores were:

2, 4, 1, 3, 4, 1, 5, 6, 6, 4, 2, 5.

What was the median score?

4   The price of a loaf of bread (in pence) at 10 shops was found to be:

44, 49, 51, 68, 62, 44, 69, 51, 44, 47.

What was the median price?

5   The weights of parcels (in kg) delivered to a library were:

7.4, 8.2, 11.1, 7.8, 2.5, 5.6, 7.1, 8.9, 2.3, 2.7, 2.9, 4.1.

Find the median weight.

18.5   The mean of a grouped distribution

Let us look first at the simpler case where the data is recorded in a frequency table:

Example 1

Find the arithmetic mean of the following scores:

Score	Frequency
1	3
2	5
3	11
4	1
5	5

The score of 2, for example, occurred five times, but instead of totalling $2 + 2 + 2 + 2 + 2$, it is quicker to multiply 2 by 5. Similarly, instead of totalling $3 + 3 + 3 + \ldots$ eleven times, it is quicker to calculate 3×11.

The table can be extended like this:

Score x	Frequency f	Score × Frequency xf
1	3	3
2	5	10
3	11	33
4	1	4
5	5	25
Totals:	$\Sigma f = 25$ (Total frequency)	$\Sigma xf = 75$ (Total of 25 scores)

$$\text{Mean score} = \frac{\Sigma xf}{\Sigma f} \left(\text{i.e. } \frac{\text{Total of 25 scores}}{\text{Total frequency}} \right)$$

$$= \frac{75}{25}$$

$$= 3$$

The mean is often denoted by $\bar{x}$.

Example 2

Thirty households were surveyed, and the number of children in each household was recorded. Find the mean and the mode.

No. of children in each family	0	1	2	3	4	5
Frequency	4	6	13	4	2	1

No. of children in each family x	Frequency f	xf
0	4	0
1	6	6
2	13	26
3	4	12
4	2	8
5	1	5
Totals:	30	57

$$\text{Mean} = \frac{\Sigma xf}{\Sigma f}$$

$$= \frac{57}{30}$$

$$= 1.9 \text{ children}$$

Mode = 2

Note that the mean need not be any of the original figures, nor indeed even a result which is possible. You cannot have 1.9 children in a family! The mean only indicates the average or 'central tendency'.

Calculating the mean of a grouped distribution is more complicated. How to deal with the values when they are in classes is shown in the following example.

Example 3

Thirty bushes were measured. Their heights (in cm) were grouped as shown.
Find the mean length and the modal class.

Height	5–15	15–25	25–35	35–45	45–55
Frequency	6	4	15	3	2

Since there is a spread in each group, it is impossible to determine the exact mean. We can only find an approximation to the mean.

It is assumed that each bush has a height equal to the middle of the range in which it lies; for instance, the 4 bushes with height 15–25 cm are all assumed to have height 20 cm, which is the **mid-interval** of the range 15–25 cm.

Length (cm)	Mid-interval (cm) x	Frequency f	Mid-interval × Frequency xf
5–15	10	6	60
15–25	20	4	80
25–35	30	15	450
35–45	40	3	120
45–55	50	2	100
Totals:		30	810

$$\text{Mean} = \frac{\text{Total of 'mid-interval} \times \text{frequency'}}{\text{Total frequency}}$$

$$= \frac{\Sigma xf}{\Sigma f}$$

$$= \frac{810}{30}$$

$$= 27\,\text{cm}$$

The modal class is 25–35 cm. (There are 15 bushes in this class, more than in any other class.)

Example 4

In one hour, twenty planes arriving at Heathrow were early. The number of minutes early (to the nearest minute) were grouped as shown. Find the mean number of minutes early.

Time (min)	0–3	4–8	9–13	14–18
Frequency	4	7	8	1

As in Example 3 there is a spread in each group. The lower and upper class boundaries need to be found carefully. The group 0–3 (minutes) will range from 0 minutes (early) to 3.5 minutes (early). Therefore, the four planes in this group are all assumed to be 1.75 minutes early, which is the mid-interval of the range 0–3.5.

The group 4–8 ranges from 3.5 to 8.5 minutes early. Hence, these seven planes are all assumed to be 6 minutes early. (6 is the mid-interval of 3.5 and 8.5.)

No. of minutes early	Mid-interval (in minutes) x	Frequency f	Mid-interval × Frequency xf
0–3	1.75	4	7
4–8	6	7	42
9–13	11	8	88
14–18	16	1	16
Totals:		20	153

$$\therefore \text{Mean} = \frac{\Sigma xf}{\Sigma f}$$

$$= \frac{153}{20}$$

$$= 7.65\,\text{minutes}$$

Note. In Examples 3 and 4 the distributions are continuous.

No problems are caused in Example 3 by including the height 15 cm in the group 5–15 and in the group 15–25 for heights of bushes since the probability of finding a bush exactly 15.000 . . . m high is nil.

When a distribution is discrete, you should *not* use 5–15 and 15–25 etc., since if an item is exactly 15, it is not clear into which group it should be placed.

If you were considering the number of people on a bus, you could use 5–14, 15–24, 25–34 etc.

The range 5–14 would have a mid-interval of 9.5, which would be used to find the mean.

Exercise 72

1 The numbers of children per family on a housing estate were recorded as follows:

No. of children	0	1	2	3	4
No. of families	12	15	5	2	1

Find the mean number of children per family.

2 An agricultural researcher counted the numbers of peas in a pod in a certain strain as follows:

No. of peas	3	4	5	6	7	8
No. of pods	5	5	20	35	25	10

Find the mean number of peas per pod.

3 The Ace Bus Company went through a bad patch when its buses always left the city centre late. This grouped distribution table shows how late:

Minutes late	0–10	10–20	20–30	30–40	40–60
Frequency	5	8	21	14	5

Find the mean number of minutes late.

4 The numbers of words per sentence on a page of book were:

No. of words	1–3	4–6	7–9	10–12	13–15
Frequency	3	38	59	27	4

Find the mean length of a sentence.

5 The weekly wages of a firm's employees were:

Wage in £	50–69.99	70–89.99	90–99.99	100–149.99
Frequency	7	9	15	25
Wage in £	150–199.99	200–249.99	250–299.99	300–350
Frequency	38	41	7	2

Find the mean weekly wage.

*18.6 Moving averages

Moving averages are a way of smoothing out irregularities in a graph so that the variations and trends over a period of time become clearer.

Before you can find a moving average, you have to decide how many individual results to average. It is necessary to decide on the length of a cycle (by common sense) and then to see how many results fit into each cycle.

We can explain how moving averages work, using an example.

Example

The sales in a shop (in £) were as shown:

	Week 1	Week 2
Tue.	225	230
Wed.	310	315
Thur.	320	315
Fri.	460	440
Sat.	700	880

a Choose the most suitable number of points for the moving average.

b Find an appropriate moving average.

c Plot the original data and the moving averages on a graph.

a Since the shop works a five-day week, and the sales pattern varies according to the day of the week, each cycle covers five results. Hence, a five-point moving average is required.

b At the end of week 1, you can clearly find the average sales per day.

At the close of business on Tuesday in week 2, you also have five previous results to average, from Wednesday (week 1) to Tuesday (week 2).

At the end of Wednesday, we can similarly find a new five-point moving average.

It may be found easier to evaluate moving averages by listing the data given in one column, and working out successive totals:

	Tues.	225						
	Wed.	310						
Week 1	Thur.	320	2015					
	Fri.	460		2020				
	Sat.	700			2025			
	Tues.	230				2020		
	Wed.	315					2000	
Week 2	Thur.	315						2180
	Fri.	440						
	Sat.	880						

Since you are finding five-point moving averages, you divide each of the totals by 5 to obtain 403, 404, 405, 404, 400, 436.

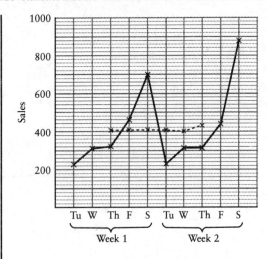

c The moving average is plotted against the middle of the points which
have been averaged. In this case, it is plotted at the third day of each cycle
(as shown).

Exercise 73

In questions 1 – 8,

a plot the data given on a graph

b decide on the number of points which the moving average should involve

c calculate the moving averages

d add these points to the graph

e state what trend can be observed.

1 The number of people having meals in a restaurant over a period of two
weeks was:

M	T	Th	F	S	Su	M	T	Th	F	S	Su
17	21	12	34	41	48	19	20	11	36	44	51

2 The average number of students absent from a college over a period of
3 years was:

1988			1989			1990		
Spr.	Sum.	Aut.	Spr.	Sum.	Aut.	Spr.	Sum.	Aut.
38	50	41	36	58	42	39	59	44

What other information should be found before the real trend could be
obtained?

3 The attendance at matinées of a play which had just opened was:

Week 1				Week 2				Week 3			
Tue.	Wed.	Fri.	Sat.	Tue.	Wed.	Fri.	Sat.	Tue.	Wed.	Fri.	Sat.
310	250	590	650	280	220	560	650	250	230	530	620

4 The number of houses completed each quarter in a planning district was:

	1989				1990			
	Spr.	Sum.	Aut.	Win.	Spr.	Sum.	Aut.	Win.
	85	88	92	38	88	93	87	43

5 The numbers of kilograms of sausages sold by a butcher's shop over a period of three weeks were:

Day	M	T	Th	F	S	M	T	Th	F	S	M	T	Th	F	S
Sales (kg)	18	9	4	11	27	15	14	6	13	24	19	12	7	14	26

6 The numbers of people (in hundreds) visiting a country house in three weeks in July were:

M	Tu	W	Th	F	Sa	Su	M	Tu	W	Th	F	Sa	Su
11	7	8	13	22	28	31	12	6	9	14	23	31	33

M	Tu	W	Th	F	Sa	Su
13	5	8	13	22	30	33

*18.7 Weighted averages

In previous examples, when calculating the mean of a set of values we have counted each value equally, but it is not always appropriate to do this.

For example, when candidates are assessed for the Statistics module, the examination mark counts for 60% of the final mark, the project for 30%, and the oral assessment for 10%. In other words, the marks are **weighted** so that the final mark is affected most by the performance in the written examination and least by the oral mark.

Example 1

On a Statistics course, a candidate receives 74% for the written exam, 64% for the coursework, and 48% for the oral. The weightings are 60, 30 and 10 respectively. Find the final mark.

Since the weightings are 60, 30 and 10, the written exam component is more important than the coursework component, and far more important than the oral component. Hence these components must be multiplied by their appropriate weightings, as follows:

Value	Weight	Value × Weight
74%	60	4440
64%	30	1920
48%	10	480
Total:	100	6840

Weighted average $= \dfrac{6840}{100} = 68.4\%$

$$\text{Weighted average} = \frac{\text{Total of (Value} \times \text{Weight)}}{\text{Total weight}} \quad \text{or} \quad \frac{\Sigma\,(\text{Value} \times \text{Weight})}{\Sigma\,\text{Weight}}$$

Example 2

A motoring correspondent gives two cars scores out of 10 on each of five attributes.

Calculate the weighted averages and say which is the better car.

Quality	Weight	Car A	Car B	Weight × Car A	Weight × Car B
Economy	2	8	4	16	8
Performance	3	5	9	15	27
Comfort	3	4	5	12	15
Reliability	4	9	6	36	24
Luggage capacity	1	9	7	9	7
Total	13			88	81

Weighted average for car A is $\frac{88}{13} = 6.77$

Weighted average for car B is $\frac{81}{13} = 6.23$ $\therefore$ Car A is the better car on the criteria given.

Exercise 74

1 In a motor treasure hunt, the time taken and the clues not found are weighted in the ratio 4:9. John takes 3.2 hours, and solves 29 of the 40 clues. Sarah takes 4.8 hours, and solves 34 clues.

What are their weighted averages?

2 In a Maths exam, the three components (written exam, oral, and coursework), are weighted 60, 10 and 30 respectively.

Jane scores 54%, 90% and 42%.
Helen scores 84%, 70% and 32%.

Calculate the weighted average mark for each girl, and state which girl obtains the better result.

3 A wine blender buys four wines:

- wine A costing 80p per litre
- wine B costing £2.20 per litre
- wine C costing £1.30 per litre
- wine D costing £4.20 per litre.

He blends them in the ratio 7 parts of A to 4 parts of B to 3 parts of C to 1 part of D.

What is the cost of the blend produced?

If the blender did not add wine D, but kept the other ratios the same, how much would the new blend cost?

4 John and Sarah decide that their budget is composed of five main items; housing, food, transport, holidays and clothes. These amounts are weighted 6:2:3:5:1. They estimate that these have increased in cost by 12% (housing), 9% (food), 8% (transport), 3% (holidays), and 21% (clothes).

What is the percentage increase in their expenditure?

19 Cumulative Frequency

The cumulative frequency is the total frequency up to a particular class boundary. It is a 'running total', and the cumulative frequency is found by adding each frequency to the sum of the previous ones.

19.1 The cumulative frequency curve (or ogive)

Virtually all cumulative frequency curves (or ogives) have an 'S' shape. How an ogive is built up will be seen in the following example.

Example

The marks obtained by 100 students in an examination were as shown below.

Marks	No. of students (frequency)
0–10	1
11–20	2
21–30	13
31–40	21
41–50	35
51–60	16
61–70	11
71–80	1

Draw the cumulative frequency curve for this data.

First, we construct a new table with an extra column. This keeps a running total of the frequencies in the second column.

Marks	No. of students (frequency)	Cumulative frequency
0–10	1	1
11–20	2	1 + 2 = 3
21–30	13	3 + 13 = 16
31–40	21	16 + 21 = 37
41–50	35	37 + 35 = 72
51–60	16	72 + 16 = 88
61–70	11	88 + 11 = 99
71–80	1	99 + 1 = 100

From the cumulative frequency column we can see that one student has 10 marks or less, three students have 20 marks or less, sixteen students have 30 marks or less, and so on.

Next we plot the cumulative frequencies against the upper class boundaries. We plot 3 (students) against 20 (marks) and 16 (students) against 30 (marks). The graph then looks like this:

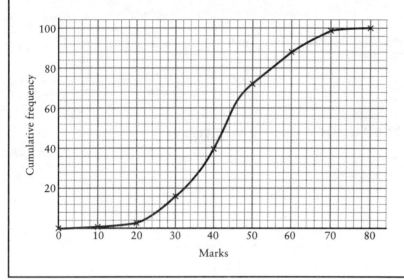

19.2   The median

The median mark is the mark obtained by the middle student. For example, if there are 100 students, the median student is the 50th (half-way point). (The strict definition would of course be the $50\frac{1}{2}$th student, but this accuracy cannot be obtained from a graph, and it is unnecessary at this stage.)

Suppose we wanted to find out the median mark in the example from section 19.1. We would look across from the 50 on the cumulative frequency axis to the curve, then read down vertically to the number of marks. In this case the median is 43 marks. This is illustrated below:

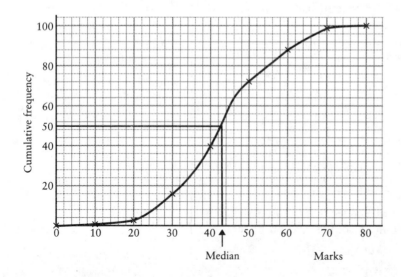

19.3 The interquartile range

(i) The **lower quartile** is the mark obtained by the student $\frac{1}{4}$ of the way along the distribution.

There are 100 students in the example on p. 131, so the 25th student is $\frac{1}{4}$ of the way up the cumulative frequency axis. From the graph, the 25th student has 35 marks, so the lower quartile is 35.

(ii) The **upper quartile** is the mark obtained by the student $\frac{3}{4}$ of the way up the cumulative frequency axis. This is the 75th student, whose mark is 52, so the upper quartile is 52.

The upper and lower quartiles are illustrated below:

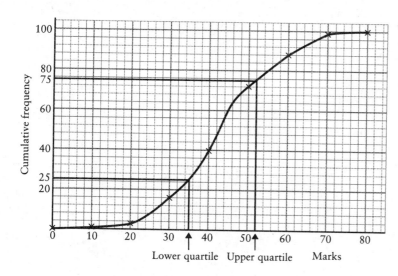

(iii) The **interquartile range** is the difference between the upper quartile and the lower quartile.
In our example, the interquartile range is:

$$52 - 36 = 16 \text{ marks.}$$

The **semi-interquartile range** is half the interquartile range, 8 in our example.

Note. Half the students have a mark between the lower and upper quartiles. Hence the interquartile range of 16 shows that half the marks obtained lie in an interval of 16. The semi-interquartile range of 8 shows that half the population lie (roughly) within 8 marks of the median mark.

*19.4 Percentiles

We can use the example on p. 131 to define a **percentile**. As the name suggests, percentiles divide the cumulative frequency distribution into 100 parts, just as the quartiles divide it into quarters. The 70th percentile, for example, is the highest mark obtained by 70% of the entry. From the cumulative frequency curve this mark is 49, so the 70th percentile is 49 marks (as shown overleaf).

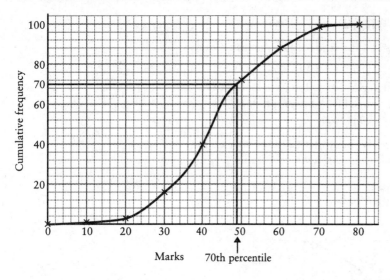

Marks   70th percentile

In questions 1 – 4, calculate the cumulative frequencies, and draw the cumulative frequency curve. Hence find:

a the median
b the lower quartile
c the upper quartile
d the interquartile range.

Part e is given separately for each question.

1   The times taken for students to complete two questions were recorded as:

Time (in minutes)	0–20	20–25	25–30	30–35	35–40	40–45	45–50
No. of students	3	12	25	49	21	7	3

   e Estimate the number of students who took more than 41 minutes.

2   The waist measurements of 100 people were recorded. The data found were:

Waist (cm)	40–50	50–60	60–70	70–80	80–90	90–100
Frequency	2	11	25	41	20	1

   e (i) Estimate the number of people with waist measurement less than 53 cm.
      (ii) Estimate the number of people with waist measurement more than 72 cm.

3   The annual gross wages of employees at a factory were:

Wage (£)	0–8000	8000–10000	10000–12000	12000–14000	14000–16000	16000–18000
Frequency	1	15	34	58	27	3

   e Estimate the 80th percentile.

4 The marks gained by 90 students on a test (out of 100) were:

Marks	0–20	21–40	41–60	61–80	81–100
Frequency	2	15	32	33	8

    e (i) What percentage of students passed the exam if the lowest pass mark was 37?

      (ii) Six students are given a distinction.
What was the lowest mark to obtain a distinction?

      (iii) Find the 90th percentile.

5 Potatoes are supplied to a greengrocer's shop in 50 kg bags. Each of the potatoes in one of these bags was weighed to the nearest gram, and the following table was drawn up:

Mass (g)	50–99	100–149	150–199	200–249	250–299	300–349
No. of potatoes	5	53	87	73	33	3

a (i) Complete the cumulative frequency for these potatoes.
   (ii) Draw the cumulative frequency graph.

b From your graph, estimate:
   (i) the median mass
   (ii) the number of potatoes weighing at least 225 g each.

c For a party, Harry requires 50 baking potatoes, each weighing at least 225 g. The greengrocer sells the potatoes without special selection.

   (i) Use your answer to b(ii) to estimate how many kilograms Harry will have to buy.

   (ii) Taking the mean mass of the 50 baking potatoes to be 260 g, estimate how many kilograms of potatoes he will have left over.

20 Dispersion

Averages are used to represent sets of data or to compare them, but an average on its own does not give sufficient information about the distributions.

Suppose we are comparing the climate of two places in Turkey. Town A has an average yearly temperature of about 12°C and town Z has an average yearly temperature of about 13°C. From this, we might suppose that the climates are similar and that town Z possibly lies south of town A.

In fact, town Z is Zonguldak, which is on the Black Sea coast, and town A is Ankara, which is further south in the interior of Turkey.

The monthly temperature distributions for the two towns are:

Month	J	F	M	A	M	J	J	A	S	O	N	D
Temperature in Ankara (°C)	−0.2	1.2	4.9	11.0	16.1	20.0	23.3	23.4	18.4	12.9	7.3	2.1
Temperature in Zonguldak (°C)	6.0	6.3	7.0	10.5	15.0	19.2	21.7	21.6	18.4	15.0	11.5	8.5

We can see that the climate in Ankara is more variable than in Zonguldak. There is a larger difference between summer and winter temperatures, whereas in Zonguldak there is a smaller spread of temperatures and a more equable climate.

We need a statistic which will measure this spread of values. The term we use to describe spread is **dispersion**, and there are several ways of measuring it.

20.1 The range

The simplest measure of spread is the **range**, which is the difference between the lowest and the highest values.

The range of temperatures for each town is

Ankara: Range = 23.4 − (−0.2) = 23.6°C
Zonguldak: Range = 21.7 − 6.0   = 15.7°C

The problems with using the range are: it only uses two values and so it can be distorted by a very high or low value; it cannot be used in further mathematical calculations.

Exercise 76

1   The numbers of cars on a hovercraft in one day were:

37, 28, 8, 5, 17, 39, 22, 10.

Find the range.

2   The goals scored by a netball team in a competition were:

10, 7, 5, 1, 12, 8, 6, 7.

Find the range.

3 At the end of a season, the bottom team in a division has 34 points. The range is 52 points. How many points has the top club?

4 The number of cars produced per year per employee by manufacturers in England were:

7.1, 12.1, 8.1, 4.7, 11.6, 9.4, 3.6, 12.5, 3.9.

Find the range.

20.2 The interquartile range

The problem of distortion by an extreme value can be overcome by calculating the range of the central part of the distribution. This is called the **interquartile range**, which we met in the previous chapter.

The interquartile and semi-interquartile ranges are always associated with the median.

To find the median and interquartile range, the values must first be written down in numerical order. The list is divided in half and then into quarters.

Ankara: −0.2 1.2 2.1 | 4.9 7.3 11.0 | 12.9 16.1 18.4 | 20.0 23.3 23.4
Zonguldak: 6.0 6.3 7.0 | 8.5 10.5 11.5 | 15.0 15.0 18.4 | 19.2 21.6 21.7

Half the values lie between the lower quartile and the upper quartile.

For Ankara:

$$\text{Median} = \frac{11.0 + 12.9}{2} = 12.0°C$$

Similarly:

Lower quartile = 3.5°C
Upper quartile = 19.2°C
Interquartile range = 19.2 − 3.5 = 15.7°C

For Zonguldak: Median = 13.3°C
Interquartile range = 11.1°C

*20.3 The standard deviation

Neither of the previous measures of spread take into account all the data. To do this, first we calculate the amount by which each value differs from the mean.

This is called the **deviation from the mean.**

For Ankara, the mean is 11.7°C.

The deviations from the mean are:
−11.9 −10.5 −6.8 −0.7 4.4 8.3 11.6 11.7 6.7 1.2 −4.4 −9.6

The total of the deviations is zero, which is not surprising because the mean is the average which 'evens out' all the differences.

Of course, we could just ignore the negative signs and find new totals. Dividing these totals would give us the average of the deviations and, in fact, this is a measure of dispersion called the mean deviation. However it is is not a very useful measure mathematically, so we solve the problem of the negative signs by *squaring* all the deviations.

The calculation for Ankara is:

Temperature (°C)	Deviations $(x - \bar{x})$	Squared deviations $(x - \bar{x})^2$
−0.2	−11.9	141.61
1.2	−10.5	110.25
4.9	−6.8	46.24
11.0	−0.7	0.49
16.1	4.4	19.36
20.0	8.3	68.89
23.3	11.6	134.56
23.4	11.7	136.89
18.4	6.7	44.89
12.9	1.2	1.44
7.3	−4.4	19.36
2.1	−9.6	92.16
Total		816.14

The total of the squared deviations $= 816.14 = \Sigma(x - \bar{x})^2$

Dividing by 12 gives:

$$\frac{816.14}{12} = 68.012 = \frac{\Sigma(x - \bar{x})^2}{n}$$

which is the average of the squared deviations and is called the **variance**.

However, the variance is in square measure.
To obtain an answer in degrees, we find the square root of the variance and this is called the **standard deviation**.

For Ankara:

Standard deviation $= \sqrt{68.012} = 8.25°C$

$$= \sqrt{\frac{\Sigma(x - \bar{x})^2}{n}}$$

Calculate the standard deviation of the temperatures for Zonguldak. You should find that:

Mean for Zonguldak $= 13.4°C$,
Standard deviation for Zonguldak $= 5.64°C$

This show, that although the mean temperatures are similar, the monthly temperatures for Zonguldak have a smaller spread than those for Ankara, i.e. they are less variable.

The standard deviation has the same advantages as the mean: it uses all the data and can be used in other mathematical calculations. It is the most widely used measure of dispersion and is always associated with the mean.

(If you have a scientific calculator, you will be able to use it to find means and standard deviations.)

1   In 10 games over a season, a rugby team scored the following number of points:

17, 41, 28, 21, 38, 45, 63, 8, 15, 34.

Find the mean number of points scored, and the standard deviation.

2   The ages of people at a youth club were as follows:

14, 15, 14, 16, 14, 14, 15, 17, 15, 18, 14, 15, 15, 16, 15, 16, 15, 14, 13, 15.

Find the mean age and the standard deviation.

3   The weight, in grams, of the marmalade in 10 jars was found to be:

456, 460, 455, 471, 453, 452, 458, 459, 465, 461.

Find the mean and the standard deviation.

There are 453 grams in 1 lb. Would the marmalade be labelled average weight 1 lb, or minimum weight 1 lb?

4   In a college the numbers of students in the various tutor sets were:

15, 13, 15, 13, 18, 14, 15, 16, 17, 21, 11, 18, 16, 17, 19, 15, 16, 19, 17, 15.

Find the mean and the standard deviation.

5   Dr Emmanuel has a choice of two routes when she drives to work. She may drive through the town centre, or take the longer route on the by-pass. She timed her journeys over a 14-day period and the results, in minutes, were:

Town route	16	20	25	26	29	17	28
By-pass route	20	25	22	25	22	26	21

a   Calculate the mean and standard deviation for each route.
b   Which route would you recommend Dr Emmanuel to use? Give a reason for your choice.

6   The speeds (in mph) of the first 12 cars passing a point on a dual carriageway after 1 pm were recorded on Friday and Saturday. The results were:

Friday: 72, 74, 81, 84, 65, 51, 52, 88, 74, 69, 75, 53
Saturday: 52, 54, 57, 72, 55, 81, 51, 56, 57, 59, 61, 51

Compare the means and standard deviations.

21 Probability

21.1 Introduction

Probability is a measure of how likely something is to occur. What occurs is an **outcome**.

This likelihood or probability can be represented on a sliding scale from the probability when an event is certain to occur to the probability when the event cannot occur.

Suppose twenty children in class 2X are studied. It is found that 19 do not wear glasses, and 1 does wear glasses. 11 of the 20 children are girls and 9 are boys.

Suppose a child is picked at random. The likelihood of the following events occurring fits into the following pattern.

Event	Likelihood
The child studied is in class 2X	Certainty
The child does not wear glasses	Highly probable
The child is a girl	Just over half
The child is a boy	Just under half
The child wears glasses	Highly unlikely
The child's age is over 70	Impossible

Since 19 out of 20 children do not wear glasses, the likelihood that a child selected at random does not wear glasses is 19 out of 20, or $\frac{19}{20}$. Therefore, the probability that a child selected at random is not wearing glasses is $\frac{19}{20}$.

If all 20 children are wearing shoes, the likelihood that a child selected at random is wearing shoes is 20 out of 20, i.e. $\frac{20}{20}$, which is 1. Therefore, the probability is 1.

None of the children have green hair. The probability that a child selected at random has green hair is 0 out of 20, i.e. $\frac{0}{20}$, which is 0.

If an event is bound to occur, then its probability is 1.

If an event cannot occur, then its probability is 0.

The probability of a child not wearing glasses is $\frac{19}{20}$.

The probability of a child wearing glasses is $\frac{1}{20}$.

The probability of a child wearing glasses or not wearing glasses is 1 (i.e. a certainty).

Note that $\frac{19}{20}$ (not glasses) $+ \frac{1}{20}$ (glasses) $= \frac{20}{20} = 1$ (certainty).

From this we can work out that:

The total probability for all possible outcomes is 1.

The events in the initial example occur with the following probabilities:

Event	Likelihood	Probability
The child studied is in class 2X	Certainty	1
The child does not wear glasses	Highly probable	$\frac{19}{20}$
The child is a girl	Just over half	$\frac{11}{20}$
The child is a boy	Just under half	$\frac{9}{20}$
The child wears glasses	Highly unlikely	$\frac{1}{20}$
The child's age is over 70	Impossible	0

21.2   Probability from theory and experiment

Probability can be found either by theory or by experiment.

The theory method relies on logical thought; the experimental method relies on the result of repetition of the event producing results which are taken to be typical.

Example

Find the probability of getting a head when tossing an unbiased coin.

Theory method
'Unbiased' means both events (head, tail) are equally likely to occur.

Heads and tails are the only outcomes,

∴ Probability of head + Probability of tail = 1

(since total of all possible outcomes is 1)

Probability of head = Probability of tail

∴ Probability of tail = $\frac{1}{2}$.

Experimental method
Toss an unbiased coin 100 times, and count the number of tails you obtain. You might obtain 49 tails.

∴ The experimental probability of getting a tail is $\frac{49}{100}$.
It is usual for the experimental probability to be close to the theoretical probability, but not to be exactly the same.

For the probability of a person being right-handed, there is no theory to use. This would be found by experiment. For example, ask 100 people and see how many are right-handed. The more people you ask, the more likely it is that your probability is accurate, but it must be appreciated that the probabilities found by experiment are not exact.

1 The probability that a street light is lit is $\frac{3}{8}$.
 What is the probability that it is not lit?

2 You toss a coin, marked with a head on both sides.
 What is the probability of getting a tail?

3 The probability of a plane being late is $\frac{2}{3}$.
 What is the probability of a plane not being late?

4 Jane has a bag containing 6 oranges and 4 apples.
 She selects a fruit from the bag at random.
 What is the probability that it is a pear?

21.3 Simple probabilities

When n different events are equally likely to occur in an experiment, then the probability of each event occurring is $\frac{1}{n}$.

The general formula when results are equally likely is:

$$\text{Probability of event} = \frac{\text{Number of results giving event}}{\text{Number of possible results}}$$

Example 1

An unbiased cubical die, marked 1 to 6, is thrown.
What is the probability of getting a 5?

There are 6 equally likely results.

∴ The probability of getting a 5 is $\frac{1}{6}$.

Example 2

A bag contains seven balls, identical in shape and size. Three balls are white and four are blue. One ball is selected from the bag at random.
What is the probability that the ball is white?

There are seven balls. The probability that any one ball is selected is $\frac{1}{7}$.

There are three white balls.

∴ The probability of selecting a white ball
$$= \frac{\text{No. of white balls}}{\text{Total no of balls}}$$
$$= \frac{3}{7}$$

1 Jane has a bag containing only 7 oranges and 3 apples. She selects a fruit from the bag at random. What is the probability that it is an apple?

2 A die, numbered 1 to 6, is thrown.
 What is the probability of getting a 6?

3 A die, numbered 1 to 6, is thrown.
 What is the probability of getting an even number?

4 From a pack of cards, one card is drawn.
 What is the probability that it is:

 a red   **b** a Jack   **c** the Queen of Spades?

5 What is the probability of picking an even number from the numbers 10 to 20 (inclusive)?

6 A bag contains 7 red balls, 8 blue balls, 3 yellow balls, and the remainder of the 20 balls are white.
 What is the probability of drawing:

 a a red ball,
 b a white ball?

7 Out of 20 lamp bulbs, three are faulty. You select a bulb at random.
 What is the probability that it works?

21.4 Possibility space

When more than one event takes place, you can either write down every possible result, or use the rules outlined on pp. 144 and 145.

The possibility space diagram is a means of identifying every possible result. It is used when two or more events occur, and where each of the outcomes is equally likely. Here is an example.

Example

Two unbiased dice, numbered 1 to 6, are thrown, and the score is found by adding the scores on the two dice.
What is the probability of obtaining a total of 10?

Second die	First die					
	1	2	3	4	5	6
1	2	3	4	5	6	7
2	3	4	5	6	7	8
3	4	5	6	7	8	9
4	5	6	7	8	9	10
5	6	7	8	9	10	11
6	7	8	9	10	11	12

The table shows each of the 36 possible results obtained by adding the score on each line. Each of these 36 is equally likely.

The total score of 10 occurs three times:
6 followed by 4; 5 followed by 5; 4 followed by 6.

$\therefore$ The probability of a total of 10 is $\frac{3}{36} = \frac{1}{12}$.

Exercise 80

1 Two dice, numbered from 1 to 6, are thrown, and the score is found by multiplying the scores on the two dice.
What is the probability of a score of:

   a 12, b 20, c 30?

2 Two coins are thrown. What is the probability of obtaining:
   a two heads, b one head and one tail?

3 Two tetrahedral dice, with faces marked 2, 4, 6 and 8, are thrown together, and the score is found by adding the scores on the two dice. What is the probability of obtaining a score of:

   a 10, b 12?

4 A die, numbered from 1 to 6, and a coin are thrown. If a head is obtained, the score is double that shown on the die. If a tail is obtained, the score is that shown on the die. What is the probability of obtaining a score of:

   a 6, b 8?

5 A card is selected from a pack of 52 cards, and a die, numbered 1 to 6, is thrown. If the card is a heart, the score is twice that shown on the die. If the card is a diamond, the score is 5. If the card is black, the score is that shown on the die. What is the probability of obtaining a score of:

   a 5, b 6?

21.5   Simple laws of addition and multiplication

Mutually exclusive events

Two events are said to be **mutually exclusive** when both cannot happen at the same time.

Suppose you toss a coin. The event 'obtain a head' and the event 'obtain a tail' are mutually exclusive, as you cannot obtain both a head and a tail at the same time.

Independent events

Two events are said to be **independent** when the result of one does not affect the result of the other.

Suppose you toss a coin *and* throw a die. Whether you obtain a head or a tail from the coin clearly does *not* affect the score on the die. These events are independent of each other.

Addition law for mutually exclusive events

If one event takes place, and you require the probability of A *or* B occurring (and *both* cannot occur), use the **addition law:**

Probability (A *or* B) = Probability (A) + Probability (B).

Example

From a pack of cards, one card is drawn. What is the probability that it is an Ace or a Jack?

$$\text{Probability (Ace)} = \frac{4}{52}$$

$$\text{Probability (Jack)} = \frac{4}{52}$$

$$\therefore \text{Probability (Ace or Jack)} = \frac{4}{52} + \frac{4}{52}$$

$$= \frac{8}{52}$$

$$= \frac{2}{13}$$

Multiplication law for independent events

If two independent events take place, and you require the probability that result A (from the first event), *and* result B (from the second event) both occur, use the **multiplication law**:

Probability (A *and* B) = Probability (A) × Probability (B)

Example

A die is thrown, and a coin is tossed. What is the probability of getting a 5 and a head?

$$\text{Probability (5 on die)} = \frac{1}{6}$$

$$\text{Probability (Head on coin)} = \frac{1}{2}$$

$$\therefore \text{Probability (5 and head)} = \frac{1}{2} \times \frac{1}{6}$$

$$= \frac{1}{12}$$

Exercise 81

1 A card is selected from a pack of 52 cards. What is the probability that the card selected is a 7 or a 10?

2 Mary drives along a road which has two sets of traffic lights. The probabilities of their being green are $\frac{3}{4}$ and $\frac{2}{5}$ respectively. What is the probability of Mary finding:

  a both sets green,
  b both sets not green?

3 The probability of Sway Wanderers winning a match is $\frac{2}{3}$. They play two matches. What is the probability that they win *both* matches?

4 A biased die has a probability of $\frac{1}{5}$ of getting a six. What is the probability that in two throws, you do not get a six?

5 A man throws a die numbered from 1 to 6. What is the probability that he gets a multiple of 5 or a multiple of 3?

6 John selects a pen at random from 6 blue pens and 5 black pens. He then selects a sheet of paper at random from 10 plain sheets and 20 lined sheets. What is the probability that he then writes with a blue pen on lined paper?

7 In a game of bingo there are 20 red, 3 black, 15 blue, and 12 white balls left. What is the probability that the next ball picked is either a red ball or a blue ball?

21.6 Tree diagrams

When two or more events take place, it is often simpler to find probabilities by means of diagrams that show all the possible events and their probabilities. A **tree diagram** is one such diagram.

Suppose a bag contains 5 blue and 7 white balls. A ball is selected at random, its colour noted, and it is then returned to the bag. The bag is shaken, and a second random selection is made.

There are two possible results of the first selection (either a blue ball or a white ball is obtained), which are represented by two branches on a tree diagram:

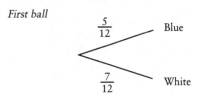

(Notice that the end-points of the branches are labelled with the event, and that the corresponding probabilities are written beside the branches.) Now consider the second selection:

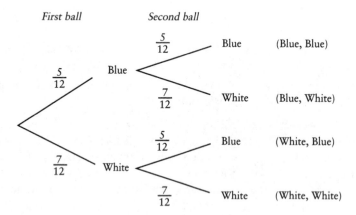

The above tree diagram shows both selections. The very top branches correspond to both selections being a blue ball, shown by (blue, blue). The probability of this is:

$$P = \frac{5}{12} \times \frac{5}{12} = \frac{25}{144}$$

This result is obtained by multiplying the probabilities on the two branches.

Similarly:

$$P\,(\text{white, blue}) \quad = \frac{5}{12} \times \frac{7}{12} = \frac{35}{144}$$

$$P\,(\text{blue, white}) \quad = \frac{7}{12} \times \frac{5}{12} = \frac{35}{144}$$

$$P\,(\text{white, white}) \; = \frac{7}{12} \times \frac{7}{12} = \frac{49}{144}$$

The completed tree diagram for this experiment is:

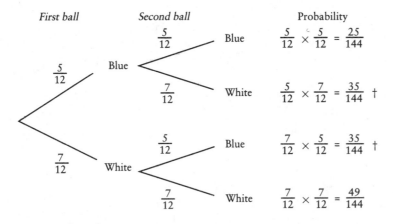

Notes

(i) The probability of any event represented by a consecutive branches of the tree is the product of the probabilities along those branches.

(ii) The probability of any event which can occur in two or more ways is the sum of the probabilities at the end of the relevant branches.

The two sets of branches marked with a dagger (†) represent picking one white ball and one blue ball.

$$\therefore \text{ The probability of one of each colour } = \frac{35}{144} + \frac{35}{144}$$

$$= \frac{70}{144}$$

$$= \frac{35}{72}$$

(For questions on this section, see Exercise 82, questions 1–5, p. 149.)

*Tree diagrams for 'without-replacement' questions

In the example above, the ball was replaced before the second selection was made. This made the probabilities for the colour of the second ball the same as those for the first ball.

When the ball is *not* replaced, it is necessary to amend each probability for the second ball. Here is an example.

Example

A bag contains 5 blue and 7 white balls. A ball is selected at random, its colour is noted, and it is *not* returned to the bag. What are the probabilities that:

a both are blue
b both are white
c one of each colour is chosen?

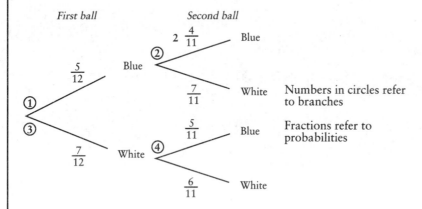

The bag contained 5 blue and 7 white balls. At branch **1**, a blue ball has been removed. This leaves only 4 blue and 7 white balls in the bag. Hence the probability of a blue ball on branch **2** is $\frac{4}{11}$. Similarly, in branch **3**, a white ball has been removed, leaving 5 blue and 6 white balls, and therefore the probability of a blue ball in branch **4** is $\frac{5}{11}$.

The required probabilities are:

a P (blue, blue) $= \dfrac{5}{12} \times \dfrac{4}{11} = \dfrac{5}{33}$

b P (white, white) $= \dfrac{7}{12} \times \dfrac{6}{11} = \dfrac{7}{22}$

c P (one of each)
$$= \left(\dfrac{5}{12} \times \dfrac{7}{11}\right) + \left(\dfrac{7}{12} \times \dfrac{5}{11}\right)$$
$$= \dfrac{35}{132} + \dfrac{35}{132}$$
$$= \dfrac{70}{132}$$
$$= \dfrac{35}{66}$$

Exercise 82

1 A bag contains 6 blue discs and 8 red discs. A disc is withdrawn, its colour is noted, and then it is replaced. A second disc is withdrawn and its colour is noted. What is the probability that the two discs are:

  a both blue
  b both red
  c one of each colour?

2 The probability of Mark oversleeping is $\frac{1}{3}$. If he oversleeps, the probability that he does not eat breakfast is $\frac{4}{5}$.
If he does not oversleep, the probability that he has breakfast is $\frac{3}{4}$.
Find the probability that Mark has breakfast.

3 Victoria travels to college either by bus (probability $\frac{3}{5}$), or by train. If she travels by bus, the probability that she is on time is $\frac{3}{4}$, but if she travels by train, the probability that she is on time is $\frac{1}{5}$.
What is the probability that Victoria is on time?

4 Two dice numbered 1 to 6 are thrown. What is the probability of obtaining:

  a two sixes
  b exactly one six?

5 A bag contains 5 white, 2 blue and 3 yellow balls. John selects a ball, notes its colour and replaces the ball. He withdraws a second ball, and notes its colour.
What is the probability that John selects:

  a two white balls
  b two balls of different colours?

*6 A bag contains 8 white and 6 black discs. Two discs are taken out and their colour noted.

What is the probability that the two discs are:

  a both white
  b of different colours?

*7 On Mothering Sunday Mrs Myers is given a box of hand-made chocolates, half of which are dark chocolate and half of which are milk chocolate. The box contains 10 chocolates. Mrs Myers eats two immediately, taking them out of the box at random.

  a Draw a tree diagram to show the possible outcomes of the type of chocolate eaten.
  b What is the probability that Mrs Myers eats two milk chocolates?
  c What is the probability that Mrs Myers eats one of each kind of chocolate?
  d Mrs Myers eats a third chocolate.
  What is the probability that she has then eaten three dark chocolates?

22 *Basic Numeracy*

Even in prehistoric times the basis of arithmetic was being developed. The earliest number system was one, two, many. Different number systems were adopted as different cultures evolved, but, in time, the **decimal system** became the most widely used.

This system, which we all use today, is based on the number 10, probably because human beings have ten fingers which are convenient for counting.

The numbers below 10 are called **digits**.

22.1   Numbers

- **Natural numbers** are the counting numbers 1, 2, 3 . . . .
  When zero is added to the set we have the
- **positive integers**: 0, 1, 2, 3, ...
  The positive integers and negative integers gives the set of
- **integers** or **directed numbers**: −4, −3, −2, −1, 0, 1, 2, ...
  When we add **fractions**, e.g. $\frac{7}{10}$, $3\frac{1}{2}$, −1.25, we obtain the set of
- **rational numbers** which are numbers that can be written in the form:

$$\frac{a}{b}$$

- **Irrational numbers** are numbers which cannot be written as fractions, e.g. π, $\sqrt{2}$, $3^{\frac{1}{4}}$.

Other types of numbers are:

Even number: integer (whole number) divisible by 2.
  e.g. 2, 4, 6, . . . or −2, −4, −6, . . .

Odd number: integer not divisible by 2,
  e.g. 1, 3, 5, ... or −1, −3, −5, . . .

Factor: a number which divides exactly into a given number leaving no remainder, e.g. 4 is a factor of 12.

Multiple: a number which is formed by multiplying a given number by an integer is a multiple of the given number, e.g. 20 is a multiple of 2, 4, 5 and 10.

Prime number: a number which has precisely two factors, the number itself and 1, e.g. 2, 3, 5, 7, 11, . . .

Square number: a number which is the result of multiplying a number by itself, e.g. 1, 4, 9, 16, ...

The **four rules of arithmetic** are:

Addition: the answer to which is called the **sum**

Subtraction: the answer to which is called the **difference**

Multiplication: the answer to which is called the **product**

Division: the answer to which is called the **quotient**

CORE MATHEMATICS

Example

Describe how the following sequences of numbers are formed and write down the next two numbers in each sequence:

a 243, 81, 27, 9, –, –.

b 1, 2, 4, 7, 11, –, –.

a Each term in the sequence is three times smaller than the previous term e.g. $81 = 243 \div 3$ and $27 = 81 \div 3$.

The next two terms are $9 \div 3 = 3$
and $3 \div 3 = 1$

b To each term, add the order number of the term to produce the next term.

To term no. 1 add 1: $1 + 1 = 2$
To term no. 2 add 2: $2 + 2 = 4$
To term no. 3 add 3: $4 + 3 = 7$

The next two terms are therefore:

$$11 + 5 = 16$$
$$\text{and } 16 + 6 = 22$$

Exercise 83

1 From the set of numbers, 4, 17, 24, 41, 49, write down:

  a a multiple of 8

  b two square numbers

  c a prime number

  d a factor of 8

  e a number which is the sum of two numbers in the set.

2 Write down the next two numbers in each sequence and describe how the sequence is formed:

  a $-3, 9, -27, 81, -, -$

  b 29, 24, 19, 14, –, –

  c 1, 8, 27, 64, –, –

  d 2, 5, 10, 17, –, –

  e 1, 1, 2, 3, 5, 8, –, –.

3 Find the sum of the first four odd numbers.

4 **a** Find the product of 100 and 6013.

  b Write your answer to **a** in words.

5 What is the remainder when 1538 is divided by 39?

6 **a** Which number between 1 and 25 has the largest number of factors?

  b Which number between 1 and 25 has the smallest number of factors?

7 What is the difference between five thousand and twenty two and eight thousand, two hundred and five?
Give your answer in words.

8 Write down the nearest integer to each of the following numbers:

  a -3.75 **b** $5\frac{1}{3}$ **c** $\sqrt{17}$ **d** π **e** $(1.92)^2$

9 From the set of numbers, $\sqrt{16}, \sqrt{19}, \sqrt{7}, \sqrt{49}, \sqrt{81}$, write down any which are:

   a irrational numbers

   b square numbers

   c even numbers

   d odd numbers

   e prime numbers.

   A calculator should *not* be used in the following questions:

10 Add 17, 33 and 12.

11 Write down the seventh prime number.

12 Subtract 583 from 1000.

13 From the set of numbers, $-8, 4, \sqrt{5}, \pi, 19$, write down those which are:

   a integers

   b natural numbers

   c irrational numbers

   d prime numbers

   e multiples of 2.

14 The attendance at a football match was eighteen thousand and forty three. Write this number in figures.

15 Write down the next two numbers in the sequence
   $1, 3, 6, 10, 15, -, -$.
   (The numbers in this sequence are called the triangular numbers.)

16 Write the number 130 402 in words.

17 List, in ascending order, all the factors of 24.

18 From the set of numbers, $2, \sqrt{16}, 9, \sqrt{101}, 15, 16, 19, 45$, write down:

   a two prime numbers

   b two square numbers

   c an irrational number

   d a factor of 30

   e a multiple of 9.

19 List the numbers $(2.9)^2, 9.34, 3^2, \sqrt{63}, \sqrt{65}$, in descending order of size.

20 Find:
   a 1.05×10 f 0.0356×0
   b $1.05 \div 10$ g 45.7×100
   c 0.003×10 h $45.7 \div 0$
   d $0.12 \div 20$ i 0.6×200
   e 67.03×1000

22.2 Directed numbers

The negative sign has two distinct uses in mathematics:

 (i) as a **subtraction** operation, e.g. $6 - 4 = 2$,
(ii) as a **direction** symbol, e.g. $-7°C$.

If we wish to show a temperature which is 7°C *below* zero, we can write $^-7°C$ *or* $-7°C$.

If a car travels 20 miles in one direction and then 15 miles in the reverse direction, we can write the distances travelled as $^+20$ **miles** and $^-15$ **miles**.

The numbers $^-7$, $^+20$ and $^-15$ are called **directed numbers**.

On your calculator you will see that there are two keys with negative symbols:

$\boxed{-}$ for subtraction

$\boxed{+/-}$ for direction.

Directed numbers can be represented on a horizontal or a vertical number line.

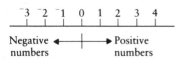

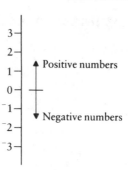

Addition and subtraction

Think of additions and subtractions of numbers as movements *up* and *down* the vertical number line.

Start at zero, move up 7 and then down 3.
Your position is now 4 *above* zero.

$$\text{i.e. } 7 + (-3) = +4$$

Subtracting two numbers is the same as finding the difference between them:

$$5 - (-3)$$

is the difference between being 3 *below* zero and being 5 *above* zero on the number line.

From 3 below to 5 above you must move up 8. Therefore

$$5 - (-3) = +8 \ or \ 5 - (-3) = 5 + 3 = 8$$

(Note that $-(-3) = +3$.)

$$-4 - (-2)$$

is the difference between being 2 below zero and being 4 below zero.

From 2 *below* to 4 *below* you must move DOWN 2, i.e.

$$-4 - (-2) = -2$$

or use the rule $- \times - = +$ to give

and $\qquad -4 - (-2) = -4 + 2 = -2$

Exercise 84

Find the answers to the following, without the aid of a calculator:

1 **a** $-5 + 7$ **c** $-2 + -8$ **e** $-21 + 27$ **g** $-19 + 19$

   b $3 + (-7)$ **d** $16 + (-12)$ **f** $35 + 5$ **h** $-35 + (-14)$

2 **a** $7 - 8$ **c** $-9 - 4$ **e** $-10 - (-7)$ **g** $25 - (-25)$

   b $12 - (-5)$ **d** $11 - (-9)$ **f** $-15 - (-15)$ **h** $100 - 64.$

For the following question you may use a calculator:

3 **a** $306 - (-219) + (-513)$ **d** $-697 + (-518) - (-29)$

   b $1427 + (-472) - 318$ **e** $-6724 + 3410 + (-148)$

   c $5061 - (-725) + (-491)$ **f** $5803 - 4190 - (-153).$

Multiplication and division

Going *up* 2 three times would mean you were now 6 *above* zero. So

$$(+2) \times 3 = +6$$
$$\text{or} \quad 2 \times 3 = 6 \qquad \text{also} \quad 3 \times 2 = 6$$

Going *down* 2 three times would mean you were now 6 *below* zero. So

$$(-2) \times 3 = -6$$
$$\text{or} \quad -2 \times 3 = -6 \qquad \text{also} \quad 3 \times -2 = -6$$

But what is $(-2) \times (-3)$?

Remember that $\quad -(-6) = +6$
and $\qquad (-2) \times 3 = -6 = -(2 \times 3)$
$\therefore \qquad (-2) \times (-3) = -(2 \times -3) = -(-6) = +6$

i.e. the reverse of moving 6 *down* is moving 6 *up*.

The rules for multiplication and division are similar:

Two numbers with like signs give a positive answer.
Two numbers with unlike signs give a negative answer.

Example 1

Find the product of $-4 \times -7 \times -3$

$$(-4 \times -7) \times -3 = +28 \times -3$$
$$= -84$$

Example 2

Evaluate $(-12 \div 4) \times -6$

$$(-12 \div 4) \times -6 = -3 \times -6$$
$$= +18$$

Exercise 85

Do *not* use a calculator in the following questions:

1.
   a $\ -3 \times 6$

   b $\ 15 \div -5$

   c $\ -7 \times -4$

   d $\ -7 \div -4$

   e $\ 10 \times -3 \div 2$

   f $\ -3 \times -3 \times -3$

   g $\ 8 \div -\frac{1}{2} \div -2$

   h $\ (-3 + -6) \times -4$

   i $\ -7 \div (3 - (-4))$

   j $\ \dfrac{20 \times -6}{-4 \times -3}$

You *may* use a calculator for the following questions:

2.
   a $\ 42 \div (-8)$

   b $\ (-47) \times (-73)$

   c $\ (-26) \times 15$

   d $\ \dfrac{252 \times (-20)}{(-30)}$

   e $\ \dfrac{(-16) \times (-27)}{(-36) \times (-10)}$

   f $\ \dfrac{(-19) + (-17)}{46 - (-26)}$

3. Given that $X = \dfrac{-2Y + 3Z}{3Y - 2Z}$, find X when

   a $\ Y = 6, Z = 3$

   b $\ Y = -3, Z = \frac{1}{2}$

   c $\ Y = 6, Z = -1$.

Order of operations

The order in which arithmetic operations are performed is as follows:

Brackets
Of
Division
Multiplication
Addition
Subtraction

(Taking the first letter of each word gives the **BODMAS** rule.)

Example 1

Evaluate $[5 - (-3)] \times [4 + (-1)]$

$$[5 - (-3)] \times [4 + (-1)] = 8 \times 3$$
$$= 24$$

Example 2

Evaluate $5 - (-3) \times 4 + (-1)$

$$5 - (-3) \times 4 + (-1) = 5 - (-12) + (-1)$$
$$= 5 + 12 - 1$$
$$= 16$$

Example 3

Evaluate $\frac{1}{2}$ of $18 - 7$

$$\frac{1}{2} \text{ of } 18 - 7 = 9 - 7$$
$$= 2$$

Example 4

Evaluate $\frac{1}{2}$ of $(18 - 7)$

$$\frac{1}{2} \text{ of } (18 - 7) = \frac{1}{2} \text{ of } 11$$
$$= 5\frac{1}{2}$$

Exercise 86

1 Use the **BODMAS** rule to evaluate the following:

   a $-4(2 - 6) + (-6) - (-3)$

   b $-3(16 - 12) - (-1)(12 - 16)$

   c $\frac{1}{2}$ of $(-35 - 21) - (-72)$

   d $12 - (7 - 3) - 4 \times 6 + 2.$

2 The formula for converting degrees Fahrenheit (°F) to degrees Celsius (°C) is

$$°C = \frac{5}{9}(°F - 32)$$

Calculate the number of degrees Celsius (to 1 decimal point) which is equivalent to:

   a 70°F   b 28°F   c ⁻4°F   d ⁻10°F

3 Given that $P = -Q(R - 2S)$, find P when

   a $Q = 4, R = 3, S = 6$

   b $Q = -10, R = -7.5, S = 2.5$

   c $Q = 8.3, R = -4.6, S = -2.9.$

22.3   Powers, roots, and reciprocals

Powers

Patterns of dots like this

1   4   9   16

form squares, and the numbers 1, 4, 9, 16 . . . are called the **square numbers**.

The square numbers are also formed by finding the product of an integer with itself:

$$1 \times 1 = 1$$
$$2 \times 2 = 4$$
$$3 \times 3 = 9$$
$$4 \times 4 = 16, \text{ etc.}$$

The sums of odd numbers also produce the square numbers:

1	= 1
1 + 3	= 4
1 + 3 + 5	= 9
1 + 3 + 5 + 7	= 16
1 + 3 + 5 + 7 + 9	= 25 etc.

If any number is multiplied by itself, the product is called the square of the number. For example:

the square of 1.2 is 1.44
the square of $\sqrt{2}$ is 2.

The square of 3 or 3 squared = $3 \times 3 = 3^2$, and the 2 is called the **index**.

Similarly:

the cube of 4 = 4 cubed = $4 \times 4 \times 4 = 4^3$

and

the fourth power of 2 = $2 \times 2 \times 2 \times 2 = 2^4$

Roots

The square of 4 = $4 \times 4 = 16$
The number 4, in this example, is called the **square root** of 16, i.e. the number which was squared to give 16.

16 has another square root because $-4 \times -4 = 16$.

Hence the square roots of 16 are +4 and −4.

The symbol $\sqrt{}$ indicates the positive square root and hence $\sqrt{16} = +4$.

Similarly, a fourth root of a given number is the number which has a fourth power equal to the given number. For example:

the positive fourth root of 16 or $\sqrt[4]{16} = 2$
$(2^4 = 2 \times 2 \times 2 \times 2 = 16)$

Reciprocals

The **reciprocal** of a number is equal to the number one divided by that number. To find the reciprocal of a number we invert it. For example:

the reciprocal of $\frac{1}{5}$ is 5
the reciprocal of 3 or $\frac{3}{1}$ is $\frac{1}{3}$
the reciprocal of $\frac{2}{3}$ is $\frac{3}{2}$ or $1\frac{1}{2}$.

To find the reciprocal of a mixed number (e.g. $2\frac{1}{4}$) it must first be converted to an improper fraction:

$$2\frac{1}{4} = \frac{9}{4}$$

Therefore the reciprocal of $2\frac{1}{4} = \frac{4}{9}$.

Exercise 87

1   Write down the first 10 square numbers.

2   Write down the first 6 cube numbers.

3   Find the value of:

    a 2^5, **b** 3^6, **c** $(^-5)^2$, **d** $(\frac{1}{2})^3$, **e** $(^-4)^3$

4   Find the value of:

    a $2^3 \times 2^2$, **b** $3^4 \times 3^2 \times 3$, **c** $2^4 \div 2^2$, **d** $\dfrac{3^4 \times 3^2}{3 \times 3^3}$

5   Find the positive square roots of the following numbers:

    a 25, **b** 121, **c** 625, **d** 225, **e** 196

6   **a** Find the cube root of (i) 27, (ii) $^-$64, (iii) 729.

    b Find the fifth root of 32.

    c Find the fourth root of (i) 81, (ii) 625.

7   Find the reciprocals of:

    a $\frac{1}{6}$, **b** $\frac{3}{4}$, **c** 0.125, **d** $3\frac{3}{4}$, **e** 9.5.

8   **a** What is the reciprocal of $\dfrac{1}{\frac{3}{4}}$?

    b What is the value of $\dfrac{1}{\frac{3}{4}}$?

9   Write 25 as the sum of a sequence of odd numbers.

10   What is the reciprocal of:

    a 11^2, **b** $\sqrt{144}$, **c** $\dfrac{2^3}{3^2}$, **d** $\dfrac{1}{(-3)^3}$, **e** 1?

A calculator should *not* be used for the following questions.

11   What is the third power of 2?

12   What is 3 cubed?

13   Write down the ninth square number.

14   Write down the fifth cube number.

15   Find the value of

    a 2^5, **b** 3^4, **c** $(\frac{1}{4})^2$, **d** $(^-2)^3$, **e** $(0.5)^2$.

16   Write down the cube root of:

    a 8, **b** 27, **c** 216, **d** $\frac{1}{8}$, **e** $^-$64.

17   What is the fourth root of 81?

18   Find the reciprocals of the following numbers, as whole numbers or fractions:

    a 4, **b** $\frac{1}{9}$, **c** 0.5, **d** $3\frac{1}{2}$, **e** 8.5.

19   Evaluate $5^2 \times \sqrt{16}$.

20   Show that $\sqrt{(16 \times 9)} = \sqrt{16} \times \sqrt{9}$.

22.4　Factors and multiples

A **factor** is an integer which divides into a given number exactly, i.e. there is no remainder.

For example:

3 and 4 are factors of 12.

A **multiple** of a number is the number multiplied by any integer.

For example, multiples of 6 are 12, 18, 24, and so on.

12 is a multiple of 3 and of 4, since 3 and 4 are factors of 12.

> **Example 1**
>
> Find all the factors of 42 and list them in pairs, the product of which is 42.
>
> The factors of 42 are 1, 2, 3, 6, 7, 14, 21, 42.
>
> Listed in pairs they are $1 \times 42, 2 \times 21, 3 \times 14, 6 \times 7$.

> **Example 2**
>
> Which of the following numbers are multiples of 9?
>
> **a** 63　**b** 732　**c** 1945?
>
> **a** $63 \div 9 = 7$　　means that 63 is a multiple of 9.
>
> **b** $732 \div 9 = 81.3$　means that 732 is not a multiple of 9.
>
> **c** $1944 \div 9 = 216$　means that 1944 is a multiple of 9.

Exercise 88

1　Find all the factors of each of the following and list the factors in pairs, the product of which is the number itself:

　a 15,　**b** 27,　**c** 36,　**d** 64,　**e** 100.

2　**a** Write down the factors of (i) 48, (ii) 72.

　b List the factors which are common to 48 and 72.

3　Of which of the following numbers is 4 a factor?
34, 64, 144, 642, 116, 2620.

4　State which of 5, 10, 20, 25 are factors of the following:

　a 300,　**b** 1050,　**c** 470,　**d** 875,　**e** 2360.

5　Which of the following numbers are divisible by **a** 3,　**b** 6,　**c** 15?
125, 240, 87, 96, 146, 255, 1073.

Prime factors

A **prime factor** is a **factor** which is also a **prime number**.

Example 1

a What are the prime factors of 12?

b Write 12 as a product of prime factors.

a Factors of 12 are   1, 2, 3, 4, 6, 12.
   Prime factors of 12 are   1, 2, 3.

b As a product of prime factors,
$$12 = 1 \times 2 \times 2 \times 3$$
$$= 2^2 \times 3 \text{ (1 is not usually included)}.$$

Example 2

Find the prime factors of 360.

Method

Begin dividing by the lowest prime factor of the given number and continue dividing until the quotient is no longer divisible by this prime number.

Find the next lowest prime factor and continue dividing.

When the quotient is 1, all prime factors have been found.

Write your answer using index notation.

2	360
2	180
2	90
3	45
3	15
5	5
	1

$$360 = 2 \times 2 \times 2 \times 3 \times 3 \times 5$$
$$= 2^3 \times 3^2 \times 5$$

Example 3

a Find three common multiples for the set of numbers 3, 4, 6.

b Describe the set of common multiples.

a Three common multiples are 12, 24 and 36. (There are, of course, many other possible answers.)

b The common multiples are all multiples of 12.

---

1 Write the following as products of prime numbers:

   a 60, **b** 315, **c** 225, **d** 8820, **e** 1694

2 (i) List the prime factors of the following pairs of numbers.
   (ii) List the factors which each pair have in common.

   a 36 and 90   **c** 312 and 260   **e** 918 and 468

   b 168 and 72   **d** 154 and 220   **f** 735 and 1050.

3 **a** Find three common multiples for each of the following sets of numbers:

   (i) 7, 14,   (ii) 6, 15,   (iii) 4, 8, 12,   (iv) 2, 3, 5.

   b Describe the set of common multiples in each case.

22.5 Highest common factor and lowest common multiple

The **highest common factor** (HCF) is the largest number which will divide exactly into each of the given numbers.

The **lowest common multiple** (LCM) is the smallest number into which each of the given numbers will divide exactly.

Example 1

Find the highest common factor of 120 and 252.

2	120
2	60
2	30
3	15
5	5
	1

2	252
2	126
3	63
3	21
7	7
	1

$$120 = 2^3 \times 3 \times 5 \qquad 252 = 2^2 \times 3^2 \times 7$$

The factors which are common to both numbers are 2^2 and 3.

The HCF of 120 and 252 $= 2^2 \times 3$
$$= 12$$

Example 2

Find the lowest common multiple of 24 and 54.

2	24
2	12
2	6
3	3
	1

$24 = 2^3 \times 3$

2	54
3	27
3	9
3	3
	1

$54 = 2 \times 3^3$

A common multiple must include the highest power of each prime factor.

The LCM of 24 and 54 $= 2^3 \times 3^3$
$$= 216$$

Exercise 90

1 Find the HCF of:

  a 64, 96   **d** 84, 56, 98

  b 105, 75   **e** 345, 920

  c 196, 252 **f** 154, 330, 242

2 Find the LCM of:

  a 15, 25, 10   **d** 900, 4200

  b 16, 18   **e** 9, 21, 28

  c 52, 78   **f** 24, 16, 36

3 a Find the HCF of 156 and 216.

  b Hence write $\dfrac{156}{216}$ in its lowest terms.

4 a Find the HCF of 900 and 2925.

  b Hence write $\dfrac{900}{2925}$ in its lowest terms.

5 Fractions $\dfrac{5}{8}, \dfrac{7}{12}, \dfrac{4}{9}$ are to be added.

  Find the LCM of their denominators.

6 A yeti and its mate are walking through the snow. The yeti's footprints are 5 metres apart, its mate's footprints are 4 metres apart. If its mate steps in the yeti's first footprint:

  a how many steps will its mate take before the prints match again?

  b how many steps will the yeti take before the prints match again?

  c what is the distance between the first and third matching prints?

7 Two buoys have lights attached which flash at regular intervals. The light on the first buoy flashes every 8 minutes and that on the second buoy, every 12 minutes. At 1.00 am the lights flash simultaneously.
At what time will they next flash simultaneously?

8 The planet Athyr has three moons, Artemis, Khons and Soma. Artemis takes 18 days to orbit the planet, Khons takes 28 days and Soma takes 36 days.

  a If the three moons can be viewed in line on one particular evening, how many days will elapse before the moons are again in line?

  b How many orbits of the planet Athyr has each moon made in this time?

9 a A manufacturer sells plastic drinking cups which are packed in boxes.
One customer has a regular order for 3750 cups and another has a regular order for 7000 cups.
How many cups should the manufacturer pack in each box so that he can fulfil both orders with complete boxes?

  b How many boxes will each customer receive?

10 A farmer delivers boxes of six eggs which are packed into cartons. To prevent damage, each carton is completely filled with boxes of eggs. His customers require 108 boxes, 90 boxes or 144 boxes.
How many boxes should each carton be designed to hold so that he can deliver full cartons to all his customers?

23 Using a Calculator

23.1 Approximations

Most calculators display answers of up to 10 digits.

In most cases, this is too many digits, therefore, when using a calculator, it is necessary to give an *approximate* answer which contains fewer digits (but which has a sensible degree of accuracy).

There are two methods in common usage:
rounding to a number of **decimal places** (e.g. 2 decimal places or 2 d.p.) and rounding to a number of **significant figures** (e.g. 3 significant figures or 3 s.f.).

The rule for rounding to, for example, 2 decimal places is:
If the digit in the third decimal place is **5 or more, round up**, i.e. increase the digit in the second decimal place by 1.

If the digit in the third decimal place is **less than 5, round down**, i.e. the digit in the second place remains the same.

Example 1

Give the following numbers, from a calculator display, to 2 d.p.
7.92341,   25.675231,   0.06666. .,   0.9999999. .

Calculator display	Degree of accuracy required (2 d.p.)	Answer correct to 2 d.p.
7.92341	7.92|341	7.92
25.675231	25.67|5231	25.68
0.066666..	0.06|6666..	0.07
0.9999999..	0.99|99999..	1.00

The rules for significant figures are similar, but you need to take care with zeros.

Zeros at the beginning of a decimal number or at the end of an integer are not counted as significant figures, but must be included in the final result. All other zeros are significant.

For example,
70 631.9 given correct to 3 significant figures is 70 600.

The three significant figures are 7, 0 and 6. The last two zeros are not significant (i.e. do not count as fourth and fifth figures), but are essential so that 7 retains its value of 70 thousand and 6 its value of 6 hundred.

Example 2

Give the following numbers, from a calculator display to 3 s.f.
7.92341, 25.675231, 0.066666. .,24380., 0.999999. .

Calculator display	Degree of accuracy required (3 s.f.)	Answer correct to 3 s.f.
7.92341	7.92|341	7.92
25.675231	25.6|75231	25.7
0.066666..	0.0666|66..	0.0667
24380.	243|80.	24400
0.999999..	0.999|999..	1.00

In the last example, the digit 1 becomes the first significant figure, and the 2 zeros are the second and third figures.

■ **Exercise 91** ■

Give each of the following numbers to the degree of accuracy requested in brackets:

1 9.736   (3 s.f.)
2 0.362 18   (2 d.p.)
3 147.49   (1 d.p.)
4 28.613   (2 s.f.)
5 0.5252   (2 s.f.)

6 4.1983   (2 d.p.)
7 1245.4   (3 s.f.)
8 0.004 25   (3 d.p.)
9 273.6   (2 s.f.)
10 459.97314 (1 d.p.).

23.2 Estimation

The answer displayed on a calculator will be correct for the values you have entered, but a calculator cannot tell you if you have pressed the wrong key or entered your numbers in the wrong order.

Each number you enter into the calculator should be checked for accuracy and the final answer should be checked by comparing it with an *estimated* answer.

Example

Estimate the value of 31.41 × 79.6.

        31.41 is approximately 30
        79.6 is approximately 80

An estimated value is therefore 30 × 80 = 2400.

If the calculator displays, for example, 25002.36, a mistake has been made with the decimal point, and the answer should read 2500.236

Exercise 92

1 By rounding all numbers to 1 significant figure, find an estimated value of
   each calculation:

   a 52.2×67.4

   b 6143×0.0381

   c 607×1.86

   d $607 \div 1.86$

   e $48.2 \div 0.203$

   f $3784 \div 412$

   g $\dfrac{520.4 \times 8.065}{99.53}$

   h $\dfrac{807}{391.2 \times 0.38}$

2 Find an estimate for each of the following calculations, by choosing an
   appropriate approximation for each number:

   a $82.3 \div 9.1$

   b $0.364 \div 6.29$

   c $\dfrac{31.73 \times 6.282}{7.918}$

3 By finding an estimate of the answer, state which of the following
   calculations are obviously incorrect. (Do not use a calculator.)

   a $8.14 \times 49.6 = 403.74$

   b $23.79 \div 5.57 = 4.27$

   c $324 \div 196 \times 0.5 = 226$

   d $3.14 \times 9.46^2 = 882.35$

   e $23.79 \div 0.213 = 11.169$

   f $\dfrac{42.3 \times 3.97}{1635} = 10.27$

   g $\sqrt{1640} = 40.5$

   h $\sqrt{650} = 80.6$

   i $(0.038)^2 = 0.00144$

   j $(0.205)^3 = 0.0862$

23.3 Standard form

Standard form is used when dealing with very large or very small numbers.

In standard form the number is always written as a number between 1 and 10
multiplied by a power of 10, i.e. as

$$A \times 10^n \text{ where } 1 \leqslant A < 10 \text{ and } n \text{ is an integer.}$$

This form is also known as scientific notation.

Example 1

The velocity of light is $300\,000\,000$ metres per second.

Write this number in standard form.

(i) Write down A, the number between 1 and 10: 3.0

(ii) Count the number of places to the *right* which the decimal point
      must be moved to give the velocity of light: 8 places.

(iii) Write the number in standard form: 3.0×10^8

or

$$300\,000\,000 = 3.0 \times 100\,000\,000$$
$$= 3.0 \times 10^8$$

Example 2

0.000 000 000 000 000 000 000 000 001 67 kg is the mass of a hydrogen atom.

Write this number in standard form.

  (i) $A = 1.67$

 (ii) In this case the decimal point must be moved 27 places to the *left*, $n = -27$

(iii) In standard form, the mass $= 1.67 \times 10^{-27}$ kg

Example 3

Convert 3.15×10^4 to a 'normal' number.

$n = 4$ so the decimal point must be moved 4 places to the right:

$$3.15 \times 10^4 = 31\,500$$

or $3.15 \times 10^4 = 3.15 \times 10\,000$
$$= 31\,500$$

Example 4

Multiply (2.7×10^4) by (5×10^{-2})

$$(2.7 \times 10^4) \times (5 \times 10^{-2}) = (2.7 \times 5) \times (10^4 \times 10^{-2})$$
$$= 13.5 \times 10^{(4 + -2)}$$
$$= 13.5 \times 10^2$$
$$= 1.35 \times 10^1 \times 10^2$$
$$= 1.35 \times 10^3 \text{ (in standard form)}$$

or $(2.7 \times 10^4) \times (5 \times 10^{-2}) = (2.7 \times 10\,000) \times (5 \div 100)$
$$= 13.5 \times 100$$
$$= 1.35 \times 1000$$
$$= 1.35 \times 10^3$$

Using a calculator:

$$(2.7 \times 10^4) \times (5 \times 10^{-2}) = 2.7 \boxed{\text{EXP}} \; 4 \; \boxed{\text{X}} \; 5 \; \boxed{\text{EXP}} \; 2 \; \boxed{+/-}$$

(On some calculators the exponential button is $\boxed{\text{EE}}$.)

Calculator display is   1350

                                $= 1.35 \times 10^3$ (in standard form)

If it is possible to use your calculator in scientific mode, the same sequence of operations will produce the answer in standard form displaying

1.35   03 or possibly 1.35^{03}

which is then written as 1.35×10^3.

(Some calculators round calculations to 2 s.f. displaying the answer as 1.4^{03}, which is then written as 1.4×10^3.)

Example 5

Divide (2.7×10^4) by (5×10^{-2}).

$$(2.7 \times 10^4) \div (5 \times 10^{-2}) = (2.7 \div 5) \times (10^4 \div 10^{-2})$$
$$= 0.54 \times 10^{(4 - -2)}$$
$$= 0.54 \times 10^6$$
$$= 5.4 \times 10^5 \text{ (in standard form)}$$

Using your calculator gives

$$(2.7 \times 10^4) \div (5 \times 10^{-2}) = 2.7 \boxed{\text{EXP}} \, 4 \div 5 \boxed{\text{EXP}} \, 2 \boxed{+/-}$$
$$= 540\,000$$
$$= 5.4 \times 10^5 \text{ (in standard form)}$$

Working in the scientific mode will give the correct answer 5.4^{05}

which is conventionally written as 5.4×10^5.

Numbers in standard form can be added and subtracted, provided the power of 10 is the same in each number.

If the powers of 10 are *not* the same, convert each number to a normal number before adding or subtracting or convert to a common power of 10.

Example 6

Evaluate $(4.3 \times 10^4) - (8.7 \times 10^3)$

Method 1

$$(4.3 \times 10^4) - (8.7 \times 10^3) = 43\,000 - 8700$$
$$= 34\,300$$
$$= 3.43 \times 10^4$$

Method 2

$$(4.3 \times 10^4) - (8.7 \times 10^3) = (4.3 \times 10^4) - (0.87 \times 10^4)$$
$$= 3.43 \times 10^4$$

Exercise 93

1 Write the following numbers in standard form:

a 790 000

b 0.0046

c 31 300

d 0.000 094 1

e 100 000

f 0.000 282

g 15.7×1000

h $4700 \times 10\,000$

i $0.000 34 \times 100$

j $0.0027 \div 1000$

k $5000 \div 10\,000$

l $0.000 028 \div 200$

2 Write the following numbers in normal form:

a 7.53×10^3 **d** 8.37×10^{-1} **g** 3.192×10^6

b 2.4×10^2 **e** 4.51×10^{-2} **h** 9.74×10^{-4}

c 1.9×10^{-3} **f** 4.042×10^4 **i** 6.8×10^{-6}

3 Evaluate the following, giving your answers in standard form:

a $(3 \times 10^2) \times (2 \times 10^2)$ **f** $(3.6 \times 10^4) \div (9.0 \times 10^5)$

b $(1.4 \times 10^{-3}) \times (3.7 \times 10^3)$ **g** $(2.7 \times 10^{-2}) \times (1.6 \times 10^{-3})$

c $(8 \times 10^3) \div (2 \times 10^2)$ **h** $(3.3 \times 10^4) - (4.6 \times 10^3)$

d $(5.3 \times 10^3) + (8.2 \times 10^3)$ **i** $(4.5 \times 10^{-3}) - (5.0 \times 10^{-3})$

e $(4.6 \times 10^3) \times (5.9 \times 10^{-1})$ **j** $(1.32 \times 10^{-1}) \div (2.2 \times 10^{-3})$

4 The distance from the Earth to the Sun is about 9.29×10^7 miles and from the Earth to the Moon is about 2.38×10^5 miles.
How many times greater is the distance to the Sun than the distance to the Moon?

5 Light travels at a speed of 3.0×10^8 metres per second.

If the distance from the Earth to the Sun is approximately 1.5×10^8 km, how long does it take the Sun's light to reach the Earth?

23.4   The keys of a calculator

To use your calculator most effectively, you must become familiar with the keys and their functions.

The booklet that accompanies your calculator will tell you the order in which the keys are used for calculations.

For example, to find $\sqrt[4]{5}$ the keys required on most calculators are

$$\boxed{5} \quad \boxed{\text{INV}} \quad \boxed{x^y} \quad \boxed{4} \quad \boxed{=}$$

Other useful keys are

a the memory keys $\boxed{\text{Min}}$ $\boxed{\text{MR}}$ $\boxed{\text{M+}}$ $\boxed{\text{M-}}$
Can you use them correctly?

b brackets, e.g. $\dfrac{36 \times 14}{21 \times 4}$ is $36 \times 14 \div (21 \times 4)$

$\qquad$ or $36 \times 14 \div 21 \div 4$

$\qquad$ *not* $36 \times 14 \div 21 \times 4$

c The $\boxed{\text{AC}}$ key clears the calculator (except the memories) before beginning a new calculation.

d The $\boxed{C}$ or $\boxed{CE}$ key is used to correct an entry error (i.e. if the wrong key has been pressed).

The key clears the last entry made (either a figure or an operation), provided it is pressed immediately after the error has been made.

The correct entry can then be made and the sequence continued.

e.g. $6 + 3\boxed{C}2 =$ will produce the answer to $6 + 2$.

Exercise 94

1 $386.9 \div 32.87$ (to 1 d.p.)

2 $2756 \div 0.03741$ (to 3 s.f.)

3 $\sqrt{\dfrac{79.5}{10.9}}$ (to 3 s.f.)

4 0.000356×385.7 (to 3 d.p.)

5 $\sqrt[3]{0.5824}$ (to 3 d.p.)

6 $\dfrac{1}{0.0345}$ (to 3 s.f.)

7 $(4.63)^2$ (to 3 s.f.)

8 $\frac{1}{2}(246.3 + 1092.8 + 376.4 + 49.8)$

9 $(51.7 + 63.2 + 28.9) \div (71.6 + 104.2)$ (to 3 d.p.)

10 $(12.7)^2 + 2 \times 12.7 - 15.3$

11 $(1.54)^3 + (1.54)^2 + 1.54$ (to 3 d.p.)

12 Some articles of clothing were left in the school changing room. To find out what they were, evaluate

a $(41^2 - 2) \times (3 \times 10^{-3})$ **b** 5×103^2

(If you have not discovered the answer, turn your calculator upside down!)

13 Who bought a car for $£[(18 \times 4 + 1) \times 4 + 1] \times 12 + 1$? (Again, turn the calculator upside down!)

14 Evaluate each answer in turn, then use it, without approximation, in the next calculation:

a $6 - \dfrac{1}{4}$ **c** $6 - \dfrac{1}{b}$

b $6 - \dfrac{1}{a}$ **d** $6 - \dfrac{1}{c}$

Write down the answer to **d** only, giving your answer correct to 4 significant figures.

15 Evaluate each answer in turn, then use it, without approximation, in the next calculation:

a $\dfrac{1}{2}\left(2 + \dfrac{5}{2}\right)$ **b** $\dfrac{1}{2}\left(a + \dfrac{5}{a}\right)$ **c** $\dfrac{1}{2}\left(b + \dfrac{5}{b}\right)$

Write down the answer to **c** only, giving your answer correct to 4 significant figures.

16 Repeat question 15 replacing [2,5] with

(i) [1,3]   (ii) [3,10]

17 Evaluate $\sqrt{5}$, $\sqrt{3}$, $\sqrt{10}$, correct to 4 significant figures. Compare these answers with the answers to questions 15 and 16 and comment.

24 Fractions

24.1 Types of fraction

A fraction is a number which can be written as a ratio, with an integer divided by an integer, e.g. $\frac{7}{9}$ or $-\frac{2}{3}$.

If a shape is divided into a number of equal parts and some of those parts are then shaded, the shaded area can be written as a fraction of the whole.

 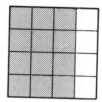

Fraction shaded $= \qquad \frac{3}{8} \qquad\qquad \frac{12}{16} \qquad\qquad \frac{2}{6}$

The **denominator** (lower integer) denotes the number of parts into which the shape was divided.

The **numerator** (upper integer) denotes the number of parts which have been shaded.

Fractions which have a smaller numerator than denominator are called **proper** fractions, e.g. $\frac{1}{2}$, $\frac{5}{12}$, $-\frac{28}{60}$

If the numerator is larger than the denominator, the fraction is an **improper** fraction, e.g. $\frac{12}{5}$, $-\frac{7}{2}$, $\frac{72}{18}$

A **mixed number** is a number composed of an integer and a proper fraction, e.g. $3\frac{1}{2}$, $-24\frac{3}{4}$, $4\frac{5}{6}$

24.2 Equivalent fractions

We could consider the larger square above to be divided into four columns instead of sixteen squares.

The shaded area is then $\frac{3}{4}$ of the whole. This means that $\frac{12}{16} = \frac{3}{4}$

$\frac{12}{16}$ and $\frac{3}{4}$ are called **equivalent fractions** because they have the same value.

Example 1

Complete $\dfrac{6}{7} = \dfrac{?}{28}$ to give equivalent fractions.

The 7 in the denominator must be increased four times to give 28.

The numerator must be treated similarly and be increased four times:

$$\frac{6}{7} = \frac{6 \times 4}{7 \times 4} = \frac{24}{28}$$

Example 2

Reduce $\dfrac{32}{72}$ to its lowest terms.

$$\frac{32}{72} = \frac{32 \div 8}{72 \div 8} = \frac{4}{9}$$

(4 and 9 have no common factor therefore the fraction is in its lowest terms.)

Example 3

Convert $13\frac{2}{5}$ to an improper fraction.

$$13\frac{2}{5} = 13 + \frac{2}{5}$$
$$= (13 \times \frac{5}{5}) + \frac{2}{5}$$
$$= \frac{65}{5} + \frac{2}{5}$$
$$= \frac{65 + 2}{5}$$
$$= \frac{67}{5}$$

Example 4

Convert $\dfrac{141}{22}$ to a mixed number.

$$22)\overline{141}$$
$$6 \text{ remainder } 9$$

$$\frac{141}{22} = 6\frac{9}{22}$$

Exercise 95

1 Complete each of the following to give equivalent fractions:

a $\dfrac{4}{5} = \dfrac{?}{10}$ e $\dfrac{21}{30} = \dfrac{?}{10}$

b $\dfrac{2}{3} = \dfrac{?}{12}$ f $\dfrac{16}{18} = \dfrac{8}{?}$

c $\dfrac{3}{4} = \dfrac{?}{24}$ g $\dfrac{65}{39} = \dfrac{?}{3}$

d $\dfrac{1}{4} = \dfrac{?}{36}$ h $\dfrac{20}{64} = \dfrac{5}{?}$

2 Reduce the following fractions to their lowest terms:

a $\dfrac{10}{15}$ b $\dfrac{18}{24}$ c $\dfrac{22}{99}$ d $\dfrac{21}{63}$

e $\dfrac{75}{100}$ f $\dfrac{40}{60}$ g $\dfrac{30}{65}$ h $\dfrac{64}{96}$

3 Convert the following mixed numbers to improper fractions:

a $2\frac{2}{9}$ c $4\frac{11}{12}$ e $8\frac{3}{4}$ g $7\frac{3}{20}$

b $5\frac{1}{6}$ d $11\frac{1}{10}$ f $12\frac{2}{3}$ h $20\frac{3}{7}$

4 Convert the following improper fractions to mixed numbers, in their lowest terms:

a $\dfrac{9}{7}$ c $\dfrac{30}{9}$ e $\dfrac{153}{11}$ g $\dfrac{132}{15}$

b $\dfrac{36}{5}$ d $\dfrac{61}{8}$ f $\dfrac{127}{12}$ h $\dfrac{94}{4}$

24.3 Operations involving fractions

Addition and subtraction

Only fractions *of the same type* can be added or subtracted, i.e. they must have the same denominator.

The method for addition or subtraction is:

(i) Find the LCM of the denominators.

(ii) Change each fraction to an equivalent fraction with the new denominator.

(iii) Add and/or subtract the fractions.

(iv) If the answer is an improper fraction, convert to a mixed number.

(v) Give the answer in its lowest terms.

Example 1

Evaluate $\dfrac{5}{8} - \dfrac{3}{4} + \dfrac{4}{5}$

The LCM is 40.

The sum in equivalent fractions is $\dfrac{25}{40} - \dfrac{30}{40} + \dfrac{32}{40} = \dfrac{25 - 30 + 32}{40}$

$$= \dfrac{27}{40} \text{ which is a fraction in its lowest terms.}$$

Example 2

Evaluate $1\dfrac{2}{3} + \dfrac{4}{5} - 2\dfrac{1}{15}$

$$1\dfrac{2}{3} + \dfrac{4}{5} - 2\dfrac{1}{15} = 1 + \dfrac{2}{3} + \dfrac{4}{5} - \left(2 + \dfrac{1}{15}\right)$$

$$= 1 + \dfrac{2}{3} + \dfrac{4}{5} - 2 - \dfrac{1}{15}$$

The LCM is 15.

The sum in equivalent fractions is $\dfrac{15}{15} + \dfrac{10}{15} + \dfrac{12}{15} - \dfrac{30}{15} - \dfrac{1}{15}$

$$= \dfrac{15 + 10 + 12 - 30 - 1}{15}$$

$$= \dfrac{6}{15} = \dfrac{2}{5} \text{ in its lowest terms}$$

Exercise 96

Evaluate the following:

1 $\dfrac{2}{5} + \dfrac{3}{4}$

2 $\dfrac{7}{8} - \dfrac{5}{6}$

3 $\dfrac{5}{12} + \dfrac{1}{4}$

4 $1\dfrac{1}{2} - \dfrac{3}{4}$

5 $4\dfrac{1}{6} - 3\dfrac{2}{3}$

6 $3\dfrac{4}{5} + 1\dfrac{1}{4}$

7 $3\dfrac{1}{2} - 4\dfrac{1}{4} + 2\dfrac{3}{4}$

8 $2\dfrac{4}{9} - 1\dfrac{2}{3} + \dfrac{5}{6}$

9 $4\dfrac{5}{9} - 1\dfrac{1}{3} - 1\dfrac{2}{3}$

10 $1\dfrac{1}{2} - 3\dfrac{1}{5} + 1\dfrac{3}{4}$

11 $3\dfrac{5}{8} - 1\dfrac{1}{4} - 1\dfrac{1}{2}$

12 $10\dfrac{1}{12} - 8\dfrac{5}{6}$

Multiplication and division

The rules for multiplication and division of fractions are very different from those for addition and subtraction.

The fractions do not have to have the same denominator, but they must not be mixed numbers.

Answers should be given in their lowest terms and as mixed numbers, if necessary.

To divide by a fraction, invert it and then multiply by the inverted fraction.

Example 1

Multiply $\dfrac{18}{25}$ by $\dfrac{5}{6}$

$$\dfrac{18}{25} \times \dfrac{5}{6} = \dfrac{18 \times 5}{25 \times 6} = \dfrac{90}{150} = \dfrac{3}{5} \qquad or \qquad \dfrac{18}{25} \times \dfrac{5}{6} = \dfrac{\overset{3}{\cancel{18}} \times \overset{1}{\cancel{5}}}{\underset{5}{\cancel{25}} \times \underset{1}{\cancel{6}}} = \dfrac{3 \times 1}{5 \times 1} = \dfrac{3}{5}$$

Example 2

Evaluate $1\dfrac{3}{4} \times 2\dfrac{2}{7}$

The mixed numbers are converted to improper fractions:

$$\dfrac{7}{4} \times \dfrac{16}{7} = \dfrac{7 \times 16}{4 \times 7} = \dfrac{112}{28} = 4 \qquad or \qquad \dfrac{7}{4} \times \dfrac{16}{7} = \dfrac{\overset{1}{\cancel{7}} \times \overset{4}{\cancel{16}}}{\underset{1}{\cancel{4}} \times \underset{1}{\cancel{7}}} = \dfrac{1 \times 4}{1 \times 1} = 4$$

Example 3

Evaluate $\dfrac{5}{9} \div \dfrac{2}{3}$

$$\frac{5}{9} \div \frac{2}{3} = \frac{5}{9} \times \frac{3}{2} = \frac{15}{18} = \frac{5}{6}$$

or

$$\frac{5}{9} \div \frac{2}{3} = \frac{5}{9} \times \frac{3}{2} = \frac{5 \times \overset{1}{\cancel{3}}}{\underset{3}{\cancel{9}} \times 2} = \frac{5 \times 1}{3 \times 2} = \frac{5}{6}$$

Example 4

Divide $3\dfrac{1}{6}$ by $\dfrac{2}{3}$

$$3\frac{1}{6} \div \frac{2}{3} = \frac{19}{6} \div \frac{2}{3} = \frac{19}{6} \times \frac{3}{2} = \frac{57}{12} = 4\frac{9}{12} = 4\frac{3}{4}$$

or

$$3\frac{1}{6} \div \frac{2}{3} = \frac{19}{6} \div \frac{2}{3} = \frac{19}{\underset{2}{\cancel{6}}} \times \frac{\overset{1}{\cancel{3}}}{2} = \frac{19 \times 1}{2 \times 2} = 4\frac{3}{4}$$

Exercise 97

Evaluate the following:

1 $\quad \dfrac{7}{8} \times \dfrac{4}{5}$

2 $\quad 1\dfrac{2}{3} \times \dfrac{4}{5}$

3 $\quad 6 \div \dfrac{1}{2}$

4 $\quad \dfrac{5}{6} \div \dfrac{3}{4}$

5 $\quad 2\dfrac{1}{5} \times 3\dfrac{1}{4}$

6 $\quad 5\dfrac{1}{7} \div 3$

7 $\quad 2\dfrac{1}{7} \times 1\dfrac{3}{5} \times 2\dfrac{1}{3}$

8 $\quad 1\dfrac{5}{9} \div 1\dfrac{1}{3} \times \dfrac{3}{4}$

9 $\quad 2\dfrac{6}{11} \div \dfrac{7}{11} \div 3\dfrac{2}{5}$

10 $\quad 4\dfrac{3}{8} \div \left(1\dfrac{5}{8} - \dfrac{3}{4}\right)$

11 $\quad 6\dfrac{2}{3} \times \left(\dfrac{1}{4} + \dfrac{1}{5}\right)$

24.4 The conversion of fractions to decimal fractions

The fraction $\frac{7}{8}$ may be stated as $7 \div 8$.

Using a calculator, $7 \div 8 = 0.875$, and this is the **decimal fraction** which is equivalent to $\frac{7}{8}$.

Example

Convert $3\frac{5}{6}$ to a decimal.

The integer part of the mixed number remains the same. Only the fractional part needs to be converted.

On the calculator $5 \div 6 = 0.833\,333\,3$ which is a recurring decimal.

$$3\frac{5}{6} = 3.8\dot{3} \text{ or } 3.83 \text{ (to 2 s.f.)}$$

All fractions convert to either a terminating or a recurring decimal. The dot above the 3 indicates that the 3 is a recurring decimal.

Exercise 98

1   Write down the shaded area as (i) a fraction, (ii) a decimal fraction of the whole area.

a 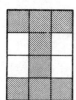   b   c

2   Convert the following fractions to decimals:

a $\frac{1}{10}$ c $\frac{3}{4}$ e $4\frac{21}{25}$ g $7\frac{4}{9}$ i $\frac{9}{11}$

b $\frac{1}{2}$ d $1\frac{9}{20}$ f $2\frac{5}{6}$ h $2\frac{8}{9}$ j $3\frac{1}{7}$

24.5   Percentages

A percentage is equivalent to a fraction with a denominator which is always 100.

They are often used for making comparisons.

For example, suppose a student received the following marks in a series of tests:

Mathematics	English	History
$\dfrac{17}{25}$	$\dfrac{32}{50}$	$\dfrac{30}{40}$

It is difficult to compare standards in each subject. The student gained most marks in English, but was this the best result?

In order to compare the marks, the fractions are all converted to equivalent fractions with the same denominator, 100.

$$\frac{17}{25} = \frac{68}{100} \qquad\qquad \frac{32}{50} = \frac{64}{100} \qquad\qquad \frac{30}{40} = \frac{75}{100}$$

and the test results are then seen as:

Mathematics	English	History
68%	64%	75%

Therefore the best result was for History.

Example 1

Convert a $\dfrac{3}{5}$,   b $\dfrac{9}{11}$ to percentages.

a $\dfrac{3}{5} = \dfrac{3}{5} \times \dfrac{20}{20} = \dfrac{60}{100} = 60\%$

or   To convert 5 to 100 multiply by $\dfrac{100}{5}$

To find the equivalent numerator multiply 3 by $\dfrac{100}{5}$:

$$3 \times \frac{100}{5} = \frac{3}{5} \times 100 = 60\%$$

b $\dfrac{9}{11}$ as a percentage $= \dfrac{9}{11} \times 100 = 81.82\%$ (to 2 d.p.)

To convert a fraction to a percentage multiply by 100.
To convert a percentage to a fraction, divide by 100.

Example 2

Convert 65% to a fraction in its lowest terms.

$$65\% = \frac{65}{100} = \frac{13}{20}$$

Exercise 99

1 Convert the following fractions to percentages:

a $\dfrac{1}{5}$ c $\dfrac{7}{10}$ e $\dfrac{2}{3}$ g $1\dfrac{3}{4}$

b $\dfrac{1}{8}$ d $\dfrac{13}{20}$ f $\dfrac{9}{25}$ h $2\dfrac{1}{2}$

2 Convert the following percentages to fractions:

a 60%   c 10%   e 15%   g $37\dfrac{1}{2}\%$

b 25%   d 85%   f 130%   h $33\dfrac{1}{3}\%$

3 Copy and complete the following table to give each quantity in its fractional, decimal and percentage form.

	Fraction	Decimal	Percentage
a	$\frac{3}{4}$		
b		0.5	
c	$\frac{1}{8}$		
d			$33\frac{1}{3}$
e		0.375	
f	$\frac{7}{10}$		
g			35
h		$0.\dot{6}$	
i	$\frac{3}{5}$		
j			62.5

4 A survey on teenage smoking found that 70% of girls of secondary school age tried smoking and that 36% of those who tried it became addicted to cigarettes.

In a secondary school with 580 female students:

a how many girls would you expect to find had tried smoking?

b how many female smokers would you expect to find?

5 a In a village with 250 on the electoral roll, 80% voted in the local elections.
How many voted?

b Of those who voted, $37\frac{1}{2}\%$ voted Labour.
How many votes did Labour receive?

c What fraction of those eligible to vote, voted Labour?

6 Linda and Mario are discussing their class.

a Linda says that 60% of the class are girls, and Mario says there are 12 boys in the class.
How many pupils are there in the class?

b Linda says that $\frac{2}{5}$ of the class eat school lunch and Mario says that $\frac{1}{3}$ of the class bring sandwiches.
If the rest of the class go home for lunch, how many go home?

7 Mr Day put $\frac{2}{3}$ of his pools winnings into shares and $\frac{1}{4}$ into a building society. He bought a new car costing £13 500 with the rest of the money.

a How much did he win?

b What percentage of his winnings did he invest?

25 Ratio and Proportion

25.1 Ratio

Nadia is 5 feet tall and her brother, Avrim, is 6 feet tall. The ratio of their heights is

$$5 : 6 \text{ or } \frac{5}{6}$$

i.e. Nadia's height is $\frac{5}{6}$ of Avrim's height

and Avrim's height is $\frac{6}{5}$ of Nadia's height.

Ratios can be written as fractions or, more usually, with a colon as $5 : 6$.

The equivalent fractions $\quad \frac{15}{20} = \frac{9}{12} = \frac{3}{4}$

can be written as equivalent ratios:

$$15 : 20 = 9 : 12 = 3 : 4.$$

Example

a Francis earns £180 per week and Gerry earns £120 per week. What, in its lowest terms, is the ratio of their weekly earnings?

b Francis and Gerry both receive a wage increase. Gerry now earns £140 per week.

If the ratio of their earnings remains the same, how much does Francis now earn?

a Francis' earnings : Gerry's earnings $= 180 : 120$
$$= 3 : 2$$

b
$$3 : 2 = ? : 140$$

$$\frac{\text{Francis' wage}}{\text{Gerry's wage}} = \frac{3}{2}$$

$$\text{Francis' wage} = \frac{3}{2} \times \text{Gerry's wage}$$

$$= \frac{3}{2} \times £140 = £210$$

Exercise 100

1 Write each pair of quantities as a ratio in its lowest terms:

   a 60 m, 40 m   f 15 cm, 10 cm

   b £2, 20p   g 750 g, 2 kg

   c 0.6 cm, 0.05 cm   h 39 litres, 26 litres

   d 0.4 m, 1.6 m   i 32 mph, 48 mph

   e 1 ft, 9 in

2 Complete the following ratios:

   a $3 : 4 = 6 : ?$ d $240 : 400 = ? : 1$

   b $18 : 9 = ? : 1$ e $20 : 1 = 64 : ?$

   c $? : 1 = 12 : 10$ f $1 : ? = 5 : 13$

3 A line is trisected (i.e. divided in the ratio 1 : 2). If the smaller length is 14 centimetres, how long is the line?

4 The circumference of a circle and its diameter have lengths which are approximately in the ratio 3 : 1.

   a What is the approximate length of the circumference of a circle with a diameter of 4.5 centimetres?

   b What is the approximate length of the diameter of a circle with a circumference of 24 inches?

5 E3 and E10 size soap packets contain powder with weight in the ratio 3 : 10.
   The E3 size contains 1.05 kilograms of powder.

   a What weight of powder does the E10 size contain?

   b What weight of powder does the E15 size contain?

6 Spring bulbs are planted in a border in the ratio of 3 yellow tulips to 2 pink tulips to 5 grape hyacinths.
   If 615 yellow tulip bulbs are planted, how many pink tulip bulbs and how many grape hyacinth bulbs are planted?

25.2   Division in a given ratio

Example

A bag contains 36 sweets in citrus flavours, orange, lemon and lime in the ratio 5 : 4 : 3 respectively.
Three children share the sweets. Paul takes all the orange sweets, Gail the lemon sweets and Ken the lime sweets. How many sweets does each child take?

Method:

(i) Find the total number of shares.

(ii) Find the amount of one share.

(iii) Find how much each receives.

(i) The total number of shares $= 5 + 4 + 3$ $= 12$

(ii) The amount of one share $= \dfrac{36}{12}$ $= 3$

(iii)   Paul's share $= 3 \times 5$ $= 15$
                  Gail's share $= 3 \times 4$ $= 12$
                  Ken's share $= 3 \times 3$ $= 9$

(Check: $15 + 12 + 9 = 36$.)

1 A medieval lord divided 5 acres of land among his serfs Alfric, Baldric and Cedric in the ratio 4 : 5 : 6 respectively.
How much land did Cedric receive?

2 A square of side 4 cm is divided into three parts whose areas are in the ratio 4 : 3 : 1.
What are the areas of the three parts?

3 Alison and Andrew share their pools winnings of £270 in the ratio of their 'stake'.
If Alison staked £3 and Andrew staked £2.40:

  a what was the ratio of the money staked?

  b how much of the £270 did they each receive?

4 a Whenever Grandma comes to visit she gives her grandchildren, Emily and David, money in the ratio of their ages.
When David is 10 years old, Emily is 6 years old.

    (i) What is the ratio of their ages, in its simplest form?

    (ii) If Grandma gives the children £12 to share, how much does each child receive?

  b The next time Grandma visits the children have both had a birthday. If they again share £12 between them, in the ratio of their ages, how much (to the nearest 1p) do they each receive?

5 The Harvey twins are baking shortbread for the cake stall at the school fete.

  Flour, butter and sugar are mixed in the ratio 3 : 2 : 1 and 12 oz of the mixture makes 15 biscuits.

  The twins decide to make $2\frac{1}{2}$ dozen shortbread biscuits.
How much of each ingredient will they require?

6 The Shang Dynasty in China was making bronze artifacts more than three thousand years ago.

  The bronze they used was an alloy of copper and zinc in the ratio (by weight) of 17 : 3.
What weights of copper and zinc were used to make a bronze bowl weighing 1.6 kilograms?

25.3 Direct proportion

If two quantities increase, or decrease, at the same rate, they are said to be in **direct proportion**.

For example, if you double your speed of walking you will travel twice as far in the same period of time.

Example

Mrs Wall usually buys 12 pints of milk each week and pays the milkman £3.72. In a week when she has visitors, she buys 3 extra pints. How much is her milk bill for that week?

    12 pints of milk cost £3.72

    1 pint of milk costs $\dfrac{£3.72}{12}$ (=31p)

    (12 + 3) pints of milk cost $\dfrac{£3.72}{12} \times 15$

                  = £4.65

Exercise 102

1 Find the cost of 6 lb of apples if 4 lb cost £2.12.

2 The baker's shop sells cheese biscuits for £1.52 per quarter pound.
How much would 7 oz cost? ($\frac{1}{4}$ lb = 4 oz).

3 Pic'n'Mix sweets cost 56p per quarter pound. How much would you be charged for 9 oz?

4 Anemone bulbs cost £1.15 for 25.
How much should be charged for 40 bulbs?

5 A shop stocks batteries in packs of five with a retail price of £2.99 per pack.
To try and sell the batteries more quickly, the shop offers them for sale in packs of three.
How much should the shop charge for the new packs?

6 It takes a hiker $3\frac{1}{2}$ hours to walk $10\frac{1}{2}$ miles.
How long will it take her to walk 12 miles, at the same speed?

7 Fruit cakes costing £1.68 per cake are each cut into ten slices to be sold at a charity tea party.
The organisers decide to make a profit of 52p on each cake.
How much will they charge for 4 slices of cake?

8 a A tin of custard is 11.5 cm high and contains 440g of custard. The manufacturer increases the height of the tin to 13.5 cm.
How many grams does a new tin hold?

b If the original tin cost 29p, how much should be charged for the new size?

c The manufacturer claims that the new tins hold 20% more.
Is he correct?

25.4   Inverse proportion

Michael Bates is slowly and carefully laying a concrete path when he suddenly notices a dark cloud approaching. Michael immediately doubles his rate of working so that he can finish the job in half the time (before it rains).

If an increase in one quantity causes a decrease, at the same rate, in another, the two quantities are **inversely** proportional.

Example

Alice, working at a rate of 20 questions per minute, completes a mental arithmetic test in 10 minutes.
Carol works at a rate of 25 questions per minute.
How long will it take Carol to complete the test?

At 20 questions per minute it takes 10 minutes

At 1 question per minute it takes 10 × 20 minutes

At 25 questions per minute it takes $\dfrac{10 \times 20}{25}$ minutes

$$= 8 \text{ minutes}$$

It takes Carol 8 minutes to finish the test.

1   Eight people share a prize and receive £125 each. How much would each have received if there had been only five prize winners?

2   It takes two combine harvesters $4\frac{1}{2}$ hours to clear a field of wheat.
How long would it take three combine harvesters to clear the same field?

3   If a job takes eight people six days to complete,

   a how long would it take five people to complete?

   b how many people would be needed if the job had to be completed in four days?

4   When eight checkout points are staffed at a supermarket, it takes an average of 54 minutes to deal with 100 customers.
If all 12 checkout points are staffed, how long does it take to deal with 100 customers?

5   It usually takes five farm workers six days to harvest the first strawberry crop. Because of wet weather, the crop needs to be harvested in two days. How many extra pickers should the farmer employ?

6   Five cousins expect to inherit £8500 each from their grandfather's estate. Unfortunately, one cousin suffers a fatal accident before he can inherit.
How much do the four surviving cousins each receive?

25.5   Measures of rate

Rate

In the previous two exercises (102 and 103), an intermediate step in the calculation was to find the **cost of one unit**: e.g., one pound, the time taken to make one article, the amount of work done in one day.

If the information given is already in this form, it is called a **rate**. For example:

A speed of 50 miles per hour is a rate of travel
French francs 9.93 per £1   is a rate of exchange
£423 per week   is a rate of pay
Time and a half   is an overtime rate

Given the basic rate, it is a simple calculation to find any other amount required.

Example 1

A firm pays an overtime rate of time and a half for work done on a Saturday.
The firm's basic rate of pay is £3.42 per hour.
How much is paid for $6\frac{1}{2}$ hours of work on a Saturday?

Basic rate of pay	= £3.42 per hour
Overtime rate of pay	= 1.5 × £3.42 per hour
	= £5.13 per hour
Saturday pay	= £5.13 × 6.5
	= £33.35

Example 2

An electric fire uses 18 units of electricity over a period of $7\frac{1}{2}$ hours.
What is the hourly rate of consumption of electricity?

$$\text{Hourly rate of consumption} = \frac{\text{Number of units used}}{\text{Time}}$$

$$= \frac{18 \text{ units}}{7.5 \text{ hours}}$$

$$= 2.4 \text{ units per hour}$$

Exercise 104

1   Carrots cost 18p per pound.
How much would $4\frac{1}{2}$ lb of carrots cost?

2   A man earns a basic wage of £10.70 per hour.
What is his basic weekly pay for a 35-hour week?

3   A car uses petrol at the rate of 36 miles per gallon.

   a How many miles can the car travel on $2\frac{1}{2}$ gallons of petrol?

   b How many gallons of petrol (to the nearest gallon) would be needed for a journey of 240 miles?

   c One gallon is approximately 4.55 litres. What is the car's rate of consumption in miles per litre?

4   The cost of bread increases from 56p per loaf to 60p per loaf.
What is the percentage rate of increase?

5   If the rate of inflation is 7%, by how much should a wage of £256 be increased to keep pace with inflation?

6   Lawn sand should be applied at a rate of 136 g per square metre.
How much lawn sand is required for a lawn 9.5 m × 7.5 m? (Give your answer to the nearest 0.1 kg.)

7   a The toll charged for a car travelling on a motorway was £33.60 for a journey of 420 kilometres.
What was the rate per kilometre?

   b Cars with trailers are charged double.
How much would it cost for a car and caravan to travel 264 kilometres?

Speed or velocity

Speed is the rate of travel.
Common measures of speed are miles per hour (mph); kilometres per hour (km/h); metres per second (m/s).

To calculate speed you need to know the distance travelled and the time taken:

$$\text{Speed} = \frac{\text{Distance travelled}}{\text{Time taken}}$$

Unless the speed is constant over the whole distance, this will be an **average speed** for the journey.

Example 1

A car travelled a distance of 306 miles in 5 hours 40 minutes.

What was the car's average speed for the journey?

$$40 \text{ minutes} = \frac{40}{60} \text{ hours} = 0.66\dot{6} \quad (\boxed{\text{Min}})$$

$$\text{Time taken} = 5.66\dot{6} \text{ hours} \quad (5 \ \boxed{\text{M+}})$$

$$\text{Speed} = \frac{\text{Distance travelled}}{\text{Time taken}}$$

$$= \frac{306 \text{ miles}}{\boxed{\text{MR}}}$$

$$= 54 \text{ mph}$$

Example 2

A car travelling at a speed of 65 kilometres per hour takes $2\frac{1}{2}$ hours for a journey.

What distance has the car covered in this time?

In 1 hour the car travels 65 km
In $2\frac{1}{2}$ hours the car travels $65 \times 2\frac{1}{2}$ km
$$= 162.5 \text{ km}$$

or Distance = Speed × Time = $65 \times 2\frac{1}{2}$ km
$$= 162.5 \text{ km}$$

Example 3

A family travelling with a car and caravan in France cover a journey of 238 kilometres travelling at an average speed of 70 kilometres per hour.

How long does the journey take?

To travel 70 km takes 1 hour

To travel 1 km takes $\frac{1}{70}$ hour

To travel 238 km takes $\frac{1}{70} \times 238$ hours

$$= 3.4 \text{ hours}$$

$$= 3 \text{ hour } 24 \text{ minutes}$$

or Time $= \dfrac{\text{Distance}}{\text{Speed}} = \dfrac{238 \text{ km}}{70 \text{ km/h}}$

$$= 3 \text{ hours } 24 \text{ minutes}$$

1 Find the average speed, in the most appropriate units, for each of the following:

   a a journey of 162 miles which took 3 hours

   b a 1500 m race which was won in a time of 2 min 54.4 s

   c a journey of 286 km which took $3\frac{1}{4}$ hours.

2 Find the distance covered:

   a by a car travelling at 85 km/h for 2.4 hours

   b by an aircraft travelling at 600 mph for 4 hours 36 minutes

   c by a manned satellite travelling at 7850 m/s for 12.5 s.

3 Find the time taken

   a to travel 186 km at an average speed of 62 km/ h

   b to travel 306 miles at an average speed of 40 mph

   c to travel 91.2 nautical miles at an average speed of 24 knots.
      (1 knot = 1 nautical mile per hour.)

4 Andy goes running every evening for 45 minutes. If he runs at a speed of 10 mph, how far does he run?

5 A cheetah is the fastest animal over short distances. It can travel at a speed of 96 km/h. At this speed, how much ground can a cheetah cover in 30 seconds?

6 a A race for small yachts has three 'legs' of distances 8 km, 6 km and 10 km. The average speed for the winning yacht was 6.2 kilometres per hour.
      How long did it take the winning yacht to complete the course?

   b The second yacht finished 8 minutes after the winner.
      What was its average speed?

Average speed

If a journey is divided into sections of different lengths travelled at different average speeds, then to calculate the average speed for the **whole** journey you need to know the **total** distance travelled and the **total** time taken.

$$\text{Average speed} = \frac{\text{Total distance travelled}}{\text{Total time taken}}$$

Example

A car travelled from Newbury to Aylesbury, a distance of 50 miles. Twenty miles of the journey was on dual carriageway and the car averaged a speed of 50 mph. The remainder of the journey was on single carriage roads and the average speed dropped to 36 mph.

What was the average speed for the whole journey?

Total distance travelled = 50 miles

To find the total time taken we need to calculate the time taken for each part of the journey.

$$\text{Time taken} = \frac{\text{Distance travelled}}{\text{Speed}}$$

On dual carriageway:

$$\text{Time taken} = \frac{20 \text{ miles}}{50 \text{ mph}}$$

$$= 0.4 \text{ hours}$$
$$= 24 \text{ minutes}$$

On single carriageway:

$$\text{Time taken} = \frac{30 \text{ miles}}{36 \text{ mph}}$$

$$= \frac{30}{36} \times 60 \text{ minutes}$$

$$= 50 \text{ minutes}$$

$$\text{Total time taken} = (50 + 24) = 74 \text{ minutes} = 1.23\dot{3} \text{ hours} \; (\boxed{\text{Min}})$$

$$\text{Average speed} = \frac{\text{Total distance travelled}}{\text{Total time taken}}$$

$$= \frac{50 \text{ miles}}{\boxed{\text{MR}}}$$

$$= 40.5 \text{ mph}$$

Exercise 106

1   Two friends, on holiday in Jersey, cycled from St Helier to St Aubin (a distance of 5 km) and back again. The ride to St Aubin took them 11 minutes and the return ride took 14 minutes.

What was their average speed for the whole journey?

2   **a** A car travelling along the M1 takes 40 minutes to travel between junction 18 and junction 21, a distance of 30 miles.
What is the car's average speed?

b Between junction 21 and junction 22, a distance of 13 miles, the car travels at a speed of 65 mph.
How long does it take?

c What is the car's average speed between junction 18 and junction 22?

3   Two friends are youth hostelling in the Northumbrian hills. At 9 o'clock one morning they set off from Rothbury to walk to Redesdale. They walk at a steady rate of $3\frac{1}{3}$ mph until noon, when they stop for lunch for 1 hour. They then

walk the remaining 14 miles and arrive at their destination at 5 pm.

a How far did they walk before lunch?

b At what average speed did they walk after lunch?

c What was their average speed for the whole journey?

4   A family, on holiday in Italy, travelled from Florence to Bologna at an average speed of 68 km/ h and from Bologna to Verona at an average speed of 81 km/h.
The journey from Florence to Bologna took them 1 hour 15 minutes and the total distance travelled was 220 km.

a How far is it from Florence to Bologna?

b How far is it from Bologna to Verona?

c How long did the journey from Bologna to Verona take?

d What was the average speed for the 220 km journey?

25.6   Models and map scales

Models and scale diagrams

A model is a replica, usually much smaller in size, of an original object. If the model is to be a faithful representation of the original, all the corresponding measurements of the original and the model must be in the same proportion.

The **scale** is the proportion by which each measurement has been reduced.

In order to fit a diagram onto paper, the measurements have to be 'scaled down'.
Once this has been done, you have a 'scale diagram'.

A scale is expressed either as a comparison between 2 lengths,
e.g. 1 cm : 2 m
or as a ratio
e.g. 1 : 200.

For example, a scale of 2 centimetres to 1 metre means that every length of 1 metre on the original is represented by a length of 2 centimetres on the model i.e. the measurements are all in the ratio of 1 : 50 and all the measurements on the model are $\frac{1}{50}$ of those of the original.

Example 1

The scale of a diagram is 1 : 5. AB in the scaled diagram is 4 cm. Find the original length AB.

$$\text{Original length} = 5 \times 4 \, \text{cm}$$
$$= 20 \, \text{cm}$$

Example 2

A model of an aircraft is made to a scale of 1 cm : 1.8 m.

a The length of the aircraft is 43 m. What is the length of the model?

b The width of the cargo hold on the model is 4.2 cm.
What is the width of the cargo hold on the aircraft?

a 1.8 m is represented by 1 cm

1 m is represented by $\dfrac{1}{1.8}$ cm

43 m is represented by $\dfrac{1}{1.8} \times 43$ cm

$$= 23.9 \, \text{cm}$$

b   1 cm represents 1.8 m
    4.2 cm represents 1.8×4.2 m
$$= 7.56 \, \text{m}$$

> **Example 3**
>
> The wing-span of a Boeing 747 is 59.6 metres. A model of the aircraft has a wing-span of 29.8 centimetres. To what scale was the model made?
>
> 29.8 cm represent 59.6 m
>
> 1 cm represents $\dfrac{59.6}{29.8}$ m = 2 m = 200 cm
>
> The scale is '1 cm represents 200 cm' or 1 : 200

Provided both lengths have the same units,

$$\text{Scale} = \frac{\text{Original length}}{\text{Scaled length}}$$

Exercise 107

1 The diagram below shows a rough plan (not drawn to scale) for a garden design.
The designer then draws an accurate scale diagram using a scale of 1 : 125.

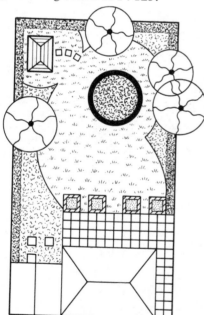

a What should the dimensions of the scale diagram be if the actual garden is 9.5 m wide by 20 m long?

b A patio is to be constructed which will measure 800 cm by 500 cm.
What dimensions should the patio have on the plan?

c The circular flower bed will have a diameter of 4 m.
What radius should the designer use to draw it on the plan?

2 A doll's house is made to a scale of 1 centimetre representing 30 centimetres.

a What are the lengths, in the doll's house, of the following:

(i) a 2.1 m high door

(ii) a carpet measuring 4.2 m × 3.6 m

(iii) a plate of diameter 24 cm?

b What would be the full-scale sizes of the following doll's house furnishings:

(i) a chair of height 3.1 cm

(ii) a candlestick 4.5 mm high

(iii) a picture 23 mm × 20 mm?

3 a When Alice ate a cake marked 'Eat me' she found herself enlarged on a scale of 1 to 1.9. If she was previously 4 ft 10 in tall, how tall was she after eating the cake?

b When Alice drank the contents of the bottle labelled 'Drink me', she shrank to a height of 10 inches.
If the miniature Alice had feet which were 3.6 cm long, how long were the feet on the enlarged Alice?

4 An 'N' gauge model railway is made to a scale of $\frac{1}{150}$. The roof of a model station is 136 mm long.

a How long would the actual station roof be?

b Taking 30 cm = 1 foot, calculate the actual length of the roof in feet.

5   Model railway scales are often quoted as the number of millimetres which represent 1 foot. The 00 gauge is a 4 mm scale, which means that 4 mm represents 1 foot.

a   The wheel on a train has a diameter of 4 ft 3 in. How many millimetres will the diameter of the wheel of a 00 gauge train be?

b   A 2 mm scale railway has passengers of height 11.5 mm. What is the actual height represented by these models:

   (i) in feet and inches

   (ii) in centimetres?

Map scales

Maps are two-dimensional representations of geographical areas. Map scales are usually given as a ratio or as a scaled line.

For example, a scale of 1 : 50 000 means that 1 cm on the map represents 50 000 cm actual distance, or 1 inch on the map represents 50 000 inches on the ground.

or   Scale:

Example 1

The scale of a map is 1 : 50 000.

a   What distance on the ground is represented by 6.3 cm on the map?

b   What distance on the map represents 23 km on the ground?

a   1 cm represents 50 000 cm

$$6.3 \text{ cm represents } 50\,000 \times 6.3 \text{ cm} = \frac{315\,000\,\text{m}}{100} = \frac{3150}{1000}\,\text{km}$$

$$= 3.15\,\text{km}$$

Actual distance = Map scale × Length on map

b $$50\,000\,\text{cm} = \frac{50\,000}{100}\,\text{m} = 500\,\text{m} = \frac{500}{1000}\,\text{km} = 0.5\,\text{km}$$

0.5 km is represented by 1 cm
1 km is represented by 2 cm
23 km is represented by 2 × 23 cm = 46 cm

or $$\textbf{Length on map} = \frac{\textbf{Actual length}}{\textbf{Map scale}}$$

$$= \frac{23 \times 1000 \times 100 \text{ cm}}{50\,000}$$

$$= 46\,\text{cm}$$

Example 2

The side of a field on the ground is 400 m. The size on the map is 5 cm. Find the scale of the map.

$$\text{Scale} = \frac{\text{Original length}}{\text{Scaled length}}$$

Converting all units to centimetres gives

$$\text{Scale} = \frac{400 \times 100}{5}$$

$$= 8000$$

The scale of the map is 1 : 8000.

Exercise 108

1 Convert the following actual lengths to map lengths using the scales given:

Actual length	Scale
a 5 metres	1 : 500
b 120 metres	1 : 2500
c 11 kilometres	1 : 10 000
d 14.4 kilometres	1 : 50 000

2 Convert the following map lengths to the original lengths using the scales given:

Map length	Scale
a 2 centimetres	1 : 1000
b 3.1 centimetres	1 : 4000
c 2.8 centimetres	1 : 6000
d 3 inches	1 : 3000
e 4.5 inches	1 : 2000

3 Calculate the scale of each of the following:

a a street map if 500 m is represented by 5 cm

b a small scale road atlas if 16 km is represented by 32 cm

c an Ordnance Survey map if 3.2 km is represented by 12.8 cm

d a town plan if 5.6 cm represents 700 m

e a large-scale road atlas if 30 mm represents 30 km.

4 An Ordnance Survey map has a scale of 1 : 25 000.

a How many metres, on the ground, are represented by 1 cm on the map?

b What is the distance on the ground, if the distance on the map is:

(i) 7.4 cm, (ii) 17.6 cm, (iii) 20 mm, (iv) 6.2 mm?

c What is the distance on the map if the distance on the ground is:

(i) 5 km, (ii) 1.8 km, (iii) 4.75 km, (iv) 10.6 km?

5 A road map has a scale of 1 : 50 000.

a How many centimetres on the map represent 1 kilometre on the ground?

b What is the distance on the ground if the distance on the map is:

(i) 4 cm, (ii) 48.8 cm, (iii) 16.3 cm, (iv) 36.9 cm?

c What is the distance on the map if the distance on the ground is:

(i) 27 km, (ii) 10.6 km, (iii) 58 m, (iv) 59.4 km?

6 On a map a reservoir covers an area of 0.6 cm². The scale of the map is 1 : 250 000.

a What distance, on the ground, is represented by 1 cm?

b What area, on the ground, is represented by 1 cm²?

c What is the actual surface area of the reservoir?

d A forest covers an area of 25 km². What area on the map does the forest cover?

7 The map below shows the centre of York (Scale: 1 cm represents 60 m)

a Mrs Hogan arrives at the station for a meeting at the careers office in Piccadilly:

(i) Find the shortest route from the station to Piccadilly.

(ii) What distance (to the nearest 50 m) will Mrs Hogan have to walk?

b By using a suitable measuring device, find the distance between the Castle and York Minster (to the nearest 50 m):
(i) as the crow flies, (ii) by road.

c Plan a guided walk around York City Centre starting and finishing at York Railway Station and taking in all the main places of interest.

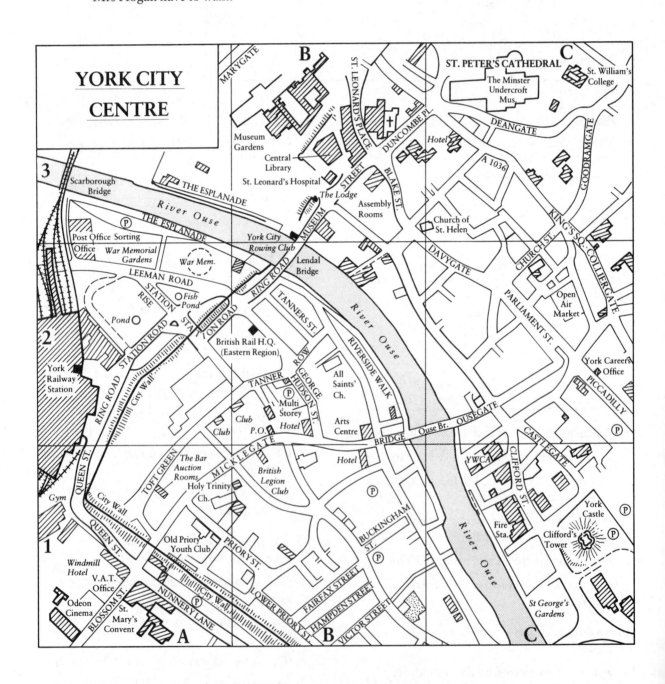

26 Measurement

26.1 Metric and imperial units

In 1971 the British monetary system was decimalised. We stopped using pounds, shillings and pence (£ s d) and started using pounds and 'new' pence (£ p). Since then, many more units have changed from imperial to metric units:

We can now buy petrol in litres not gallons.
We buy material in metres instead of yards.
The standard length of a ruler is 30 centimetres not 12 inches.
The weather forecast gives temperatures in degrees Celsius rather than in degrees Fahrenheit.

We do, however, still use imperial measures:

Distances are given in miles and speed restrictions in miles per hour.
Although most grocery items are labelled with weights in both grams and ounces, fruit, vegetables and sweets are bought in pounds and ounces.
Milk and beer are still sold in pints.
Carpets are often sold in feet and yards.

The advantage of the metric system over the imperial is the ease with which one unit can be converted to another.

Metric units

The metric system is a decimal system and units are converted by multiplying or dividing by powers of 10 (i.e. 10, 100, 1000).

Each prefix to a standard measure (eg metre, gram, litre) indicates the relative size:

kilo (k)   means 1000 ×   deci (d)   means $\frac{1}{10}$ ×

hecto (H) means 100 ×

deka (D) means 10 ×   centi (c) means $\frac{1}{100}$ ×

milli (m) means $\frac{1}{1000}$ ×

The most common metric units in everyday use are:

Length Metre (m)	Weight Gram (g)	Capacity Litre (l)
1 kilometre = 1000 metres	1 kg = 1000 g	1 litre = 100 cl
1 centimetre = 10 millimetres	1 g = 1000 mg	1 litre = 1000 ml
1 metre = 100 centimetres	1 tonne = 1000 kg	1 cl = 10 ml
1 metre = 1000 millimetres		

Example

a Convert 2.75 kg to g.

b Convert 6.3 ml to cl.

c Convert 56 cm to km.

a $1\,kg = 1000\,g$, so

$2.75\,kg = 1000 \times 2.75\,g = 2750\,g$

b $1\,ml = \dfrac{1}{10}\,cl$, so

$6.3\,ml = \dfrac{1}{10} \times 6.3\,cl = 0.63\,cl$

c $1\,cm = \dfrac{1}{100}\,m = \dfrac{1}{1000} \times \dfrac{1}{100}\,km$, so

$56\,cm = \dfrac{1}{1000} \times \dfrac{1}{100} \times 56$

$= 0.000\,56\,km$

Exercise 109

1 Change:

a 1232 m to km

b 0.032 m to mm

c 626 g to kg

d 0.731 litres to cl

e 1.62 cl to ml

f 12.7 km to m

g 59.1 ml to litres

h 3.4 t to kg

i 1.39×10^{-5} kg to mg

j 8.5×10^{7} mg to kg

2 The weight of sliced meat is marked on the packet as 0.23 kg.
What is the weight in grams?

3 A load of 2 tonnes of potatoes is to be split into 10 kg bags.
How many bags can be filled from this load?

4 The weight of a packet of crisps is 25 g.
How many milligrams is this?

5 Two pieces of wood of lengths 2.56 m and 37 cm are joined together.
What is the total length of wood

a in metres,

b in centimetres?

6 A quantity of sugar is weighed out on a scale and the reading is 250 g. A spoonful of sugar is removed and the new reading is 221.6 g.

a How many grams of sugar are removed?

b What is this weight in milligrams?

c Write your answer to b in standard form.

7 A bottle of lemonade is emptied into eight cups, each of which holds 150 ml.
How many litres of lemonade were in the bottle?

8 A bottle of weedkiller holds 1 litre of concentrate. To make up a solution, 5 capfuls are added to 2 litres of water. A capful is 15 ml.
How many litres of solution can be made from the bottle?

Imperial units

The most common imperial units in everyday use are:

Length	Weight	Capacity
1 mile = 1760 yards	1 ton = 20 hundredweight	1 gallon = 8 pints
1 yard = 3 feet (ft or ′)	(cwt)	1 pint = 20 fluid
1 foot = 12 inches (in or ″)	1 cwt = 112 pounds (lb)	ounces (fl oz)
	1 lb = 16 ounces (oz)	
	1 stone = 14 pounds	

Example

a Convert 76″ to feet and inches.

b Convert 6 lb 5 oz to ounces.

c Convert 18 pints to gallons.

a 12 in = 1 ft, so

$$76'' = \frac{76}{12} \text{ft} = 6\frac{4}{12} \text{ft} = 6' \, 4''$$

b 1 lb = 16 oz,   so

$$6 \text{ lb } 5 \text{ oz} = (6 \times 16) + 5 \text{ oz} = 101 \text{ oz}$$

c 8 pt = 1 gal,   so

$$18 \text{ pt} = \frac{18}{8} \text{gal} = 2.25 \text{ gal}$$

Exercise 110

1 Change:

    a 39 in to feet

    b 40 oz to lb and oz

    c $16\frac{1}{2}$ ft to yards

    d 2 ft 3 in to inches

    e 3 lb 7 oz to oz

    f 20 pints to gallons

    g $3\frac{1}{2}$ gallons to pints

    h 0.7 pint to fluid ounces

2 A milk churn holds 5 gallons of milk.
How many pint bottles of milk can be filled from the churn?

3 A recipe requires 15 fluid ounces of water, but the measuring jug is calibrated in pints.
How many pints of water should be measured?

4 Coal is delivered in hundredweight bags. A load of twenty-four bags is delivered to a house.
What is the weight of the load in tons?

5 A quantity of honey is weighed out in a bowl. The reading on the scale is 1 lb 5 oz. The bowl is known to weigh $12\frac{1}{4}$ oz.
How much does the honey weigh?

6 Nina measures her bedroom and finds that the length is 12′ 9″ and the breadth is 10′ 6″.
What is the perimeter of the room (i) in feet, (ii) in yards?

7 A grocery store divides a block of cheese weighing 4 lb $9\frac{1}{2}$ oz into twelve equal portions.
How many ounces does each portion weigh?

8 A bottle of concentrated orange juice holds 8 fluid ounces. To make up the juice for drinking, three times this amount of water is added.
If 1 gallon of diluted juice is required, how many bottles of concentrated juice should be used?

26.2 Conversion between metric and imperial units

It is often necessary, for comparison, to convert from imperial to metric units or vice versa.

If a rough comparison is all that is required, an approximate conversion factor can be used. For large quantities, or where a correct comparison is required, a conversion factor of the appropriate degree of accuracy should be used.

Approximate conversion		More accurate conversion	
5 miles	$\approx 8\,km$	1 mile	$= 1.609\,km$
1 yard	$\approx 1\,m$	1 yard	$= 0.914\,m$
1 foot	$\approx 30\,cm$	1 foot	$= 30.48\,cm$
1 inch	$\approx 2\frac{1}{2}\,cm$	1 inch	$= 2.54\,cm$
1 kg	$\approx 2\,lb$ or $2.2\,lb$	1 kg	$= 2.205\,lb$
1 litre	$\approx 1\frac{3}{4}$ pints	1 litre	$= 1.76\,pt$
		1 gallon	$= 4.55\,litres$

Example 1

The contents of a packet of sugar weigh $250\,g$.
What weight, in ounces, should be printed on the packet?

Merchandise must carry an accurate description, and therefore the weight given must be correct to a reasonable degree of accuracy.
The weight stated on a packet, tin, etc., is usually the minimum weight of the contents.

$$\text{The conversion used is } 1\,kg = 2.205\,lb$$
$$\therefore\ 1000\,g = 2.205 \times 16\,oz = 35.28\,oz$$
$$1\,g = 0.035\,28\,oz$$
$$250\,g = 0.035\,28 \times 250\,oz$$
$$= 8.82\,oz$$

The minimum weight of the sugar is $8.8\,oz$.

Example 2

A family, on a self-catering holiday in Europe, wish to buy the equivalent in weight of $3\,lb$ of potatoes and $\frac{1}{2}\,lb$ of butter.
How much of each item should they buy?

Exact weights are not required.

$$\text{The conversion used is } 2\,lb \approx 1\,kg$$
$$\text{or } 1\,lb \approx 0.5\,kg$$
$$\text{so } 3\,lb \approx 1.5\,kg$$
$$\text{and } \tfrac{1}{2}\,lb \approx 0.25\,kg = 250\,g$$

They should buy $1\frac{1}{2}\,kg$ of potatoes and $250\,g$ of butter.

Exercise 111

1 Vegetables are sold in tins of various sizes. What weight, in ounces to the nearest 0.1 oz, should be printed on a tin if the contents weigh:

   a 425 g   **b** 440 g   **c** 415 g?

2 A shop prepacks its groceries and labels the packets with both metric and imperial weights. For each of the following items, calculate the missing weight:

   a 1 lb of mince   **d** 0.74 lb of cheese

   b 150 g of cooked ham **e** 14.2 g of dried herbs

   c 5 lb of potatoes

3 A pair of kitchen scales is calibrated in both imperial and metric weights. Calculate the number of pounds and ounces (to the nearest ounce) which can be read alongside the metric weights of:

   a 1 kg   **b** 0.4 kg   **c** 1.5 kg   **d** 2.1 kg

4 **a** What is the equivalent of 1 litre of petrol in gallons?

   b A driver buys 20 litres of petrol. Approximately, how many gallons of petrol does he buy (correct to 1 d.p.)?

5 The specifications for a particular make of car give the petrol consumption as 40 miles per gallon. How many miles per litre is this?

6 A couple on a touring holiday in France cover a certain distance each day. Calculate the approximate number of miles which are equivalent to:

   a 120 km **b** 324 km **c** 429 km **d** 56 km

   e 370 km

7 Gary is shopping for shelving for his collection of CDs. He finds some which is 6 inches wide. His CDs are 14.1 cm wide. Is the shelving wide enough?

26.3 Perimeters of polygons

A **plane figure** is a two-dimensional shape which is bounded by lines called **sides**.

Here are some examples:

Triangle

Rhombus

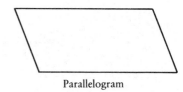
Parallelogram

Circle

Pentagon

Regular hexagon

Plane figures which are bounded by straight lines are called **polygons**.

A **regular polygon** is one which has all its sides the same length. The regular three-sided polygon (like the one above) is an **equilateral triangle**.

The **perimeter** of a plane figure is the total length of the sides.

Example

Find the perimeter of the shape given below:

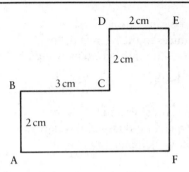

Perimeter = AB + BC + CD + DE + EF + FA
= (2 + 3 + 2 + 2 + 4 + 5) cm
= 18 cm

Exercise 112

1 Find the perimeter of each of the following:

a a rectangle of length 8 cm and width 3 cm

b a square of side 7 m

c a rhombus of side 6 inches

d a parallelogram with adjacent sides 7 cm and 9 cm

e a regular hexagon of side 7.3 cm.

2 Find the perimeter of each of the following shapes:

a

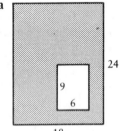

c

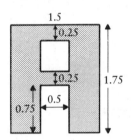

e

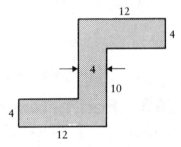

b

d

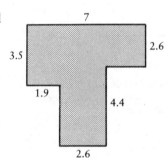

f

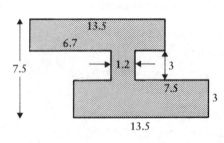

3 A mirror is cut to the shape of a regular octagon. The edges are then beaded for protection. What length of beading is required?

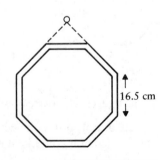

26.4 Area

Area is a measure of the surface covered by a given shape.

Example 1

Consider the shapes below and place them in ascending order according to their area.

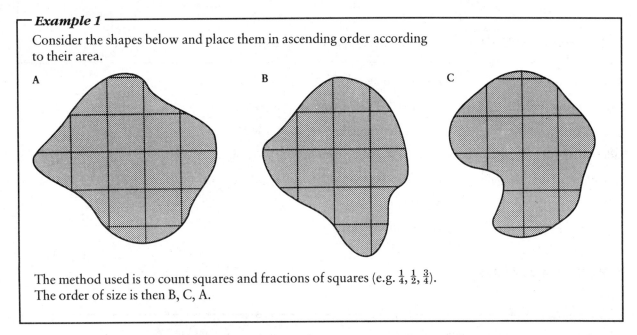

The method used is to count squares and fractions of squares (e.g. $\frac{1}{4}, \frac{1}{2}, \frac{3}{4}$).
The order of size is then B, C, A.

Figures which have straight sides (polygons) are much easier to compare.

Example 2

Consider the following polygons:

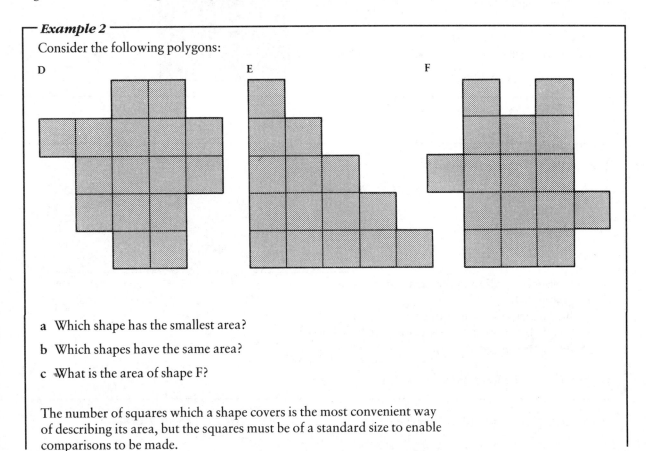

a Which shape has the smallest area?

b Which shapes have the same area?

c What is the area of shape F?

The number of squares which a shape covers is the most convenient way of describing its area, but the squares must be of a standard size to enable comparisons to be made.

The square used in the diagrams above is a 1 centimetre square.

The area of a 1 cm square = 1 square centimetre or $1\,cm^2$.

The area [ ] $= 1\,cm^2$.

The answers are: **a** E **b** D and E **c** 16 square centimetres

The most common measures of an area are based on:

side = 1 cm square,   area = 1 square centimetre = $1\,cm^2$

side = 1 m square,   area = 1 square metre   = $1\,m^2$

side = 1 mm square,   area = 1 square millimetre = $1\,mm^2$

The imperial units for area are, 1 square inch = $1\,in^2$

1 square foot = $1\,ft^2$

1 square yard = $1\,yd^2$

Area of a rectangle

Example 1

What is the area of the rectangle shown?
(Each square represents $1\,cm^2$.)

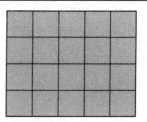

The rectangle covers 20 squares, each of area $1\,cm^2$.

∴ Area of rectangle = $20\,cm^2$

Example 2

What is the area of this rectangle?

3 cm

8 cm

It is not necessary to draw in the 1 cm squares.
It can be seen that

the number of squares fitting along the length = 8
the number of squares fitting along the breadth = 3
the number of squares required for the area   = 8×3
= 24

∴ Area of the rectangle is $24\,cm^2$

For a rectangle: **Area = Length × Breadth**

$$A = L \times B$$

For a square: Area = Length × Breadth

$$A = L^2$$

Some shapes, which are more complicated, can be split up into rectangles in order to find the area.

Example 3

Find the area of this shape.

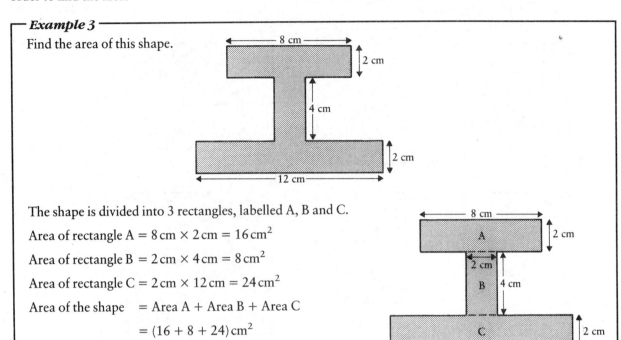

The shape is divided into 3 rectangles, labelled A, B and C.

Area of rectangle A = 8 cm × 2 cm = 16 cm^2

Area of rectangle B = 2 cm × 4 cm = 8 cm^2

Area of rectangle C = 2 cm × 12 cm = 24 cm^2

Area of the shape   = Area A + Area B + Area C

                = (16 + 8 + 24) cm^2

                = 48 cm^2.

In some cases it is quicker to subtract areas than to add.

Example 4

Find the area of this shape.

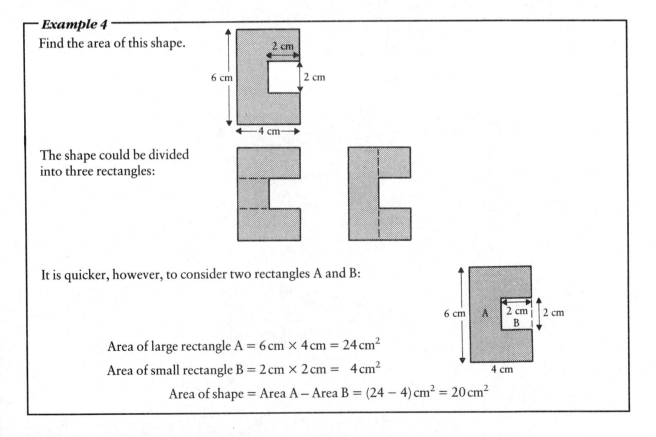

The shape could be divided into three rectangles:

It is quicker, however, to consider two rectangles A and B:

Area of large rectangle A = 6 cm × 4 cm = 24 cm^2

Area of small rectangle B = 2 cm × 2 cm = 4 cm^2

Area of shape = Area A − Area B = (24 − 4) cm^2 = 20 cm^2

1   Write down the areas of the following
    rectangles. (Each square represents 1 cm².)

a

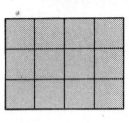

b

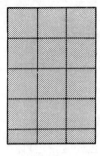

c

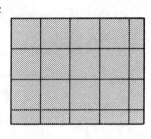

d

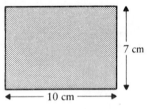

e

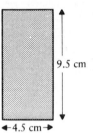

f

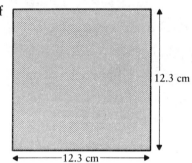

2   Calculate the area of a rectangle with
    dimensions:

a 12 cm × 5 cm   d 17.5 cm × 2.5 cm

b 15 cm × 20 cm   e 22.8 cm × 11.6 cm

c 10 cm × 7.5 cm   f 15.5 cm × 15.5 cm.

3   Calculate the areas of the following shapes:

a

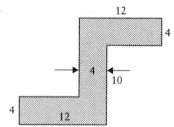

c

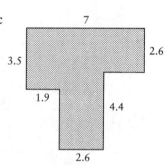

b

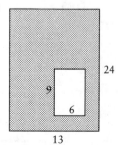

d

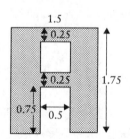

4 A window has 24 panes of glass, each measuring 26.5 cm by 21.5 cm.
What is the total area of glass in the window?

5 Carpet is bought by the metre, but is priced by the square metre.

   a What is the area of a carpet which is 4 m wide and 7.5 m long?

   b What is the cost of the carpet if the price is £9.99 per square metre?

6 A path 1 m wide is laid all round the edge of a lawn which is 6 m by 5 m.
What is the area of the path?

7 A piece of card, 20 cm square, is used to make a surround for a picture.

   a What size square is cut out of the card if the frame is 4 cm wide?

   b What is the area of the card surround?

8 A rectangular piece of land covers an area of 156 m².

   a If its length is 13 m, what is its width?

   b What is its perimeter?

9 A square has a perimeter which is 22 cm long. What is the area of the square?

10 a A rectangle has an area of 5 m². How many square centimetres is the area?

   b How many square inches are there in an area of 6 ft²?

   c A rectangular area is 12.6 cm². How many square millimetres is this area?

Area of a triangle

Draw a rectangle ABCD and mark a point E anywhere along AB.
Join ED and EC.

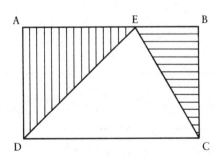

If triangles ADE and BCE are cut out, they can be placed on top of triangle DEC so that they exactly cover triangle DEC.

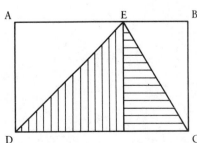

Area of triangle DEC = Half the area of rectangle ABCD
Area of rectangle ABCD = Length DC × Width BC
DC is the base of triangle DEC and BC equals its height.

Therefore: **Area of triangle DEC = ½ Base × Height**

$$A = \tfrac{1}{2}bh$$

Example 1

Show that the area of the triangle below is $\frac{1}{2}$ Base × Height.

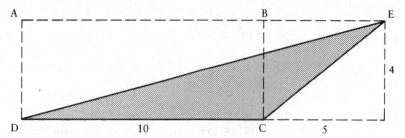

Area of the rectangle ABCD = $10 \times 4 = 40$

Area of the triangle DEC $= \frac{1}{2}(15 \times 4) - \frac{1}{2}(5 \times 4) = 30 - 10$

$\qquad\qquad\qquad\qquad = 20$

$\qquad\qquad\qquad\qquad = $ Half rectangle ABCD

$\qquad\qquad\qquad\qquad = \frac{1}{2}$ Base × Height

Example 2

Find the area of the following:

a

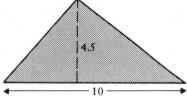

b

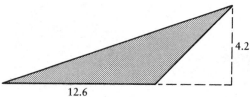

a Area $= \frac{1}{2}bh$

$\qquad = \frac{1}{2} \times 10 \times 4.5$

$\qquad = 22.5$

b Area $= \frac{1}{2}bh$

$\qquad = \frac{1}{2} \times 12.6 \times 4.2$

$\qquad = 26.46$

Example 3

A triangle has an area of $35\,\text{cm}^2$ and a height of $4\,\text{cm}$.
What is the length of its base?

$\qquad A = \frac{1}{2}bh \qquad\qquad or \qquad\qquad A = \frac{1}{2}bh$

$\qquad 35 = \frac{1}{2} \times b \times 4 \qquad\qquad\qquad\qquad b = \dfrac{2A}{h}$

$\qquad\qquad = 2b \qquad\qquad\qquad\qquad\qquad\quad = \dfrac{2 \times 35}{4} = 17.5$

$\qquad b = \dfrac{35}{2} = 17.5$

$\qquad$ Base = $17.5\,\text{cm}$ $\qquad\qquad\qquad\qquad$ Base = $17.5\,\text{cm}$

Exercise 114

1 Find the areas of the following triangles:

a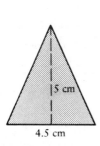

5 cm

4.5 cm

b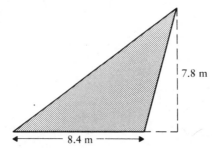

7.8 m

8.4 m

c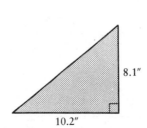

8.1″

10.2″

2 Find the area of a triangle with:

a Base = 16 cm, Height = 11 cm

c Base = 10.5 in, Height = 1 ft

b Base = 28 mm, Height = 17.5 mm

d Base = 4.6 m, Height = 8.4 m

3 a A triangle has an area of 40 m² and a height of 8 m.
What is the length of its base?

b A triangle has a base of length 12 cm and an area of 96 cm².
What is its height?

4 Calculate the missing dimension in the following triangles:

a Area = 144 cm²   Base = 18 cm   Height = ? cm

b Area = 52 cm²   Base = ? cm   Height = 13 cm

c Area = 45 mm²   Base = 7.5 mm   Height = ? mm

d Area = ? m²   Base = 7.2 m   Height = 3.4 m

e Area = ? in²   Base = 5.5 in   Height = 8.6 in

5 Find the area of each of the following shapes:

a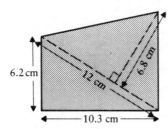

6.2 cm

12 cm

6.8 cm

10.3 cm

c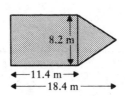

8.2 m

11.4 m

18.4 m

e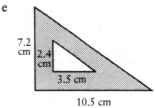

7.2 cm

2.4 cm

3.5 cm

10.5 cm

b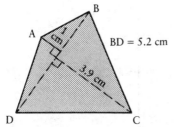

B

A

cm

BD = 5.2 cm

3.9 cm

D

C

d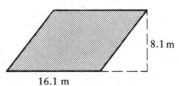

8.1 m

16.1 m

f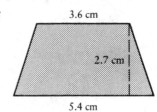

3.6 cm

2.7 cm

5.4 cm

Area of a parallelogram

A parallelogram is a quadrilateral (four-sided figure) which has both pairs of opposite sides parallel.

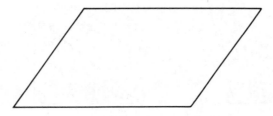

Any parallelogram can be divided into two identical triangles by drawing in a diagonal.

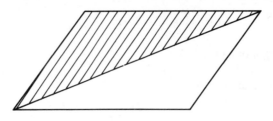

Area of the parallelogram = 2 × Area of one triangle

$$A = b \times h$$

Example 1

Find the area of a parallelogram with base 12.5 cm and height 8.2 cm.

$$A = b \times h$$
$$= 12.5 \times 8.2 \,\text{cm}^2$$
$$\text{Area} = 102.5 \,\text{cm}^2$$

Example 2

A parallelogram has a base of 14 cm and an area of 168 cm².
What is the height of the parallelogram?

$$A = b \times h \qquad or \qquad A = bh$$

$$168 = 14 \times h \qquad\qquad h = \frac{A}{b}$$

$$h = \frac{168}{14} = 12 \qquad\qquad h = \frac{168}{14} = 12$$

Height = 12 cm   Height = 12 cm

Exercise 115

1 Find the areas of the following parallelograms:

    a Base = 10 cm   Height = 12 cm

    b Base = 12.6 cm   Height = 6 cm

    c Base = 34.7 in   Height = 13 in

    d Base = 1.3 m   Height = 26 cm

    e Base = 19.2 mm   Height = 10.3 cm.

2 Find the missing dimension for the following
   parallelograms:

    a Area = 24 cm^2 Base = ?   Height = 10 cm

    b Area = 1.3 m^2 Base = 3.9 m   Height = ?

    c Area = ?   Base = 8.4 cm   Height = 16 mm

    d Area = 0.68 m^2 Base = 17 cm   Height = ?

    e Area = 2.4 yd^2 Base = ?   Height = 3 ft.

3 Calculate the area of the following shapes:

    a   **b**   **c**

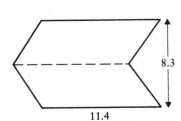
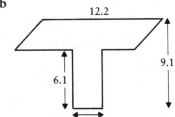
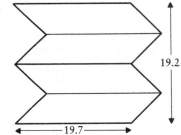

26.5 The circumference and area of a circle

Circumference of a circle

The **perimeter** of a circle is called the **circumference**.

Take a long piece of thread or thin string and cut off
a piece equal to the length of the diameter of the
circle on the right.

Cut off a second length equal to the length of the
circumference. (You will need to take care when
measuring the circumference.)

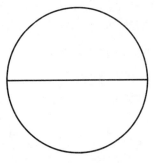

Compare the two lengths of thread and you should find that the longer piece is
just over three times the length of the shorter piece. That is:

$$\text{Circumference} \approx 3 \times \text{Diameter}$$

This formula was good enough for the mathematicians of ancient times, but we now have an accurate value for this multiple of the diameter.

The multiple is given the symbol π, the Greek letter for 'p', pronounced 'pi'.

If you have a π button on your calculator, press it and you will see that $\pi = 3.141592654$.

In fact, the number goes on for ever, but 9 decimal places are adequate for most purposes! For all circles

$$\textbf{Circumference} = \pi \times \textbf{Diameter}$$

$$C = \pi d$$

or
$$\textbf{Circumference} = 2 \times \pi \times \textbf{Radius}$$

$$C = 2\pi r$$

In calculations, you may use the 'pi' button or an appropriate approximation,

e.g. $\pi = 3$ for a rough estimate,
$\pi = 3.14$ for a more accurate answer.

You will usually be told in an examination which value of π to use.

Example 1

Find the circumference of a circle with radius 6 cm. (Take $\pi = 3.14$.)

$$\text{Circumference} = 2 \times \pi \times r$$
$$= 2 \times 3.14 \times 6 \,\text{cm}$$
$$= 37.68 \,\text{cm}$$

Example 2

A circle has a circumference of length 30 m. Find:

a a rough estimate of the diameter

b an estimate of the diameter, correct to d.p.

a For a rough estimate take

$$C \approx 3 \times d$$
$$\therefore d \approx \frac{C}{3} = \frac{30}{3} = 10$$

The diameter is 10 m

b
$$C = \pi \times d$$
$$\therefore d = \frac{C}{\pi} = \frac{30}{3.14} \text{ or } \frac{30}{\pi}$$

and use the π button on your calculator.

The diameter = 9.55 m (to 2 d.p.).

Exercise 116

1 Use the π button on your calculator (or take π = 3.14) to calculate the missing lengths below. Give your answers correct to 1 d.p.

    a Diameter = 10 cm   Radius = ?   Circumference = ?

    b Diameter = ?   Radius = ?   Circumference = 27″

    c Diameter = ?   Radius = 3.5 mm   Circumference = ?

    d Diameter = ?   Radius = ?   Circumference = 51 cm

    e Diameter = ?   Radius = ?   Circumference = 6.12 m

2 The diameter of a 10p coin is 3 cm.
   What is the length of its circumference, correct to 2 d.p.?

3 The radius of a 1p coin is 5.5 mm.
   What is the length of its circumference in centimetres, correct to 2 d.p.?

4 A bicycle wheel has a radius of 34 cm.
   How far does it travel in 5 revolutions?

5 A gardener wishes to put an edging around his circular rose border.
   If the diameter of the border is 3.5 m, what length of edging will he need?

6 Find the perimeters of the following shapes:

    a   **b**   **c**

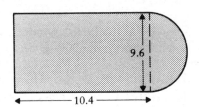
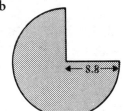
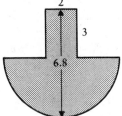

Area of a circle

A circle with radius l has an area = π

A circle with radius r has an area = $\pi \times r^2$.

For all circles:

$$A = \pi r^2$$

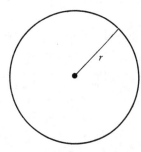

Example 1

A circle is drawn which has a diameter of 6 cm.
What is the area of the circle?

$$\text{Diameter} = 6\,\text{cm}$$
$$\text{Radius} = 3\,\text{cm}$$
$$\text{Area} = \pi \times r^2$$
$$= \pi \times 3 \times 3\,\text{cm}^2$$
$$= 28.3\,\text{cm}^2 \text{ (to 3 s.f.)}$$

Example 2

A circle has an area of $10\,\text{m}^2$.
What is the radius of the circle?

$$\text{Area} = \pi r^2$$
$$\pi \times r^2 = 10$$
$$r^2 = \frac{10}{\pi}$$
$$r = \sqrt{\frac{10}{\pi}} \qquad = 1.78\,\text{m} \text{ (to 3 s.f.)}$$

Exercise 117

1 Use the π button on your calculator, or take $\pi = 3.14$, to calculate the area of:

  a a circle of radius 7.1 cm

  b a circle of radius 29.5 in

  c a circle of diameter 13.6 mm

  d a semicircle of radius 4.9 cm

  e a semicircle of diameter 9.28 m

2 By taking an approximate value for π of 3, estimate the length of the radius of a circle with area:

  a $27\,\text{cm}^2$ b $150\,\text{in}^2$ c $108\,\text{cm}^2$ d $48\,\text{m}^2$
  e $300\,\text{ft}^2$

3 Calculate the length of the radius, correct to 3 s.f., for each of the circles in question 2 above.

4 A discotheque has a circular dance floor which covers $50\,\text{m}^2$.
What is the diameter of the dance floor, correct to 1 d.p.?

5 The Costain's buy a semicircular rug to fit their hearth which is 1.2 m wide.
If the carpet costs £12.99 per square metre, how much does the rug cost?

6 Mr Swain is making a circular fish pond. Each fish requires a surface area of 1000 square centimetres.
If Mr Swain intends to keep ten fish, what is the smallest radius, correct to the nearest centimetre, he must use for his pond?

26.6 Composite areas

Example

A luggage label is made of thin cardboard with the dimensions shown in the figure.
Calculate the area of cardboard which the label covers.

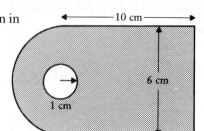

Area of the rectangle $= L \times B = 10 \times 6 = 60\,\text{cm}^2$

Area of the semicircle $= \frac{1}{2}\pi r^2 = \frac{1}{2} \times \pi \times \left(\frac{6}{2}\right)^2 = 14.14\,\text{cm}^2$

Area of the cut-out circle $= \pi r^2 = \pi \times (1)^2 = 3.14\,\text{cm}^2$

Area of cardboard = Area of rectangle + Area of
semicircle − Area of circle
$$= (60 + 14.14 - 3.14)\,\text{cm}^2$$
$$= 71\,\text{cm}^2 \text{ (to the nearest cm}^2\text{)}$$

Exercise 118

1   A running track of length 400 m is the shape of a rectangle with 2
    semicircular ends of radius 40 m.

    a What is the length of each straight?

    b What is the total area enclosed by the running track?

2   Find (i) the perimeter and (ii) the area of each of the following shapes:

a   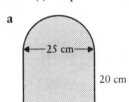   **b** 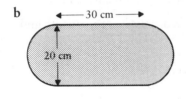   **c** 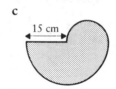   **d**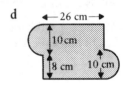

3   Calculate the shaded area in each of the following:

a 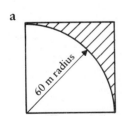   **b** 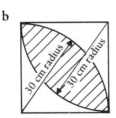   **c** 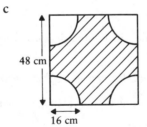   **d**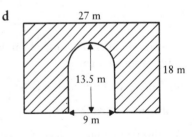

4   **a** Calculate the area of each semicircle.

    b What is the sum of the areas of the two smaller semicircles?

    c How are the areas of the three semicircles connected?

    d What is the total area of the figure?

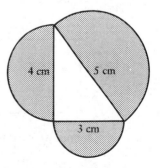

5   The diagram shows the cross-section of a drainage pipe.

The radius of the outer circle
is R and of the inner circle is r.
(The shaded area is called an annulus.)

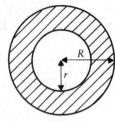

   a Write down an expression for the area of the outer circle in terms of R.

   b Write down an expression for the area of the inner circle in terms of r.

   c Write down an expression for the shaded area.

   d If $R = 13.2$ cm and $r = 10.2$ cm, find the area of the cross-section of the pipe.

6   Find the area of the cross-section of a pipe with an outer diameter of 16 mm and an inner diameter of 12 mm.

26.7   Volume

Volume is a measure of the amount of space which is taken up by a solid shape.

Solids are three-dimensional shapes, i.e. they have length, breadth and height.

A **cube** is a solid with six square faces (or sides), hence the length, breadth and height of a cube are all equal.

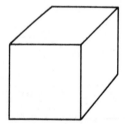

A cube which is 1 cm long, 1 cm wide and 1 cm high has a volume of 1 cubic centimetre ($= 1$ cm^3).

Similarly, the volume of a cube of side 1 metre is 1 cubic metre ($= 1$ m^3).

In imperial units the measures of volume are cubic feet, cubic inches, etc.

A **cuboid** is a solid with six rectangular faces.

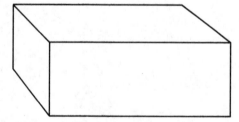

To find the volume of a cuboid we need to find out how many unit cubes it contains.

Example 1

A cuboid which is 6 cm × 4 cm × 3 cm can be divided as shown:

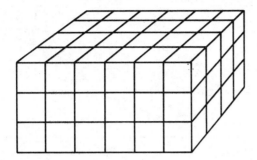

6 one centimetre cubes can be along the length
4 rows (each of 6 cubes) can be fitted into the base = 6 × 4 cubes
3 layers (of 6 × 4 cubes) are needed for the height = 6 × 4 × 3 cubes
= 72 cubes

The volume is therefore 72 cubic centimetres.

Volume of a cuboid = Length × Breadth × Height

$$V = L \times B \times H$$

For a cube, *Length = Breadth = Height*, and

$$V = L^3$$

A **prism** is a solid which has a constant cross-section (i.e. the cross-section of the top is exactly the same as the cross-section of the base).

Some common prisms are:

(i) a **cylinder** (with a cross-section which is a circle)

(ii) a **triangular prism**

(iii) a **rectangular prism** or cuboid

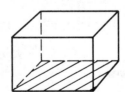

For a prism:

Volume = Area of cross-section × Height

For a cylinder:

Volume = Area of circle × Height

$$V = \pi r^2 h$$

Example 1

A Toblerone packet is a triangular prism.
The triangular cross-section has an area of 2.25 cm² and the length of the
packet is 16.8 cm.
What is the volume of the Toblerone packet?

$$\text{Volume} = \text{Area of cross-section} \times \text{Length}$$
$$= 2.25 \, \text{cm}^2 \times 16.8 \, \text{cm}$$
$$= 37.8 \, \text{cm}^3$$

Example 2

Find the volume of a cylinder with a base radius of 3.5 cm and a height of
8 cm.

$$\text{Volume of cylinder} = \text{Area of base} \times \text{Height}$$
$$= \pi r^2 h$$
$$= \pi \times 3.5 \times 3.5 \times 8 \, \text{cm}^3$$
$$= 307.9 \, \text{cm}^3$$

Example 3

Find the volume of a block of wood of length 12 cm which has a constant
cross-section as shown in the diagram.

3.2 cm

7.4 cm

$$\text{Area of triangle} = \tfrac{1}{2}bh$$
$$= \tfrac{1}{2} \times 6.4 \times 4.2 \, \text{cm}^2$$
$$= 13.44 \, \text{cm}^2$$
$$\text{Area of semicircle} = \tfrac{1}{2}\pi r^2$$
$$= \tfrac{1}{2} \times \pi \times (3.2)^2 \, \text{cm}^2$$
$$= 16.08 \, \text{cm}^2$$
$$\text{Area of cross-section} = 13.44 + 16.08 \, \text{cm}^2$$
$$= 29.52 \, \text{cm}^2$$
$$\text{Volume} = \text{Area of cross-section} \times \text{Height}$$
$$= 29.52 \times 12 \, \text{cm}^3$$
$$= 354 \, \text{cm}^3$$

▨ *Exercise 119* ▨

1 Calculate the volume of the following cuboids:

    a Length = 6 cm   Breadth = 5 cm   Height = 3 cm

    b Length = 3.4 m   Breadth = 2.6 m   Height = 5.8 m

    c Length = 0.7 m   Breadth = 0.6 m   Height = 0.8 m

    d Length = 30 cm   Breadth = 22 cm   Height = 22 mm

    e Length = 2 m   Breadth = 60 cm   Height = 1 m

2 The base of a cuboid has an area of 12 cm^2. The volume of the cuboid is 40 cm^3.
    What is the height of the cuboid?

3 Calculate the volume of a cube of side:

    a 8 cm   **b** 11 cm   **c** 4.1 cm

4 The volume of a cube is 63 cm^3.
    What is the length of a side, correct to 1 d.p.?

5 A child has 27 wooden blocks, each a cube of edge 1 inch. She builds one large cube using all the blocks.

    a What is the volume of the large cube?

    b What is the length of an edge of the large cube?

6 An oil tank is 6 ft × 4 ft × 4 ft high.

    a How many cubic feet of oil does the tank hold when full?

    b When 54.5 cubic feet of oil has been used, by how much has the level in the tank fallen?

7 Find the volume of the following solids which have a length as given and a uniform cross-section as shown in the diagram.

    a Length = 15 cm   **b** Length = 5 cm   **c** Length = 10.5 cm

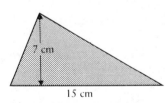

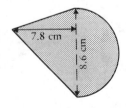

8 A greenhouse is 10 m long and has a cross-section which is a square of side 5 m on top of which is a triangle, as shown in the diagram.

    The overall height of the greenhouse is 7.5 m.
    Calculate the volume of air inside the greenhouse.

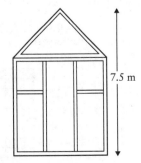

9 Calculate the volume of a cylinder with:

    a Radius = 4 cm   Height = 12 cm

    b Radius = 9 cm   Length = 12 cm

    c Diameter = 14 in   Height = 11 in

    d Diameter = 8.4 cm   Length = 7 cm.

10 Calculate the radius of a cylinder with

    a Volume = 32 cm^3 Height = 5 cm

    b Volume = 49 cm^3 Length = 2.3 cm

    c Volume = 84 in^3 Height = 9 in.

11 A tunnel is excavated from a hillside. The length of the tunnel is 300 m and its cross-section is as shown:

    Calculate the volume of earth which is removed to make the tunnel.

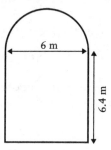

12 A section of metal pipe has an outer diameter of 5 cm and the metal is 3 mm thick. The pipe is 18 cm long.
What volume of metal is used in making the section of pipe?

26.8   Areas and volumes of similar shapes

Areas of similar shapes

If a plane figure is **enlarged** to form a new figure, all the corresponding lengths of the two figures will be in the same ratio.

The two figures are said to be **similar**.

The areas of two similar shapes are in the ratio of the squares of the corresponding dimensions.

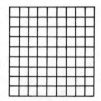

The larger square has sides which are 3 times the length of the sides of the smaller square.

The area of the larger square is 9 times the area of the smaller square:

$$\text{Larger area} = 3^2 \times \text{Smaller area}$$

The smaller circle has a radius r; the larger circle has a radius R.

 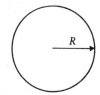

$$\text{Area of smaller circle : area of larger circle} = \pi\, r^2 : \pi\, R^2$$
$$= r^2 : R^2$$

For all similar shapes:

$$\textbf{Larger area : Smaller area} = L^2 : l^2$$

where L and l are the lengths of corresponding dimensions.

Example

The triangle PQR has every length three times the corresponding lengths in triangle ABC. The area of ABC is $21\,\text{cm}^2$. Find the area of PQR.

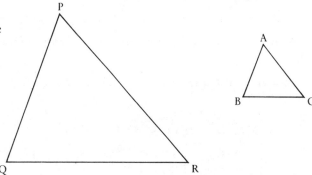

Area PQR = $3^2 \times$ Area ABC

Area PQR = $3^2 \times 21$

$\qquad\quad = 189\,\text{cm}^2$

Exercise 120

1 The quadrilaterals ABCD and PQRS are similar. The area of quadrilateral ABCD is $7\,\text{cm}^2$. PQ = 2AB. Calculate the area of quadrilateral PQRS.

2 The area of a pentagon is $38\,\text{cm}^2$. The pentagon is enlarged by scale factor 5. Find the area of the new shape.

3 The area of quadrilateral PQRS is $31\,\text{cm}^2$. The quadrilateral is enlarged by scale factor $\frac{1}{2}$. Find the area of the new shape.

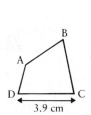 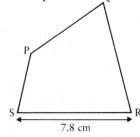

4 The radii of two circles are in the ratio 1:6. The area of the larger circle is $306\,\text{cm}^2$. Calculate the area of the smaller circle.

5 The area of a circle is $74\,\text{cm}^2$. After an enlargement, the area is $296\,\text{cm}^2$. Find the scale factor of the enlargement.

6 The area of a triangle is $24\,\text{in}^2$. After an enlargement, the area is $384\,\text{in}^2$. Find the scale factor of the enlargement.

7 The area of a triangle is $38\,\text{cm}^2$. After an enlargement, the area is $9.5\,\text{cm}^2$. Find the scale factor of the enlargement.

8 A semicircle has an area of $108\,\text{m}^2$. It is enlarged to a semicircle of area $192\,\text{m}^2$. In what ratio are the diameters of the two semicircles?

Volumes of similar solids

When a Russian doll, like the ones shown, is lifted, inside is another doll looking just the same, but smaller and inside that doll is another and then another.

All the dolls in the set are similar to each other, but each one is a scale model of another.

Just as areas can be enlarged, so can volumes.

Two solids are **similar** if the ratios between their corresponding dimensions are equal.

The larger cube has sides which are three times the length of the sides of the smaller cube.

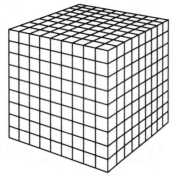

The volume of the larger cube is 27 times the volume of the smaller cube.

$$\text{Larger volume} = 3^3 \times \text{Smaller volume}$$

For all similar shapes:

$$\text{Larger volume} : \text{Smaller volume} = L^3 : l^3$$

where L and l are the lengths of corresponding dimensions.

Example

A small tin of cocoa has a height of 8 cm. A similar tin is an enlargement of scale factor $1\frac{1}{4}$.
The volume of the smaller tin is 320 cm^3.

a What is the height of the larger tin?

b What is the volume of the larger tin?

c If the larger tin contains 244 g of cocoa, what weight does the smaller tin contain?

a Ht. of larger tin $= 1\frac{1}{4} \times$ Ht. of smaller tin

$= 1.25 \times 8$ cm

$= 10$ cm

b Vol. of larger tin $= (1\frac{1}{4})^3 \times$ Vol. of smaller tin

$= (1.25)^3 \times 320$ cm^3

$= 625$ cm^3

c As weight depends on volume, the weights of the contents of the two tins are in the same ratio as their volumes.

Wt. in smaller tin $= \dfrac{\text{Wt. in larger tin}}{(1\frac{1}{4})^3}$

$= 125$ g

Exercise 121

1 A child's set of 64 building blocks are packed into a box which is similar in shape to the blocks.

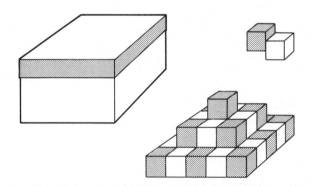

a The volume of a block is 3 in^3.
What is the volume of the box?

b The height of a block is 2 in.
What is the height of the box?

2 Two cylinders are similar in shape. Their dimensions are in the ratio of 1:5.

a If the volume of the larger cylinder is 875 cm^3, what is the volume of the smaller cylinder?

b If the area of the base of the smaller cylinder is 1.4 cm^2, what is the area of the base of the larger cylinder?

3 A small bottle of medicine has a capacity of 20 ml. A similar bottle has a capacity of 160 ml.
If the height of the smaller bottle is 7.5 cm, what is the height of the larger bottle?

4 Two buckets are similar in shape.
Their volumes are in the ratio 27:8.

a In what ratio are their heights?

b In what ratio are the areas of their circular bases?

c If the height of the smaller bucket is 28 cm, what is the height of the larger bucket?

5 The dimensions of a giant ice cream cone are $1\frac{1}{2}$ times those of a small cone.

a The volume of ice cream in the small cone is 112 cm^3. How much ice cream does the giant cone hold?

b If the price of a small cone is 60p, what should the ice cream vendor charge for a giant cone?

27 Algebra

27.1 The basics of algebra

From words to symbols

Every Saturday the milkman delivers the following to 16 Cromarty St:
3 loaves of bread, 4 pints of milk and a dozen eggs.

He then presents his bill.

A sentence such as 'the cost of three loaves of bread, four pints of milk and a dozen eggs' is difficult to deal with mathematically.

Algebra is a mathematical language which enables us to deal with problems more easily.

Suppose we let:

b stand for 'the cost of a loaf of bread'
m stand for 'the cost of a pint of milk'
and e stand for 'the cost of a dozen eggs'.

The milkman's bill for 3 loaves of bread, 4 pints of milk and 1 dozen eggs becomes

$$3b + 4m + 1e \quad \text{or} \quad 3b + 4m + e$$

where $3b$ means 3 times b (or $3 \times b$).
We can miss out the 1 in front of the e
(e means $1 \times e$).

The foundation of algebra, as a science of equations, was laid down in the year 825 by an Islamic mathematician, Muhammad Ibn Musa al-Khwarizmi.

Example 1

A holiday has two prices, one for adults and the other for children.
Write down an algebraic expression for the cost of a holiday for two adults and three children.

Let the cost for one adult be £A and the cost for one child be £C:

Cost for 2 adults   = £$2A$

Cost for 3 children = £$3C$

Cost of the holiday = £$2A$ + £$3C$ = £$(2A + 3C)$

Example 2

Don thinks of a number, multiplies it by 3, adds 4 and then divides this answer by the original number.
Write his final number in algebraic terms.

Let his first number be n

Multiply by 3: $3n$

Add 4: $3n + 4$

Divide by the original number: $\dfrac{3n + 4}{n}$

Final number $= \dfrac{3n + 4}{n}$

Simplify the following phrases and sentences by translating them into algebraic expressions:

1 The cost of 3 pounds of apples, 4 pounds of bananas, and 2 pounds of cherries.

2 The entrance fee to the leisure centre for 3 adults and 5 children.

3 The weight of 6 tablespoonfuls of flour and 3 of sugar.

4 The sum of seven times x and five times y.

5 The difference between q and twice p.

6 Double the number a and divide by the number b.

7 A number is formed by multiplying a number x by five and then subtracting eight.

8 A number is formed by dividing a number y by four and then adding six.

9 Think of a number, multiply by two, add three and divide by eight.

10 Think of a number, multiply by four, subtract one and then divide by twice the original number.

Collecting like terms

Mrs Goodman empties her son's pockets on washing day and finds two sweets, one dirty hanky, two pencils, two more sweets and another pencil.

The contents of the pocket can be listed in algebraic terms as

$$2s + 1h + 2p + 2s + 1p$$

The list can be simplified if articles of the same kind are added together. This is called **collecting like terms**. If we collect the like terms from Master Goodman's pocket, the list becomes $h + 3p + 4s$ (which cannot be simplified further as each term is different).

Addition and subtraction

Only like terms may be added or subtracted.

Like terms are those which are multiples of the same algebraic variable.

For example, $3a$, $7a$ and $-8a$ are all like terms.

The expression $3a + 7a - 8a$ can be simplified to $2a$.

The terms $7a$, $3b$ and $4c$ are unlike terms and the expression $7a + 3b + 4c$ cannot be simplified any further.

Example 1

Simplify the following expression:

$$2a + 3c + a + 4b + 2c + b$$

$$2a + 3c + a + 4b + 2c + b = (2a + a) + (4b + b) + (3c + 2c)$$
$$= 3a + 5b + 5c$$

Example 2

Simplify the following by collecting like terms:

$$5x + 2y - 4z + 3x - y + z$$

$$5x + 2y - 4z + 3x - y + z = (5x + 3x) + (2y - y) + (z - 4z)$$
$$= 8x + y - 3z$$

Exercise 123

Simplify the following by collecting terms:

1   $2x + 3x$ 6 $x + 2y + z + 3y + 8z$

2   $8n - n$ 7 $5p + 3q + 2r - q + 7p - 3p - r$

3   $3a - a + 6a$ 8 $2x - y + 3x - 2y + z$

4   $5y - 6y + 3y$ 9 $a + b - 2a + c - 2b + 3c$

5   $3a + 2a + c + b + 2a + 3c + 5b$ 10 $3x - y + \frac{1}{2}y - 2\frac{1}{2}y + 8y$

Substitution

The letters in an algebraic expression stand for numbers or amounts.

For example, in the expression $3b + 4m + e$, b stands for the cost of a loaf of bread, m for the cost of a pint of milk and e for the cost of a dozen eggs.

The advantage of the letters is that the amounts are not fixed, they can change or vary.

The letters are called **variables**. The numbers, which are constant, are called **coefficients**.

If a loaf of bread costs 54p, a pint of milk 29p and a box of eggs 63p (i.e. $b = 54, m = 29, e = 63$), then

Weekly bill $= 3b + 4m + e$

$$= (3 \times 54) + (4 \times 29) + 63$$
$$= 341 \text{ pence}$$

If, however, prices rise and $b = 56$, $m = 31$, $e = 66$, then

Weekly bill $= 3b + 4m + e$

$$= (3 \times 56) + (4 \times 31) + 66$$

$$= 358 \text{ pence}$$

Example 1

Evaluate $\dfrac{3x - y}{x + y}$ when $x = 3$, $y = -1$

$$\frac{3x - y}{x + y} = \frac{(3 \times 3) - (-1)}{3 + (-1)} = \frac{9 + 1}{3 - 1} = \frac{10}{2} = 5$$

Exercise 124

1   Evaluate the following expressions when $x = 3$, $y = 5$, $z = 2$.

   a $2x + y$ **d** $\dfrac{12}{z}$ **g** $3z - 12x$ **j** $xy + z$

   b $3x - 2y$ **e** $\dfrac{y}{10}$ **h** $2x - 5z$ **k** $y - xz$

   c $y + 4z$ **f** $4z - 6$ **i** $x + y - 4z$ **l** $yz - 4x$

2   Evaluate each of the expressions in question 1 when $x = -1$, $y = 3$, $z = -2$.

3   Evaluate the following expressions if $x = 4$, $y = 3$, $z = -2$:

   a y^2 **d** $\dfrac{x}{y}$ **g** $2x^2 - yz$ **j** $\dfrac{2y}{x - z}$

   b xyz **e** $xz + y$ **h** $\dfrac{x + y}{x - y}$ **k** $\dfrac{x + 2z}{y}$

   c $2z^2$ **f** $\dfrac{y}{2z}$ **i** $x(x + y)$ **l** $(x + y)(x + z)$

4   Evaluate each of the expressions in question 3 when $x = -2$, $y = \frac{1}{2}$, $z = -1$.

Multiplication and division

Multiplication of algebraic terms is usually easier than multiplication in arithmetic. For example:

$$a \text{ multiplied by } 4 = a \times 4 = 4a$$

$$p \text{ multiplied by } q = p \times q = pq$$

$$x \text{ divided by } y = x \div y = \frac{x}{y}$$

a multiplied by $a = a \times a = a^2$

(Algebraic terms are usually listed in alphabetical order with the constant first.)

The rules for directed numbers are the same as on p. 154. For example:

$$(+x) \times (+y) = xy$$

$$(-x) \times (+y) = -xy$$

$$(+x) \div (-y) = -\frac{x}{y}$$

$$(-x) \div (-y) = \frac{x}{y}$$

Example 1

Simplify $5 \times p \times 6 \times q \div 3 \div r$

Collect numbers and letters separately:

$$5 \times p \times 6 \times q \div 3 \div r = (5 \times 6 \div 3) \times (p \times q \div r)$$

$$= \frac{10pq}{r}$$

Example 2

Simplify $(3x) \times (-2y) \times (-z)$

$$(3x) \times (-2y) \times (-z) = (3 \times -2 \times -1) \times (x \times y \times z)$$

$$= 6xyz$$

Example 3

Simplify $3a \times (2b)^2 \times (-a^2)$

$$3a \times (2b)^2 \times (-a^2) = (3 \times 2^2 \times -1) \times a \times b^2 \times a^2$$

$$= -12a^3b^2$$

Exercise 125

Simplify:

1   $5 \times x$

2   $3 \times m \times n$

3   $2 \times y \div z$

4   $a \times c \times b$

5   $2 \times p \times 3 \times r \times q$

6   $c \div 2d \times 8b$

7   $(-2x) \times (-3y)$

8   $(4p) \div (-2q)$

9   $(6x) \times (-3y) \div (-2z)$

10   $a \times a \times a$

11   $a \times a \times b \times b$

12   $3a \times a^2$

13 $3a \times b^2 \times a^2$ **15** $3ab^2 \times 2ab \times bc^2$ **17** $(-2ab) \times (3b)^2$ **19** $x^2y \times xy^2$

14 $2a^2b \times -3bc$ **16** $(-3a)^2 \times b$ **18** $(-2a) \times (-b) \div (-b)$ **20** $12ab \div (-4c^2) \times (3ac)$

Indices

$$a \times a \times a = a^3$$

The number 3 is called an **index**. It shows, or indicates, the number of *a*s which have been multiplied together to give the third power of *a*.

$$a^1 \times a^2 = a \times a \times a \qquad = a^3 \text{ i.e. } a^{(1+2)}$$

$$a^3 \times a^2 = a \times a \times a \times a \times a = a^5 \text{ i.e. } a^{(3+2)}$$

In general: $a^x \times a^y = a^{(x+y)}$

$$a^3 \div a^2 = \frac{a \times a \times a}{a \times a} = a \text{ i.e. } a^{(3-2)}$$

$$a^5 \div a^3 = \frac{a \times a \times a \times a \times a}{a \times a \times a} = a \times a = a^2 \text{ i.e. } a^{(5-3)}$$

In general: $a^x \div a^y = a^{x-y}$

$$a^3 \div a^5 = \frac{a \times a \times a}{a \times a \times a \times a \times a} \qquad = \frac{1}{a \times a} \quad = \frac{1}{a^2}$$

but $a^3 \div a^5 = a^{(3-5)} = a^{-2}$ i.e. $\frac{1}{a^2} = a^{-2}$

In general: $a^{-x} = \dfrac{1}{a^x}$

$$a^3 \div a^3 = \frac{a \times a \times a}{a \times a \times a} = 1$$

but $a^3 \div a^3 = a^{(3-3)} = a^0$

In general: $a^0 = 1$

The rules for indices can be summarised as follows:

$$a^x \times a^y = a^{(x+y)}$$

$$a^x \div a^y = a^{(x-y)}$$

$$a^{-x} = \frac{1}{a^x}$$

$$a^0 = 1$$

Example 1

Simplify $x^5 \div x^3 \times x^2$

$$x^5 \div x^3 \times x^2 = \frac{x \times x \times x \times x \times x \times x \times x \times x \times x \times x}{x \times x \times x \times x} = x^4$$

or $\quad x^5 \div x^3 \times x^2 = x^{(5-3+2)} = x^4$

Example 2

Simplify $\dfrac{5x^2y \times 3y^3}{6xy^2}$

$$\frac{5x^2y \times 3y^3}{6xy^2} = \frac{5 \times x \times x \times y \times 3 \times y \times y \times y}{6 \times x \times y \times y} = \frac{5xy^2}{2}$$

or $\quad \dfrac{5x^2y \times 3y^3}{6xy^2} = \dfrac{5 \times 3}{6} \times x^{(2-1)} \times y^{(1+3-2)} \qquad = \dfrac{5xy^2}{2}$

Exercise 126

Simplify the following expressions:

1 $x^5 \times x^3$

2 $x^3 \times x \times x^2$

3 $x^7 \div x^4$

4 $x^4 \div x^5$

5 $x^4 \div x^4$

6 x^{-2}

7 x^0

8 $3x^3 \times 2x \div 4x^2$

9 $2x^3 \div 4x^5$

10 $(-2x)^2$

11 $(3x)^2 \times (-2x^2)$

12 $\dfrac{7x^3 \times (-3x)}{6x^4}$

13 $5xy \times 2x$

14 $\dfrac{x^2y \times y^2x}{xy}$

15 $\dfrac{3ab \times 2b^2 \times a}{3b^2c^2}$

16 $\dfrac{12x^2y^2}{4z^2xy}$

17 $\dfrac{6y^2 \times 8x^2z}{2xy \times 3y^2x}$

18 $\dfrac{18a^2b \times 3}{6b^2 \times 9a}$

19 $\dfrac{6x^2}{5} \div \dfrac{3x}{10}$

20 $\dfrac{2x^3}{y^2} \div \dfrac{4x^2}{yz}$

27.2 Brackets

The milkman's order for 16 Cromarty Street is three loaves of bread, four pints of milk and one dozen eggs per week. After five weeks the milkman will have delivered five times this amount.

Suppose the cost of a loaf of bread is b, the cost of a pint of milk is m, and that a dozen eggs costs e. Then suppose that we want to work out the total cost after five weeks.

The neatest way to write this cost in algebraic terms is to use a bracket:

Cost after 5 weeks $= 5(3b + 4m + e)$

To 'remove' the bracket from the expression, each term must be multiplied by 5:

$$5(3b + 4m + e) = 15b + 20m + 5e$$

If the number outside the bracket is a negative number, take care: the rules for multiplication of directed numbers must be applied.

Example 1

Remove the bracket from: **a** $4(3x - 2y)$ **b** $-2(x - 3y)$

a $4(3x - 2y) = 4 \times 3x - 4 \times 2y$
$= 12x - 8y$

b $-2(x - 3y) = -2 \times x - (-2) \times 3y$
$= -2x + 6y$

Example 2

Remove the brackets and simplify:
a $3(2x - y) - 2(x - 4y)$ **b** $a(a + b) - b(a + 2b)$

a $3(2x - y) - 2(x - 4y) = 6x - 3y - 2x + 8y$
$= 4x + 5y$

b $a(a + b) - b(a + 2b) = a^2 + ab - ba - 2b^2$
$= a^2 - 2b^2$ (since $ba = ab$)

Exercise 127

Remove the brackets and simplify:

1 $4(5x + 2y)$

2 $3(2a - 4b)$

3 $-2(3p - 6q)$

4 $5(x - 3y + 2z)$

5 $-(a - b - c)$

6 $x(3x - y + 2z)$

7 $-a(a + b - c)$

8 $x(2x^2 + 3x + 2)$

9 $4y(y^2 - 3y + 1)$

10 $-5x(1 - 2x + x^2)$

11 $2(x - 2y) + 3(x + 2y)$

12 $y(y + 4) - 3(y + 4)$

13 $5(2a + 3b) - (a - 2b)$

14 $2(p - 6q) + 3(2p - q)$

15 $5x(x - y) - 2x(2x + 3y)$

16 $-x^2y(x - xy + 2y^2)$

17 $3(p - 2q + r) + 2(p + q + 2r)$

18 $(x + y - z) - 3(x - y + z)$

19 $4(x + y) - 3(x - y) + 2(x - y)$

20 $3xy(x + y) + 2x(xy - y^2)$

27.3   Common factors

The process which is the reverse of multiplying out brackets is called
factorising.

Factorising is a very important technique in algebra. It enables expressions to
be simplified and hence makes the solving of problems easier.

Factors have already been met in section 22.4 on p. 158 and a factor in algebra
is the same as a factor in arithmetic. Remember: a factor is a number which
will divide exactly into a given number.

Consider the number $2x + 6y$.

2 is a factor of each part (or term) of this number and therefore of the whole
number.

$$\therefore 2x + 6y = 2 \times x + 2 \times 3 \times y$$
$$= 2 \times (x + 3 \times y)$$
$$= 2(x + 3y)$$

$(x + 3y)$ and 2 are both factors of $2x + 6y$.

Example 1

Factorise $6p + 3q + 9r$.

The factor which is common to each term is 3.

$$\therefore 6p + 3q + 9r = 3(2p + q + 3r)$$

Example 2

Factorise $x^2 + xy + 6x$.

The factor which is common to each term is x

$$\therefore x^2 + xy + 6x = x(x + y + 6)$$

Example 3

Factorise $2x^2 - 4xy$.

This expression has more than one common factor.
Both 2 and x are common factors.

$$\therefore 2x^2 - 4xy = 2x(x - 2y)$$

$2x^2 - 4xy$ therefore has three factors: 2, x and $(x - 2y)$.

Factorise the following expressions:

1 $4x + 12y$	6 $12s + 20t$	11 $6xy + 3x$	16 $7p - 14q + 7r$
2 $3p - 6q$	7 $xy + xz$	12 $5x - 10x^2$	17 $2a^2 + 4ab - 8a$
3 $5a + 10$	8 $xy + x^2$	13 $2a + 8b - 4c$	18 $3p^2 + 6pq - 9p$
4 $10b - 5$	9 $y^2 - 2y$	14 $9x - 3y - 6z$	19 $x^2y + xyz + xy^2$
5 $14m - 21n$	10 $2y^2 - 4y$	15 $15x - 5y + 10$	20 $4pqr - 12p^2q$

27.4 The addition and subtraction of fractions

The rules for adding and subtracting algebraic fractions are exactly the same as for fractions in arithmetic:

Method	Arithmetical fraction	Algebraic fraction
To add	$\dfrac{2}{3} + \dfrac{2}{5}$	$\dfrac{a}{3} + \dfrac{a}{5}$
(i) Find the LCM of the denominators	LCM = 15	LCM = 15
(ii) Convert each fraction to an equivalent fraction with this denominator	$\dfrac{10}{15} + \dfrac{6}{15}$	$\dfrac{5a}{15} + \dfrac{3a}{15}$
(iii) Collect like terms, i.e. add the numerators	$\dfrac{16}{15}$	$\dfrac{8a}{15}$

Example 1

Express as a single fraction $\dfrac{4}{a} + \dfrac{3}{a}$

The fractions have a common denominator a:

$$\frac{4}{a} + \frac{3}{a} = \frac{4 + 3}{a} = \frac{7}{a}$$

Example 2

Express as a single fraction $\dfrac{2}{a} + \dfrac{3}{b} - \dfrac{1}{c}$

The common denominator is abc and the equivalent fractions are:

$$\frac{2}{a} \times \frac{bc}{bc} = \frac{2bc}{abc} \qquad \frac{3}{b} \times \frac{ac}{ac} = \frac{3ac}{abc} \qquad \frac{1}{c} \times \frac{ab}{ab} = \frac{ab}{abc}$$

In equivalent fractions:

$$\frac{2}{a}+\frac{3}{b}-\frac{1}{c}=\frac{2bc}{abc}+\frac{3ac}{abc}-\frac{ab}{abc}=\frac{2bc+3ac-ab}{abc}$$ as a single fraction.

As there are no 'like terms' the answer cannot be simplified further.

Example 3

Express as a single fraction $\dfrac{(x-1)}{2}-\dfrac{(2x+1)}{6}$

The LCM of 2 and 6 is 6:

$$\frac{(x-1)}{2}-\frac{(2x+1)}{6}$$

$$=\frac{3(x-1)}{6}-\frac{1(2x+1)}{6} \quad \text{(as equivalent fractions)}$$

$$=\frac{3(x-1)-(2x+1)}{6} \quad \text{(as a single fraction)}$$

$$=\frac{3x-3-2x-1}{6} \quad \text{(multiplying out brackets)}$$

$$=\frac{x-4}{6} \quad \text{(collecting like terms)}$$

Exercise 129

Express each of the following as a single fraction and simplify where possible:

1 $\dfrac{2x}{5}+\dfrac{x}{5}$

6 $\dfrac{2x}{5}-\dfrac{x}{3}$

11 $\dfrac{1}{x}+\dfrac{2}{x}$

16 $1+\dfrac{2}{x}+\dfrac{3}{2x}$

2 $\dfrac{2a}{3}+\dfrac{a}{3}$

7 $\dfrac{x}{2}+\dfrac{x}{3}+\dfrac{x}{4}$

12 $\dfrac{3}{5x}-\dfrac{1}{2x}$

17 $\dfrac{(x+2)}{3}-\dfrac{x}{4}$

3 $\dfrac{x}{2}-\dfrac{x}{4}$

8 $\dfrac{3a}{4}+\dfrac{a}{3}-\dfrac{5a}{6}$

13 $\dfrac{4}{y}-\dfrac{5}{2y}$

18 $\dfrac{(2x-1)}{2}+\dfrac{(x+3)}{5}$

4 $x+\dfrac{3x}{4}$

9 $\dfrac{y}{6}+\dfrac{y}{2}-\dfrac{y}{3}$

14 $\dfrac{2}{3y}+\dfrac{5}{6y}-\dfrac{1}{y}$

19 $\dfrac{(5-2x)}{6}+\dfrac{(4x-1)}{3}$

5 $\dfrac{x}{4}+\dfrac{x}{3}$

10 $\dfrac{2y}{5}-\dfrac{y}{3}$

15 $\dfrac{1}{4a}+\dfrac{2}{5a}-\dfrac{1}{2a}$

20 $\dfrac{4x}{5}-\dfrac{(x+3)}{2}$

27.5   Equations

Forming equations

Many problems are solved more quickly if they are first written in algebraic terms.

An equation is, in algebra, the equivalent of a sentence. All equations must contain an equals sign.

The following problem is solved by first translating into algebra.

Example

My brother is twice as old as I am and the sum of our ages is 42.
How old am I?

English	Algebra
My age	x (years)
My brother's age	$2x$
The sum of our ages	$x + 2x$
The sum of our ages is forty-two	$x + 2x = 42$

In algebra the problem becomes:

$$x + 2x = 42$$
$$3x = 42 \text{ (collecting like terms)}$$
$$x = 14 \text{ (dividing by 3)}$$

Which translates back into English as:
My age is 14 years.

Example 2

I think of a number, treble it, add seven and the answer is 19. Form an equation using this information.

Let the number thought of be n.
Then

treble the number add seven is nineteen becomes
$$3n \quad + \quad 7 \quad = \quad 19$$

Exercise 130

For each of the following, rewrite the problem as an algebraic equation:

1   Five added to a number gives the answer twelve.

2   Seven subtracted from a number leaves thirteen.

3   If 4 is added to twice a number, the answer is equal to 10.

4   Seven times a number is twenty-one.

5   Six subtracted from five times a number gives the answer twenty-nine.

6   Ten added to half of a number gives the answer twenty-two.

7   Think of a number, double it, subtract three and the answer is three.

8 Think of a number, divide by four, subtract three and the answer is three.

9 A rectangle has a length which is double its width. The perimeter of the rectangle is eighteen.

10 The length of a rectangle is 4 cm more than its width. Its perimeter is 28 cm.

Solving equations

Once a problem has been written as an algebraic equation, the problem can be solved by solving the equation.

To solve $3n + 7 = 19$ means finding the value of n which makes the equation true.

To do this, the 7 and 3 must be eliminated from the LHS (left-hand side) of the equation to leave $n =$ the solution.

An equation must always be balanced, i.e. the LHS must always equal the RHS (right-hand side).

In the equation above

$3n + 7$ balances 19

If 7 is deducted from the LHS, the equation will no longer be balanced.

To maintain the balance, 7 must also be deducted from the RHS, giving $3n = 12$.

$3n$ must now be reduced to $1n$ by dividing both sides by 3, giving $n = 4$.

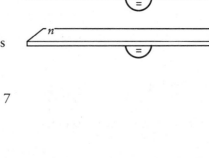

The stages are: $3n + 7 = 19$

Subtract 7 from both sides: $3n + 7 - 7 = 19 - 7$

giving $3n = 12$

Divide both sides by 3: $\dfrac{3n}{3} = \dfrac{12}{3}$

giving $n = 4$

$\therefore$ The number was 4.

Example 1

Solve the equation $\frac{1}{2}x - 1 = 6$.

$$\frac{1}{2}x - 1 = 6$$

Multiply through by 2 to remove fraction: $x - 2 = 12$

Add 2 to each side: $x - 2 + 2 = 12 + 2$

$$x = 14$$

Example 2

Solve the equation $2(x - 3) = 5$.

$$2(x - 3) = 5$$

Multiply out the bracket: $2x - 6 = 5$

Add 6 to each side: $2x = 11$

Halve each side: $x = 5\frac{1}{2}$

Exercise 131

1–8   Solve the equations formed in questions 1–8 of Exercise 130 to find the number.

9   Solve the equation formed in question 9 of Exercise 130 to find the width of the rectangle.

10   Solve the equation formed in question 10 of Exercise 130 to find the length of the rectangle.

In the following questions, solve the equations to find the value of x:

11 $x + 7 = 11$ **15** $12 - x = 9$ **19** $\dfrac{x}{3} - 2 = 1$

12 $5x = 15$ **16** $17 - 2x = 5$ **20** $\dfrac{2x}{5} + 2 = 10$

13 $4x - 5 = 19$ **17** $2(x - 6) = 8$

14 $\dfrac{3x}{4} = 6$ **18** $\dfrac{x}{2} + 3 = 7$

Harder equations

The method used above can be very time-consuming and, when solving more difficult equations, a quicker method is used.

For the equation

$$3n + 7 = 19$$

when 7 is subtracted from both sides, we know it will disappear from the LHS; therefore it is only necessary to write it on the RHS:

$$3n = 19 - 7$$

i.e. the 7 has *changed sides* and has *changed sign* from + to −.

We now have:

$$3n = 12$$

Similarly, when both sides are divided by 3 we know that the LHS will be reduced to n.

Showing the division on the RHS only we have:

$$n = \frac{12}{3} = 4$$

i.e. the 3 has *changed sides* and *changed sign* from $\times$ to $\div$.

The rule for eliminating a quantity from one side of an equation is

change to the opposite side and change to the opposite sign.

This process is called **transposing**.

Example 1

Solve the equation $5x - 4 = 3x + 12$

In this type of equation, the x terms should be collected on one side of the equation and the numerical terms on the other, i.e. $3x$ must be eliminated from the RHS and -4 from the LHS.

$$5x - 4 = 3x + 12$$

Transposing $3x$ and -4 gives:
$$5x - 3x = 12 + 4$$
$$2x = 16$$
$$x = 8$$

Answers to algebraic equations should *always* be checked by substituting back into the LHS and RHS of the original equation, as shown below:

$$\text{LHS} = 5x - 4 = 5 \times 8 - 4 = 36$$
$$\text{RHS} = 3x + 12 = 3 \times 8 + 12 = 36$$
$$\text{LHS} = \text{RHS, so the solution is correct.}$$

Example 2

Solve $2(2x - 5) = 3(x - 4)$

$$2(2x - 5) = 3(x - 4)$$

Multiply out the brackets:
$$4x - 10 = 3x - 12$$

Transpose $3x$ and -10:
$$4x - 3x = 10 - 12$$
$$x = -2$$

Example 3

Solve $4(x + 3) - 2(x - 5) = 46$

$$4(x + 3) - 2(x - 5) = 46$$

Multiply out the brackets:
$$4x + 12 - 2x + 10 = 46$$

Collect terms:
$$2x + 22 = 46$$

Transpose 22:
$$2x = 24$$

Divide by 2:
$$x = 12$$

Exercise 132

Solve the following equations to find the value of x:

1 $7x + 3 = 5x + 11$

2 $5x - 2 = 2x + 7$

3 $6x - 4 = 10 - x$

4 $3 - 2x = 4 - 5x$

5 $3(x - 5) = 12$

6 $5 + 2(x + 1) = 11$

7 $2(x + 7) + 4 = 18$

8 $2(x - 4) = (x + 2)$

9 $3(2x - 1) = 7(x - 1)$

10 $(x + 5) + 2(3x - 2) = 8$

11 $4(x + 2) + 2(x + 3) = 32$

12 $2(2x + 1) - 3(3x - 4) = 40$

Further problems

Example 1

I buy a pizza and cut it into three pieces. When I weigh the pieces, I find that one piece is 8 g lighter than the largest piece and 5 g heavier than the smallest piece.
If the whole pizza weighs 360 g, how much does the smallest piece weigh?

Let the weight of the smallest piece be x grams
The weights of the other two pieces are $(x + 5)$ grams
and $(x + 5 + 8)$ grams

Total weight of the three pieces $= x + (x + 5) + (x + 13)$

$$\therefore x + (x + 5) + (x + 13) = 360$$
$$3x + 18 = 360$$
$$3x = 342$$
$$x = 114$$

The weight of the smallest piece is 114 grams.

Example 2

This year, Dawn is three times as old as her brother Marcus, but in four years' time she will be twice as old. How old are Dawn and Marcus now?

Let Marcus' age now be x years
Then Dawn's age now is $3x$ years
In four years' time:
Marcus will be $(x + 4)$ years
Dawn will be $(3x + 4)$ years

Dawn will then be twice as old as Marcus

$$\therefore (3x + 4) = 2(x + 4)$$
$$3x + 4 = 2x + 8$$
$$3x - 2x = 8 - 4$$
$$x = 4$$

Marcus is 4 years old and Dawn is 12 years old.

Exercise 133

1 A rubber costs 25p more than a pencil. Twelve pencils and ten rubbers are bought for a bran tub. The cost of a pencil is x pence.

  a Write down, in terms of x:

    (i) the cost of a rubber

    (ii) the cost of 12 pencils

    (iii) the cost of 10 rubbers.

  b The total cost of the 12 pencils and 10 rubbers is £6.90. Using this information:

    (i) write down an equation in terms of x

    (ii) solve the equation to find x

    (iii) find the cost of one rubber.

2 Mr Wilson regularly attends football matches when his team plays at home.

  a For x number of games he buys a seat in the stands. Each seat costs £9.
  Write down an expression for the cost of these x games.

  b For the remaining matches he buys a ticket for the terraces. Each ticket costs £5.
  If he attends 20 games in a season, write down an expression in x for the cost of tickets for the terraces.

  c Using your answers to a and b, write down and simplify an expression for the total cost for the season.

  d The cost for the season was £148.
  Write down an equation in terms of x.

  e Solve the equation to find the number of matches he watched from the stands.

3 Eighteen theatre seats were bought for a school party. Seats in the circle cost £x, seats in the stalls cost £1.80 less. Eight of the party sat in the circle, the remainder in the stalls.

  a What was the cost of a seat in the stalls in terms of x?

  b Write down and simplify an expression for the total cost of the seats, in terms of x.

  c The total cost of the seats was £136.80. Write down an equation in terms of x and solve it to find the cost of:

    (i) a seat in the circle

    (ii) a seat in the stalls.

4 A pound of apples costs 5p more than a pound of pears. The cost of 5 pounds of apples and 3 pounds of pears is £4.65.
What is the cost of one pound of pears?

5 In a mathematics test, Ann scored 3 more marks than Brian who scored 5 marks more than Carol.

The total of their three scores was 139.
How many marks did they each score?

6 Alan weighs 3 kg less than Barry who weighs 4 kg less than Colin.

Barry weighs x kilograms and the total weight of the three boys is 193 kg.
How much does Colin weigh?

7 Apples cost 30p per pound more than bananas and 72p per pound less than cherries.

The cost of 2 pounds of apples and 3 pounds of bananas is the same as the cost of 2 pounds of cherries.
What is the cost of a pound of cherries?

27.6 Formulae

Substitution in formulae

A second important use of algebra is to enable a relationship between two or more quantities to be expressed in a short but easily understood form. This is called a **formula**.

The volume of a cylinder is found by multiplying the area of the circular cross-section of the cylinder by its height. This is more neatly expressed by the formula

$$V = \pi r^2 h$$

In order to calculate the volume, the given values of r (radius) and h (height) are substituted into the formula.

Example

The distance travelled by a vehicle in a given time can be found by using the formula:

$$s = ut + \tfrac{1}{2}at^2$$

where s is the distance travelled in metres, u is the starting velocity in metres per second, a is the acceleration in metres per second per second, and t is the time in seconds.

Find the distance travelled by a sports car:

a in 4 s if $u = 20$ m/s and $a = 2.5$ m/s^2

b in 10 s starting from rest with an acceleration of 3 m/s^2.

a $t = 4$, $u = 20$, $a = 2.5$

Substituting in $s = ut + \tfrac{1}{2}at^2$

gives $s = (20 \times 4) + (\tfrac{1}{2} \times 2.5 \times 4^2)$
$= 80 + 20$
$= 100$

The distance travelled was 100 m.

b $t = 10$, $u = 0$, $a = 3$

Substituting in $s = ut + \tfrac{1}{2}at^2$

gives $s = 0 + \tfrac{1}{2} \times 3 \times 10^2$
$= 150$

The distance travelled was 150 m.

Exercise 134

1 a The speed, or velocity, at which the sports car in the example above is travelling after a certain time, is given by the formula

$$v = u + at$$

Calculate the speed, in m/s, when
(i) $t = 4$, $u = 20$, $a = 2.5$
(ii) $t = 10$, $u = 0$, $a = 3.6$.

b The formula $V = 2.25v$ converts v m/s to V mph.
Calculate the sports car's speed in miles per hour for each of parts (i) and (ii) above.

2 The formula for a straight line graph is

$$y = mx + c$$

Find the coordinate y if $m = 2$, $c = -3$ and $x = 4$.

3 To convert degrees Fahrenheit F to degrees Celsius C, the following formula is used:

$$C = \frac{5}{9}(F - 32)$$

Convert to degrees Celsius:

a 50°F, **b** 77°F, **c** 14°F.

4 The surface area of a cylinder is given by the formula

$$S = 2\pi r(r + h)$$

Find the surface area of a cylinder if $r = 6$ cm and $h = 14$ cm.

5   The focal length of a lens is given by

$$\frac{1}{f} = \frac{1}{v} + \frac{1}{u}$$

Find f when:

a $u = 12$, $v = 18$ **b** $u = 10.5$, $v = 7$.

6   When an amount of money, £P, is invested at a compound interest rate of $r\%$ per annum, the amount in the account after n years is given by

$$A = P\left(1 + \frac{r}{100}\right)^n$$

Find the amount in an account after 2 years if £1600 was invested at 9.5% per annum.

7   The formula for the area of a trapezium is

$$A = \frac{h}{2}(a + b)$$

where a and b are the lengths of the parallel sides of the trapezium and h is its height.

Find the area of a trapezium when

a $a = 5$ cm $b = 7$ cm $h = 4$ cm

b $a = 4.3$ cm, $b = 10.5$ cm, $h = 5.6$ cm.

8   The time, in seconds, for a pendulum to complete a full swing is given by

$$T = 2\pi\sqrt{\frac{l}{10}}$$

where l metres is the length of the pendulum.

How long, to the nearest second, will it take a pendulum of length 0.5 m to complete five full swings?

Rearranging formulae

The formula which converts degrees Fahrenheit to degrees Celsius is

$$C = \frac{5}{9}(F - 32)$$

It may, however, be necessary to convert degrees Celsius to degrees Fahrenheit. In this case, the formula needs to be **rearranged**, or **transposed**, to give F in terms of C, i.e. F is made the subject of the formula.

Example

a Rearrange the formula $C = \frac{5}{9}(F - 32)$ to give F in terms of C.

b Use the transposed formula to convert 15°C to °F.

a The method is the same as for solving equations.

$$C = \frac{5}{9}(F - 32)$$

(i)   Multiply through by 9: $9C = 5(F - 32)$

(ii) Remove the bracket: $9C = 5F - 160$

(iii) Transpose 160: $9C + 160 = 5F$

(iv) Dividing through by 5 gives the formula for F: $F = \dfrac{9C + 160}{5}$

b When $C = 15$

$$F = \frac{9 \times 15 + 160}{5}$$

$$= \frac{135 + 160}{5}$$

$$= \frac{295}{5}$$

$$= 59$$

$$\left(\text{Check.} \quad C = \frac{5}{9}(F - 32) = \frac{5}{9}(59 - 32) = \frac{5}{9} \times 27 = 15\right)$$

Exercise 135

1 Rearrange

  a $P = 2(L + B)$ to make B the subject

  b $S = 2\pi rh$ to make h the subject

  c $A = \pi r^2$ to make r the subject

  d $S = \left(\dfrac{u + y}{2}\right)t$ to make t the subject

  e $v = u + at$ to make a the subject.

2 a Rearrange the straight line equation

$$y = mx + c$$

  to give x in terms of y.

  b Find x if:

   (i) $m = 2$, $c = 1$, $y = 5$

   (ii) $m = -1$, $c = 4$, $y = -3$.

3 a The mean of three numbers a, b and c is given by

$$M = \frac{a + b + c}{3}$$

  Find the mean of the numbers 12, 8 and 16.

  b Rearrange the formula to find a in terms of b, c and M.

  c If the mean of three numbers is 7 and two of the numbers are 8 and 9, what is the third number?

4 The area of a trapezium is given by

$$A = \frac{h(a + b)}{2}$$

  a Make h the subject of the formula and find h when $A = 40$, $a = 10$, $b = 6$.

  b Make a the subject of the formula and find a when $A = 63$, $h = 9$, $b = 6$.

5 Any term in the sequence of numbers such as

$$3, 7, 11, 15 \ldots$$

can be calculated using the formula

$$N = a + (n - 1)d$$

where:
a is the first term,
d is the difference between successive terms, and
n is the number of the term to be calculated
(i.e. $n = 3$ gives the third term).

  a By taking $n = 4$, show that the formula is correct for the sequence given above.

  b Find the eleventh term in this sequence.

  c Rearrange the formula to make d the subject.

  d The first term of a sequence is 2 and the eighth term is 23:

   (i) What is the difference between successive terms?

   (ii) Write down the first five terms in the sequence.

6 The deposit required when booking a self-catering holiday consists of a fixed amount A, which is non-returnable, plus $\frac{1}{20}$ of the cost of the accommodation.

$$D = A + \frac{NC}{20}$$

where C is the cost of the accommodation per person and N is the number of people booking the holiday.

a Find the deposit payable for six people, if the holiday costs £210 per person and there is a non-returnable amount of £50.

b Rearrange the formula to make C the subject.

c Calculate C if $D = £133$, $A = £65$ and $N = 4$.

Exercise 136: Miscellaneous questions

1 Simplify the following:

   a $9x - 2y + 5y - 4x$

   b $6p - 7q + 4r + 3p + 5q - 6r$

2 If $x = -2$ and $y = 4$, find the value of:

   a $4x + 3y$ **b** $\dfrac{2x - 4y}{5}$ **c** $x^2y + x^3$

3 Simplify:

   a $3a \times 4b$ **d** $2a^2 \times 5a^4$

   b $(-2a)^2 \times 2b$ **e** x^0

   c $x^{-4} \times x^6$ **f** $3pq \times (-4q^2r) \times (-pr^2)$

4 Simplify:

   a $x^7 \div x^4$ **d** $\dfrac{5}{3a^2} \times \dfrac{9a^5}{10}$

   b $x^2 \div x^5$

   c $x^3 \div x^5 \times x^2$ **e** $\dfrac{3x^2}{2} \div \dfrac{x^3}{4}$

5 Remove brackets and simplify:

   a $3x + 2(3x - 2)$

   b $2(4x + 5) + (x - 7)$

   c $4(x - 1) - 2(3x - 2)$

6 Factorise:

   a $6ab + 9ac + 3a$

   b $xyz - xy^2$

   c $2abc^2 - 6a^2bc + 4ab^2c$

7 Express as a single fraction, simplifying where possible:

   a $\dfrac{x}{2} + \dfrac{x}{4} - \dfrac{2x}{3}$

   b $\dfrac{4}{5x} - \dfrac{3}{10x}$

   c $\dfrac{2(x - 1)}{3} - \dfrac{(x - 3)}{4}$

8 Solve the following equations:

   a $9x - 5 = 67$

   b $\dfrac{3x}{5} + 2 = 20$

   c $4(x + 1) - 3(2 - x) = 12$

9 $S = \frac{1}{2}n(n + 1)$ is a formula which gives the sum of the positive integers from 1 up to n.

   a Find the sum of the numbers from 1 to 20.

   b Find the sum of the first 50 positive integers.

10 **a** Find y when $x = -4$, given that

$$y = \frac{3x + 2}{x}$$

   b Rearrange the above expression to give x in terms of y.

   c Find x when   (i) $y = 2.5$,   (ii) $y = 7$.

28 Geometry

Geometry is one of the oldest branches of mathematics. The name is derived from the Greek words 'ge', meaning earth, and 'metrein', the verb to measure. The Ancient Babylonians, Egyptians and Greeks all contributed to the development of geometry.

28.1   Lines and angles

Lines

Lines can be curved or straight.

There are very few straight lines in nature, but much of geometry is concerned with straight lines.

On a flat surface – and most ancient civilisations believed the earth to be flat – a straight line is the shortest distance between two points.

A **point** marks a position and has no size.

The line below joins the two points A and B:

A   B

To indicate a point in a line, it is usual to mark the point with a small 'dash':

A

The next diagram shows two points A and B and a line passing between the two points.
The line extends infinitely in both directions.

If we wish to refer to the part of the line between the points A and B, we should, strictly speaking, refer to the **line segment** AB. It is, however, usual to use **'the line AB'** for the line segment.

A   B

Straight lines are drawn using a ruler, a rule or a straight edge.

Angles

When two straight lines meet they form an angle. The angle between AO and OB is called the angle AOB (or it could be called BOA).

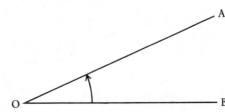

The size of the angle AOB on p. 239 is a measure of how far the line OA has been rotated from a starting position along OB. If we continue to rotate OA in a clockwise direction until it once again lies along OB, it will have been rotated through a full circle or **360 degrees** (also written as **360°**).

The reason there are 360 degrees in a full circle is because the Babylonians believed there were 360 days in the year. An angular measure was derived in order to chart the movements of the stars, which took one year for a full rotation about the earth.

A rotation through half of a circle, or half-turn, is a turn through 180°.

Angles which add together to make 180° form a straight line and are called **supplementary angles**. COB and AOC are supplementary angles.

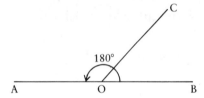

A rotation through a quarter of a circle, or a quarter-turn, is a turn through 90°.

An angle of 90° is usually called a **right angle** and angles which add together to make 90° are called **complementary angles**. Angles AOC and BOC shown below are complementary angles.

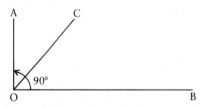

Right angles occur a great deal in the man-made world. If you look around any room, you will see many examples of right angles.

Draughtsmen use a set square for drawing and checking right angles.

Two lines which intersect at right angles are called **perpendicular lines**. The diagram on the right shows two perpendicular lines.

The special symbol for a right angle has been used in the diagram.

Other types of angle are:

acute angles, of size between 0° and 90°

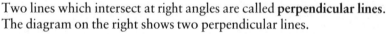

obtuse angles, of size between 90° and 180°

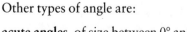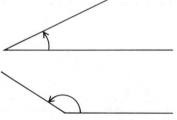

*reflex angles, of size greater than 180°

vertically opposite angles, which are formed when two straight lines cross.

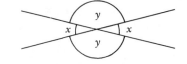

In the diagram alongside, the angles marked *x* are vertically opposite angles and they are equal in size. The angles marked *y* are also equal.

From what you have read in this chapter you should be able to work out the angles in the following exercise.

Exercise 137

1 Calculate the size of the missing angle in each case:

a   b   c   d

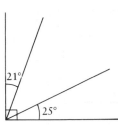 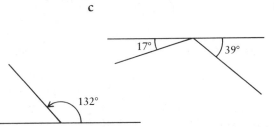 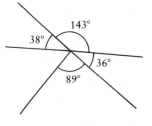

2 State whether the following angles are acute, obtuse or reflex:

a   c   e

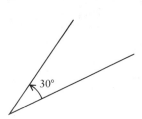 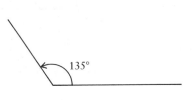 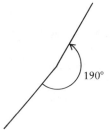

b   d   f

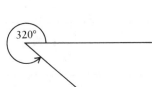 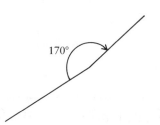 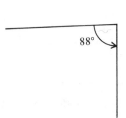

3 Find the size of the unknown angles:

a   b   c

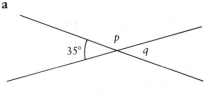 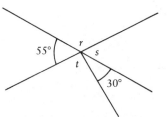 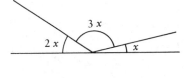

Angles and parallel lines

Lines which do not intersect, no matter how far they are extended, are called **parallel lines**.

There are many examples of parallel lines around us.

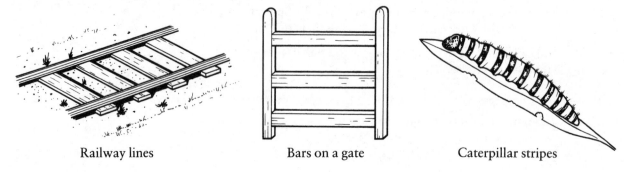

Railway lines   Bars on a gate   Caterpillar stripes

On a wide gate, an extra bar is added at an angle to act as a brace or support.

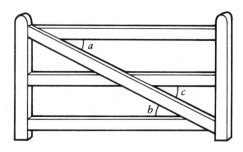

Several angles are formed.

Angles such as *a* and *b*, which are on opposite sides of the bracing bar, are called **alternate angles**.

Angles such as *a* and *c*, which are in similar positions on the same side of the bar, are called **corresponding angles**.

Angles *a*, *b* and *c* are all equal in size.

The arrows indicate that the lines are parallel.
All the angles marked *a* are equal in size.
All the angles marked *b* are equal in size.

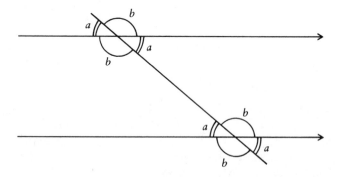

Example

Find the size of the angles lettered *a*, *b* and *c* in the diagram below:

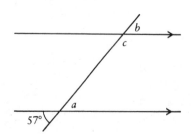

$a = 57°$ (vertically opposite angles)

$b = a$ (corresponding angles)
$ = 57°$

$b + c = 180°$ (angles on a straight line)
$ c = 123°$

(*Note.* *a* and *c* are also supplementary angles)

Exercise 138

Find the size of each lettered angle in the following diagrams:

1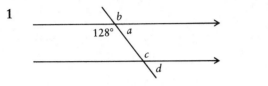

2

3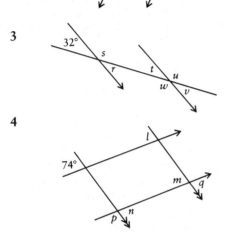

4

5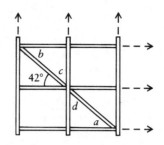

The diagram shows part of a fence panel.

6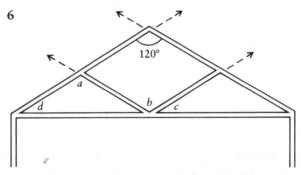

The diagram shows a cross-section through the roof of a barn.

28.2 Symmetry

Symmetry implies balance and harmony.
Shapes which are symmetrical are balanced because there is a repeated pattern.
There are two types of symmetry, **reflexive** (or line) symmetry and **rotational** symmetry.

Reflexive symmetry

A shape is said to have **reflexive symmetry** if a line can be drawn which cuts the shape into two halves and each half is the mirror image of the other.

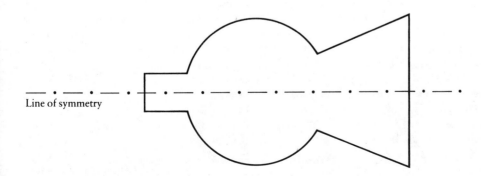

Line of symmetry

In the example below, AB is *not* a line of symmetry, even though the two halves PQBA and ABRS are identical.

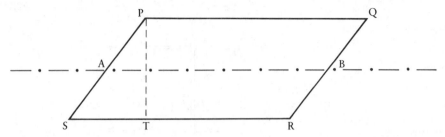

The mirror image of point P is point T, therefore ABRS cannot be the mirror image of PQBA.

Shapes may have several lines of symmetry.
For example, the hexagon on the right has six lines of symmetry, shown by the dotted lines.

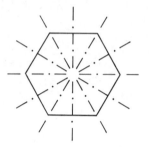

Exercise 139

1   State the number of lines of symmetry which could be drawn through the following shapes:

a

c

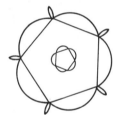

b

d

2   Copy the letters of the alphabet, shown below, and draw the maximum number of lines of symmetry on to each letter.

A H X Z D

3 Lines of symmetry are indicated on the diagrams below, but the shapes are incomplete.
Copy and complete each shape.

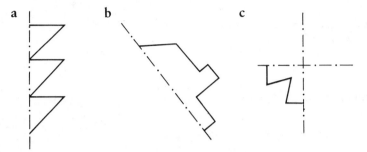

Rotational symmetry

The parallelogram shown below is not symmetrical about any line. If, however, the parallelogram is rotated about point O until PQ is in the position previously occupied by RS, the shape will look exactly the same as the original:

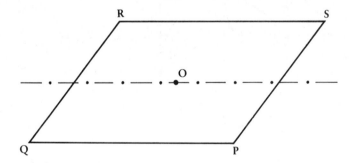

Parallelograms have **rotational symmetry**.

Because there are *two* positions in which the shape looks identical, the parallelogram has rotational symmetry of **order 2**.

The equilateral triangle ABC can be rotated about O, through 120° anticlockwise, to the position CAB, which is identical to ABC:

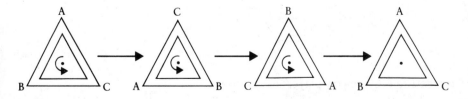

Repeating this rotation twice more will bring the triangle back to its original position.

The triangle has rotational symmetry of **order 3**.

1 List the letters of the alphabet which have rotational symmetry and state the order of symmetry.

2 **a** Which of the shapes in question 1 of Exercise 139 have rotational symmetry?

   b What is the order of symmetry of each?

3 For each of the shapes shown below, state (where appropriate):

  (i) the number of lines of symmetry

  (ii) the order of rotational symmetry.

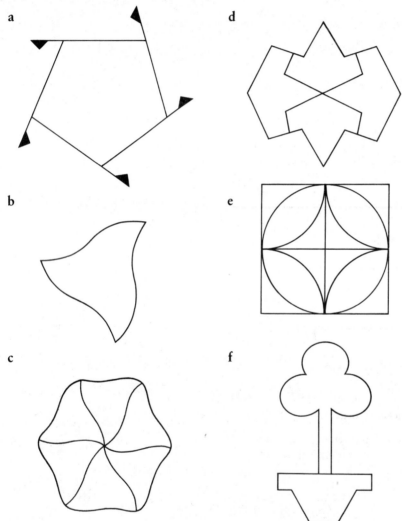

a

d

b

e

c

f

Historic note. Patterns **b, c** and **d** are Arabic. Symmetry played an important part in Arabic art and mathematics.

Pattern **e** was given to students in Babylonian times for the calculation of areas.

All these patterns tessellate.

28.3 Triangles

A **triangle** is a three-sided polygon.

All triangles belong to one of three categories:

Scalene triangles

The three sides are of different length.

The three angles are of different size.

A scalene triangle has no reflexive or rotational symmetry.

Isosceles triangles

Two sides are equal in length.

Two angles are equal in size.

An isosceles triangle has one line of symmetry.

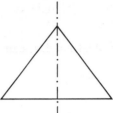

Equilateral triangles

All three sides are equal in length.

All three angles are equal in size.

An equilateral triangle has three lines of symmetry and rotational symmetry of order 3.

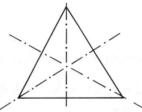

The triangles above are all **acute-angled** triangles, i.e. all three angles are less than 90°.

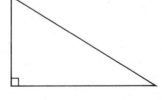

Triangles can also be:

Right-angled

if one angle is 90°.

Obtuse-angled

if one angle is greater than 90°.

Properties of triangles

The basic property of all triangles is this:

> **The three interior angles of a triangle always total 180°.**
>
> $a + b + c = 180°.$

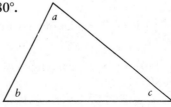

It follows from this that, in the triangle below, $d = a + b$

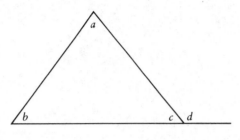

$$a + b + c = 180° \qquad \text{(angles of a triangle)}$$
$$d + c = 180° \qquad \text{(angles on a straight line)}$$
$$d = a + b$$

In other words: **The exterior angle of a triangle is equal to the sum of the opposite interior angles.**

Equilateral triangles

Because all three angles are equal in size, the angles of an equilaterial triangle are all 60°.

Isosceles triangles

An isosceles triangle has two sides equal in length and the angles opposite these sides equal in size.
In the diagram below, the equal sides are marked with dashes and the equal angles with arcs.

$$AB = AC$$

$$\text{Angle B} = \text{Angle C}$$

which we can also write as

$$\text{Angle ABC} = \text{Angle ACB}$$

$$\text{or } \angle \text{ABC} = \angle \text{ACB}$$

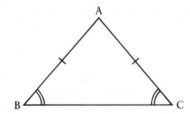

Example 1

In triangle PQR, PQ = PR and the angle PQR = 75°.

Find: **a** angle PRQ, **b** angle QPR.

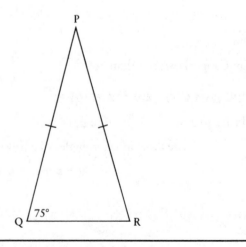

Angle PRQ = Angle PQR (angles of an isosceles triangle)
      = 75°

Angle QPR = 180° − (75° + 75°) (angles of a triangle)
      = 30°

Example 2

In the diagram, find angle XZY.

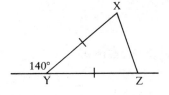

Angle XZY + Angle ZXY = 140° (exterior angle of a △ = sum of interior opposite angles)

But

Angle XZY = Angle ZYX (angles of an isosceles △)
= 70°

Exercise 141

1 Find angles P and R.

4 Find angles *x*, *y* and *z*.

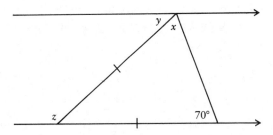

2 Find angles A and B.

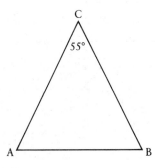

5 Find angles *x* and *y*.

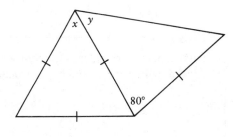

3 Find angles ACB, CAB, DCA, ADC.

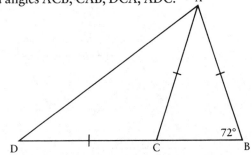

Congruent triangles

Two triangles are congruent if they are identical.
This means that:

(i) the angles in the first triangle are equal to the corresponding angles in the second triangle.

(ii) the lengths of the sides in the first triangle are equal to the corresponding lengths of the sides in the second triangle.

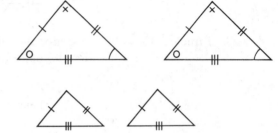

Triangles are known to be congruent if:

(i) **three sides** in the first triangle are equal to three sides in the second triangle

(ii) **two sides and the angle between them** in the first triangle are equal to two sides and the angle between them in the second triangle (the only exception is a pair of right-angled triangles)

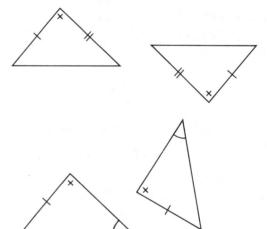

(iii) **two angles and one side** in the first triangle are equal to two angles and the corresponding side in the second triangle.

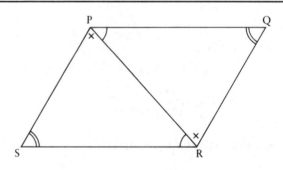

Example

Name a pair of triangles in the following diagram which are definitely congruent.

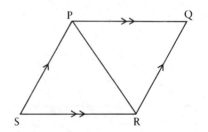

Marking the equal sides and equal angles in the triangles PQR and PRS, we can see that

∠ QPR = ∠ PRS,
∠ PQR = ∠ PSR,
∠ PRQ = ∠ RPS

Side PR is in both triangles.

∴ Triangle PQR is congruent to triangle RSP.

In each of the following questions, where possible, name a pair of triangles which are definitely congruent:

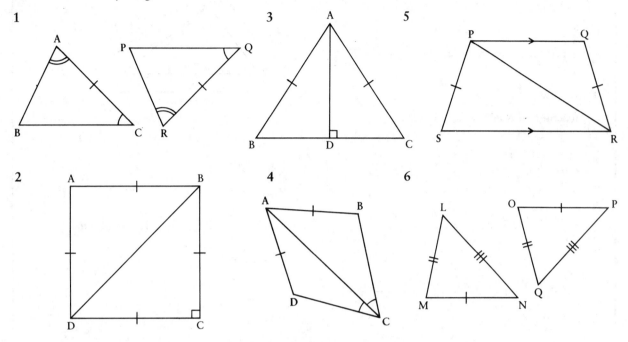

Similar triangles

Two triangles are similar if the angles in the first triangle are the same size as the angles in the second triangle, i.e. they have the same shape but are of different size.

If two triangles are similar, their corresponding sides are in the same ratio.

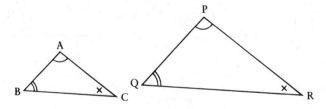

$\angle A = \angle P, \angle B = \angle Q, \angle C = \angle R,$
∴ Triangles ABC and PQR are similar

and $\dfrac{AB}{PQ} = \dfrac{BC}{QR} = \dfrac{CA}{RP}$

If $AB = 2\,cm$ and $PQ = 5\,cm$,

$\dfrac{PQ}{AB} = \dfrac{5}{2}$

i.e. triangle PQR is an **enlargement** of triangle ABC and the scale factor of the enlargement is $\frac{5}{2}$ or 2.5.

Example

Triangles PQR and XYZ are similar.
QR = 2.5 cm, PR = 2.0 cm, PQ = 1.5 cm.
YZ = 1.5 cm

Find:

a the ratio between the sides of triangles PQR and XYZ

b the length of XZ

c the length of XY.

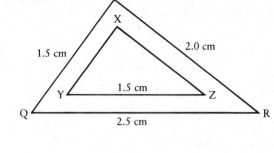

a $\dfrac{QR}{YZ} = \dfrac{2.5}{1.5} = \dfrac{5}{3}$

∴ The ratio is 5:3

b $\dfrac{XZ}{PR} = \dfrac{3}{5}$

∴ $XZ = \dfrac{3}{5} \times PR = \dfrac{3}{5} \times 2.0 = 1.2\,cm$

$XZ = 1.2\,cm$

c $\dfrac{XY}{PQ} = \dfrac{3}{5}$

∴ $XY = \dfrac{3}{5} \times PQ = \dfrac{3}{5} \times 1.5 = 0.9\,cm$

$XY = 0.9\,cm$

Exercise 143

1 Triangles ABC and PQR are equiangular.

Find PQ and PR.

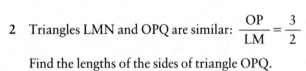

2 Triangles LMN and OPQ are similar: $\dfrac{OP}{LM} = \dfrac{3}{2}$

Find the lengths of the sides of triangle OPQ.

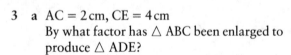

3 **a** AC = 2 cm, CE = 4 cm
By what factor has △ ABC been enlarged to produce △ ADE?

b If BC = 2.7 cm, what is the length of DE?

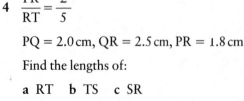

4 $\dfrac{PR}{RT} = \dfrac{2}{5}$

PQ = 2.0 cm, QR = 2.5 cm, PR = 1.8 cm

Find the lengths of:

a RT **b** TS **c** SR

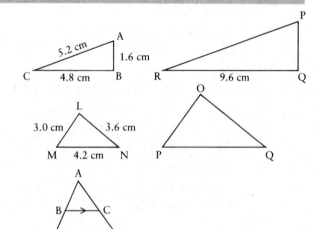

28.4 Quadrilaterals

As the name suggests, **quadrilaterals** are four-sided polygons.

A line joining two vertices of a polygon is called a **diagonal**.

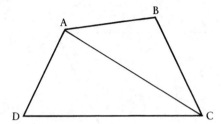

In the quadrilateral ABCD, above, the diagonal AC divides the quadrilateral into two triangles.

The sum of the angles of each triangle is 180°.

From this we deduce that:

The sum of the angles of a quadrilateral is 360°

Some special quadrilaterals are described below:

A trapezium is a quadrilateral which has one pair of opposite sides parallel.

A parallelogram is a quadrilateral which has both pairs of opposite sides parallel.

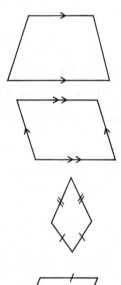

A kite is a quadrilateral which has two pairs of adjacent sides equal in length.

A rhombus is a parallelogram which has four sides of equal length.

A rectangle is a parallelogram which has angles of 90°.

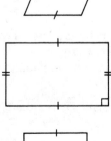

A square is a rectangle which has four sides of equal length, (or a rhombus with angles of 90°).

Properties of parallelograms

Applying a knowledge of symmetry, the properties of parallelograms can be listed.

All parallelograms. The parallelogram ABCD has rotational symmetry about O. Therefore:

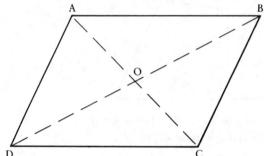

1   AB = DC and AD = BC

    i.e. **opposite sides are equal**

2   ∠ A = ∠ C and ∠ B = ∠ D

    i.e. **opposite angles are equal**

3   AO = OC and BO = OD

    i.e. **diagonals bisect each other.**

Rhombuses. A rhombus is a parallelogram and therefore all the properties listed above are true for a rhombus.
In addition, EG and FH are lines of symmetry. Therefore:

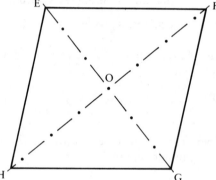

1   ∠ HEG = ∠ FEG,
    ∠ EFH = ∠ GFH,
    ∠ FGE = ∠ HGE,
    ∠ GHF = ∠ EHF

    i.e. **Diagonals bisect opposite angles.**

2   ∠ EOF = ∠ EOH = ∠ GOH = ∠ GOF = 90°.

    i.e. **The diagonals bisect each other at 90°.**

Rectangles. A rectangle is a parallelogram which has angles of 90°. Rectangles have two lines of symmetry and rotational symmetry.
Therefore:

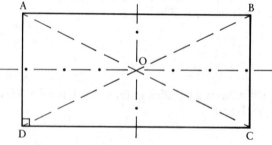

1   AO = OD = OC = OB

    i.e. **The diagonals are equal.**

Squares. A square is a rectangle which has all its sides equal. All the properties of a rectangle and a rhombus apply to a square.
Squares have four lines of symmetry and rotational symmetry of order 4.
Therefore:

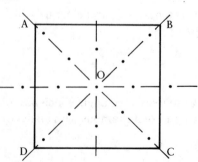

All the angles formed in the square ABCD are either 90° or 45°.

Example

ABCD is a parallelogram.

∠ B = 63°, ∠ BAC = 47°

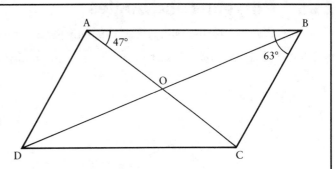

Find **a** ∠ A   **b** ∠ ACB

a   ∠ A + ∠ B = 180° (supplementary angles)

∠ A = 117°

b   ∠ DAC = 117° − 47°
= 70°

∠ ACB = ∠ DAC (alternate angles)
= 70°

Exercise 144

1   Use your knowledge of symmetry to list the
properties of the kite ABCD.

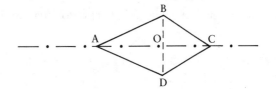

2   The trapezium PQRS is an isosceles trapezium,
i.e. PS = QR.

∠ S = 44°

Find the size of angles P, Q and R.

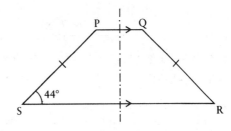

3   ABCD is a rectangle with BC = CX.
Find the angle BXC.

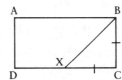

4   In rhombus PQRS, ∠ PSQ = 32°

Find: **a** angle PQS, **b** angle PRQ.

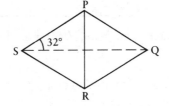

5   In parallelogram EFGH, ∠ EFH = 29°,
∠ EOH = 57°, ∠ EHG = 110°

Find: **a** ∠FOG, **b** ∠HFG, **c** ∠EGH.

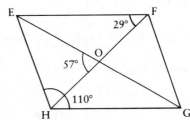

28.5 Polygons and angles

A regular polygon has all its sides equal in length and all its angles equal in size.

In the polygon below, the marked angles are the **exterior angles** of the polygon.

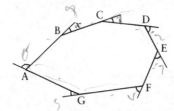

If you face along the direction AB and turn through the angle marked x, you will face along direction BC.
Turning through y will face you in the direction CD.
When you have turned through all of the exterior angles, you will again be facing the direction AB, having made a complete rotation of 360°.

The sum of the exterior angles of any polygon is 360°.

In the regular pentagon shown below, x is an exterior angle and y is an interior angle.

Interior angle + Exterior angle = 180° = 2 right angles

For a polygon with n sides,

$$\text{Interior angles} + \text{Exterior angles} = 180° \times n = 2n \text{ right angles}$$
$$\text{Exterior angles} = 360° = 4 \text{ right angles}$$
$$\therefore \quad \text{Interior angles} = (2n - 4) \text{ right angles.}$$

Sum of the interior angles of any polygon = $(2n - 4)$ right angles.

(*Note.* For regular polygons, it is sufficient to remember that the sum of the exterior angles is always 360°.)

Example 1

Find the size of an interior angle of a regular hexagon.

A regular hexagon has six exterior angles.

The sum of the exterior angles = 360°

$$\therefore \text{Each exterior angle} = \frac{360°}{6} = 60°$$

Interior angle + Exterior angle = 180°

$$\therefore \text{Each interior angle} = 180° - 60°$$

$$= 120°$$

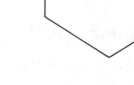

Example 2

The sum of the interior angles of a regular polygon is 1260°.

How many sides has the polygon?

Interior angles + Exterior angles = 1260° + 360°
$$= 1620°$$

1 interior angle + 1 exterior angle = 180°

$$\therefore \quad \text{No. of sides of the polygon} = \frac{1620°}{180°} = 9$$

The polygon has 9 sides (a nonagon).

Exercise 145

1 A regular polygon has 8 sides. What is the size of:

 a each exterior angle

 b each interior angle?

2 Repeat question 1 for a regular decagon, which is a polygon with 10 sides.

3 The exterior angle of a regular polygon is 20°. How many sides does it have?

4 The interior angle of a regular polygon is 140°. How many sides does it have?

5 What is the sum of the interior angles of a regular pentagon?

6 What is the sum of the interior angles of a regular decagon?

7 How many sides has a polygon if the sum of the interior angles is:

 a 1620° b 2340°?

28.6 Circles and angles

Definitions

A circle is a set of points which are a fixed distance from a given point.
The given point is the **centre** of the circle and the fixed distance is the **radius**.

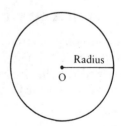

The **circumference** is the perimeter of the circle.

An **arc** is a section of the circumference.

A **chord** is a line joining two points on the circumference.

A **diameter** is a chord which passes through the centre.

A **tangent** is a line which touches the circle at one point only.

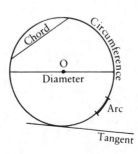

A **segment** is part of a circle cut off by a straight line.
A **sector** is part of a circle cut off by the radii.

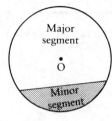

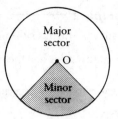

Angle ACB is an **angle subtended** at a point, C, on the circumference **by the arc** AB.

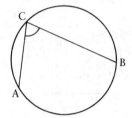

Circles and right angles

(i) The angle subtended at the circumference by a diameter is a right angle:

$\angle$ ACB = 90°

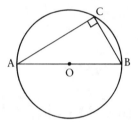

(ii) The angle formed by a radius and a tangent to the circle is a right angle:

$\angle$ OAT = 90°

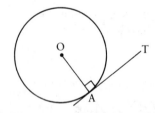

*Angles in a circle

There are several properties relating to angles in a circle which derive from the fact that the angle subtended at the centre of the circle, by an arc, is twice the angle subtended at the circumference, by the same arc. (In each of the circles below the angle indicated by the double arc is twice the angle indicated by the single arc.)

(i)

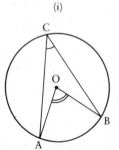

(ii)

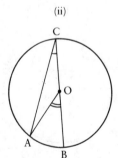

(iii)

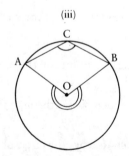

Theorems

1 The angle at the centre is twice the angle at the circumference, subtended by the same arc.

$$\angle\, AOB = 2\, \angle\, ACB$$

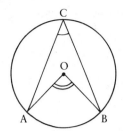

2 The angle in a semicircle is a right angle.

$$\angle\, ACB = 90°$$

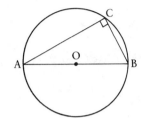

3 Angles in the same segment of a circle are equal (i.e. angles subtended at the circumference by the same arc).

$$\angle\, ACB = \angle\, ADB$$
$$\angle\, CAD = \angle\, CBD$$

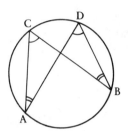

4 The opposite angles of a cyclic quadrilateral are supplementary.
(A cyclic quadrilateral is any quadrilateral whose vertices are points on the circumference of a circle.)

$$\angle\, ABC + \angle\, ADC = 180°$$

$$\angle\, BAD + \angle\, BCD = 180°$$

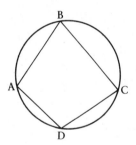

Exercise 146

1 Find: **a** angle OPQ   **b** angle BCD.

3 Find **a** angle BCA, **b** angle CAD.

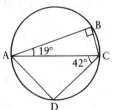

2 Find the angles of triangle DEB.

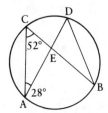

4 PQ and SR are parallel.
   Angle PQS = 32°, angle SPQ = 74°.

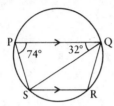

Find angles: **a** QSR, **b** QRS, **c** SQR.

5 AB and BC are equal in length. Angle CAB = 41°.

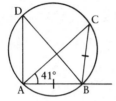

Find angles: **a** ACB, **b** ADB.

6 Angle PTQ = 25°, RT is a diameter, and
   TS = SR.

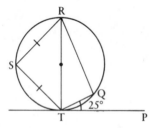

Find angles: **a** QTR, **b** TRQ, **c** SRQ.

7 O is the centre of the circle.
   ∠ AOB = 108°, ∠ ACO = 33°

   Find angle OBC.

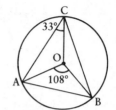

8 AB is a diameter of the circle.

   Find: **a** angle DCB
          b angle DAB
          c angle AXD.

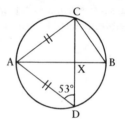

28.7   The use of drawing instruments

Equipment

You need:

a **ruler** or **rule** for measuring lengths, both in centimetres and inches

a **pencil** with a sharp point

a **protractor** for measuring angles, (either circular or semicircular)

a **pair of compasses** for drawing arcs of circles

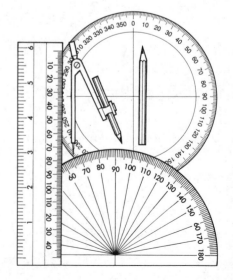

Construction of triangles

To construct a triangle, you need to know three facts about it, one of which must be the length of one side.

You need to know:
either (i) the lengths of all three sides
or (ii) the lengths of two sides, and the angle between them
or (iii) the length of one side, and the size of two angles.

It is conventional to represent the length of the side of a triangle by the lower case letter of the angle opposite to that side.

In triangle ABC on the right the lengths of the sides opposite to angles A, B and C are represented by a, b and c respectively.

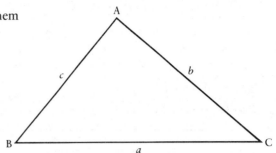

Example 1

Construct a triangle ABC given the three lengths:

$a = 7$ cm, $b = 9$ cm, $c = 5$ cm.

Select the largest side to draw first.

Draw AC = 9 cm.

With centre A and radius 5 cm draw an arc.

Since AB = 5 cm, B must lie somewhere on this arc.
With centre C and radius 7 cm draw an arc on which B also must lie, since CB = 7 cm.

Therefore this arc cuts the first arc at B.

Join AB and BC.

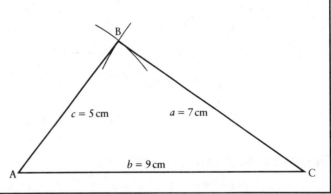

Example 2

Construct a triangle ABC given $a = 7$ cm, $b = 9$ cm, $\angle C = 70°$.

Once again, select the largest side first.
Draw AC = 9 cm.

With your protractor, centre at C, mark a point at 70° with AC.

This point is on BC.

Draw BC = 7 cm.

Join AB.

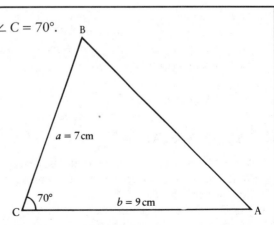

Example 3

Construct triangle ABC given $a = 7$ cm, $\angle B = 80°$, $\angle A = 48°$.

Draw BC = 7 cm.
With your protractor, centre at B, mark a point at 80° with BC.
This point is on AB.
Join B to this point and extend the line as appropriate.

($\angle C$ is required but was not given. However
$$\angle A + \angle B + \angle C = 180°$$
$$\therefore \ \angle C = 52°.)$$

Repeat at C with 52°.

Where the two lines meet is point A.

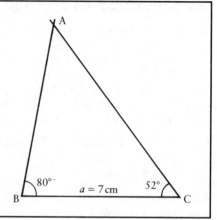

When you construct a diagram, you should try to ensure that your accuracy is within 1 mm (for length) and 1° (for angles).

Exercise 147

Draw the following triangles accurately and make the specified measurement.

1   AB = 6 cm   Angle A = 50°   Angle B = 70°   Measure AC

2   PQ = 4.3 cm   Angle P = 70°   Angle Q = 48°   Measure PR

3   PQ = 2.5 in   QR = 3.7 in   RP = 2.8 in   Measure angle P

4   XY = 6.4 cm   YZ = 7.1 cm   ZX = 4.9 cm   Measure angle Z

5   PQ = 4.9 cm   QR = 3.1 cm   Angle Q = 71°   Measure PR

Geometrical constructions

The following constructions require only the use of a ruler and a pair of compasses.

(i) To bisect the line AB

With A as centre, draw arcs above and below AB.
(The radius must be more than $\frac{1}{2}$AB.)
With B as centre, draw arcs to intersect the first arcs.
Join the points of intersection.
This lines bisects AB.
(It is also perpendicular to AB.)

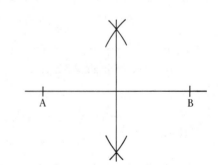

(ii) To bisect angle BAC

With A as centre, draw arcs to cut AB and AC.
With these two points as centres draw arcs to intersect each other.
Join A to the point of intersection.
This line bisects $\angle$ BAC.

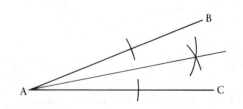

(iii) To draw a perpendicular from point P to line AB

With P as centre, draw an arc to cut AB in two points.
With these points as centres, draw arcs to intersect each other above, or below AB.
Join P to the point of intersection, producing the line if necessary, to cut AB.
This line is perpendicular to AB.

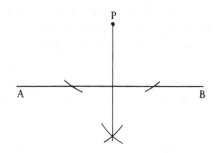

▨ *Exercise 148* ▨

1 Construct an equilateral triangle of side 5 cm, using ruler and compass only.

2 Triangle ABC is isosceles with BC = 6.4 cm, AB = AC = 5.2 cm.
Construct the perpendicular bisector of the base BC.
Measure:
    a angle ABC
    b the altitude of triangle ABC.
(The **altitude** of the triangle is its height, i.e. the perpendicular line from A to BC.)

3 Draw any triangle. Bisect the three angles of the triangle to show that the angle bisectors are concurrent (all meet in the same point).

4 Show, by construction, that the three altitudes of a triangle are concurrent.

5 The position of a boat entering a harbour protected by two piers with lighthouses is shown in the diagram.

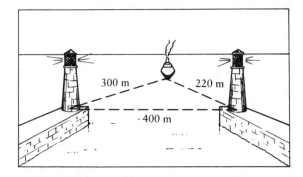

Construct the triangle formed by the boat and the two lighthouses.
By constructing the perpendicular from the position of the boat to the line between the lighthouses, find the closest distance the boat will sail to a pier.

6 A helicopter is on a mercy flight with food supplies to an isolated farm which has been cut off by snow drifts.

It flies a horizontal path at an altitude of 650 m above the ground.

Parcels dropped by the helicopter will be carried by the wind in a direction which is 30° to the vertical.

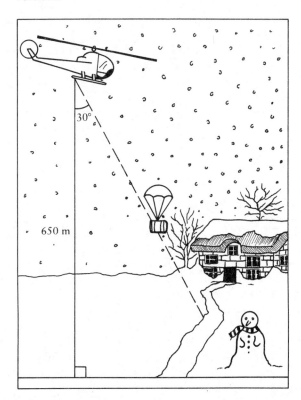

Using ruler and compasses only, construct a diagram to show the position of the helicopter, relative to the dropping point, when the parcels are released.

What is the horizontal distance of the helicopter from the dropping point?

7   A bridge, 200 metres long, spans a river. The
    bridge is suspended by steel cables from arches
    which are arcs of a circle, as shown in the
    diagram.

    Using ruler and compasses only, find the height of
    an arch at its centre.

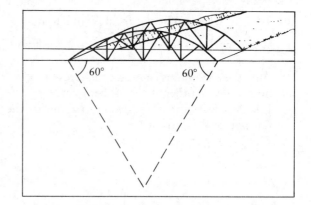

29 Trigonometry

29.1 Pythagoras' theorem

The name of Pythagoras must be the most widely recognised of all the famous Greek mathematicians.

The theorem attributed to his name appears in all school geometry and trigonometry textbooks.

He was born in 582BC on the island of Samos. He later lived in Crotona where he founded a secret society known as the Pythagoreans. The main purpose of this 'brotherhood' was political, but they were also interested in mathematics and astronomy. They believed that everything could be explained in terms of numbers and number patterns.

The theorem for which they are best known, concerning the sides of a right-angled triangle, had been known in a limited form for many hundreds of years.

Egyptian surveyors, or rope-stretchers, knew that a triangle with sides of lengths 3, 4 and 5 units always contained a right angle.
They used a rope with 12 equally spaced knots.

The ends could then be joined together and the rope stretched to form a right-angled triangle which could be used for marking out fields after the Nile floods or for building.

Even before the Egyptians, the Babylonians knew of hundreds more of these triangles.

For example, triangles with sides of 5, 12 and 13 units or 8, 15 and 17 units are always right-angled triangles.

It was Pythagoras, or one of the Pythagoreans, however, who discovered the relationship which connects the sides of these triangles.

Pythagoras' theorem states that in any right-angled triangle:

The square of the hypotenuse is equal to the sum of the squares of the other two sides

(The **hypotenuse** is the side of a triangle which is opposite to a right angle.)

The theorem can easily be illustrated for the Egyptians' 3, 4, 5 triangle:

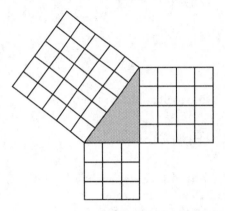

It was about two hundred years later that the theorem was proved, by Euclid when he was writing his textbook of geometry theorems.

(The theorem is in fact true for any shapes which are constructed on the three sides of a right-angled triangle: see Exercise 118, question 4, p. 209.)

For a triangle ABC with sides of length a, b and c, as shown:

$$c^2 = a^2 + b^2$$

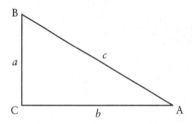

Example 1

John is building a garage with a rectangular base, 6.5 m long and 3.4 m wide.
John knows that to ensure he marks out right angles at each corner he must also mark out one of the diagonals.
Calculate the length of this diagonal.

By Pythagoras' theorem:

$$x^2 = 6.5^2 + 3.4^2$$
$$= 42.25 + 11.56$$
$$= 53.81$$
$$\therefore x = 7.34$$

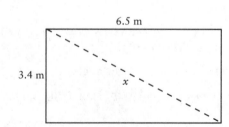

∴ The diagonal has a length of 7.34 m.

Example 2

In PQR,   R = 90°, PQ = 17 in, QR = 10 in. Find PR.

By Pythagoras' theorem:
$$r^2 = p^2 + q^2$$
$$\therefore 17^2 = 10^2 + q^2$$
$$\therefore q^2 = 17^2 - 10^2$$
$$= 289 - 100$$
$$= 189$$
$$\therefore q = \sqrt{189}$$
$$= 13.7$$
$$\therefore \text{PR is } 13.7 \text{ in.}$$

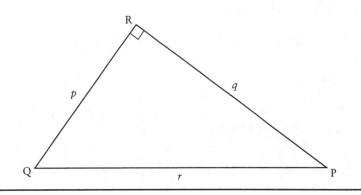

Example 3

In the diagram below, $\angle CAB = \angle CBD = 90°$. AC = 8 cm, AB = 6 cm, and CD = 12 cm.

Find BC and BD.

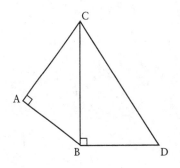

Insert all the known lengths and angles on the diagram.

Two lengths are known in triangle ABC. Therefore use triangle ABC first.

In △ ABC
$$x^2 = 6^2 + 8^2$$
$$= 100$$
$$\therefore x = 10$$

x is 10 cm.

In △ BCD
$$x^2 + y^2 = 12^2$$
$$\therefore 10^2 + y^2 = 12^2$$
$$\therefore y^2 = 144 - 100$$
$$= 44$$
$$\therefore y = \sqrt{44}$$
$$\therefore y = 6.633$$

y is 6.63 cm.

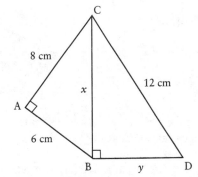

Exercise 149

1   **a** Construct a triangle with sides 6 cm, 8 cm, 10 cm and check that it is right-angled by measuring the largest angle.
An accuracy of between 89° and 91° should be obtained.

  b Measure the two remaining acute angles in the triangle.

2 The following questions refer to this diagram:

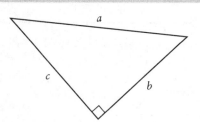

   a $b = 5$ cm, $c = 12$ cm. Find a.

   b $b = 2.7$ mm, $a = 3.8$ mm. Find c.

   c $a = 3.5$ in, $c = 2.1$ in. Find b.

3 A triangle has sides $p = 10$ cm, $q = 17$ cm, $r = 13$ cm.
  Is it a right-angled triangle?

4 Find the unknown lengths:

a

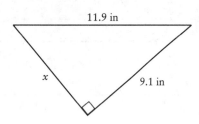

c

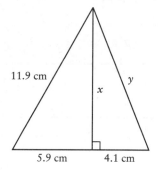

b

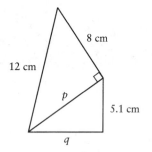

d

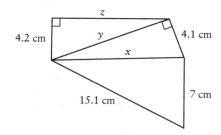

5 A square tin of sweets has a side of length 17 cm. It is decorated by a ribbon
  across one diagonal.
  What is the length of the ribbon?

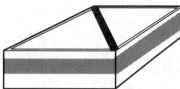

6 From corner to corner across the floor of a square room measures 182 feet.
  What are the dimensions of the room?

7 A concrete fence post, TY, is supported by two wire struts, XY and YZ, of
  equal length.
  Calculate the length of a strut.

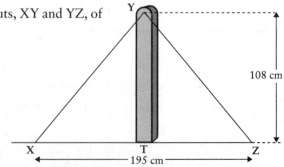

8   A house has a dormer window, as shown in the diagram. Calculate the length of the roof, over the dormer window, from the ridge to the gutter edge.

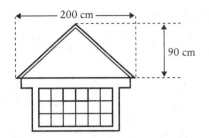

9   A side-view of the dormer window in question 8 is shown here. Calculate how far the window projects from the roof of the house.

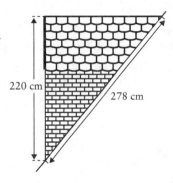

29.2   Trigonometrical ratios

The tangent ratio

The word **trigonometry** means triangle measurement.
The surveyors of ancient times were able to find the height of a tall object by comparing its shadow with the shadow of an object (e.g. a stick) of known height.

Because the angle of the Sun's rays was the same for both objects, two similar triangles were formed, in which the lengths of the shadows and the heights were in the same ratio.

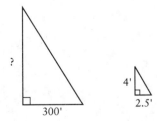

$$\frac{\text{Height of pyramid}}{\text{Length of pyramid's shadow}} = \frac{\text{Height of stick}}{\text{Length of stick's shadow}} = \frac{4.0}{2.5}$$

The unknown height could then be calculated.

$$\text{Height of pyramid} = \frac{4.0}{2.5} \times \text{Length of its shadow}$$

$$= 1.6 \times 300\,\text{ft}$$

$$= 480\,\text{ft}$$

(The original height of the Great Pyramid was 481.4 ft, although 31 feet are now missing.)

Any other triangle with equal angles will also have its sides in the same ratio and a height can quickly be calculated if the ratio is known.

For example, in the triangles below, the angle at the base is 58°. When the base of the triangle is 2.5 units, the height of the triangle is 4.0 units. Therefore, when the base is 1 unit the height is 1.6 units.

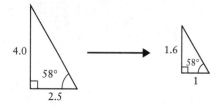

The height of any similar triangle can be found by multiplying the length of its base by 1.6.

Height of tree = 5.5 m × 1.6
          = 8.8 m

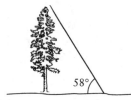

The ratio 1.6 is called the **tangent of 58°**.

In about 150 BC, a Greek mathematician working in Alexandria, called Hipparchus, realized that a great deal of time could be saved if a table of values for different angles was constructed.

In a triangle ABC, if A is the angle involved in a calculation, the sides are named as follows:

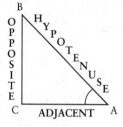

The tangent ratio is therefore:

$$\text{Tangent of angle A} = \frac{\text{Side opposite to angle A}}{\text{Side adjacent to angle A}} = \frac{\text{BC}}{\text{CA}}$$

or
$$\text{Tan A} = \frac{\textbf{Opposite}}{\textbf{Adjacent}}$$

Example 1

In the triangle shown, find angle A.

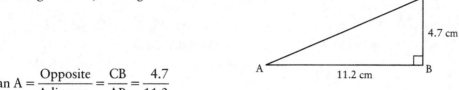

$$\text{Tan A} = \frac{\text{Opposite}}{\text{Adjacent}} = \frac{\text{CB}}{\text{AB}} = \frac{4.7}{11.2}$$

$$= 0.4196428$$

$$\angle \text{A} = \boxed{\text{INV}} \ \boxed{\text{tan}} \ 0.4196428 \left(\text{or} \ \boxed{\text{tan}^{-1}} \ \text{or} \ \boxed{\text{arc tan}} \ \text{key} \right)$$

$$\angle \text{A} = 22.8°.$$

Example 2

In the triangle below, find PQ.

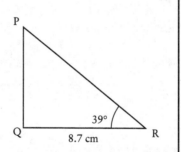

$$\text{Tan R} = \frac{\text{Opposite}}{\text{Adjacent}}$$

$$\text{Tan 39°} = \frac{\text{PQ}}{8.7}$$

$$\therefore \text{PQ} = \text{Tan 39°} \times 8.7$$

$$= 7.05 \, \text{cm}.$$

Example 3

In the triangle shown, find XY.

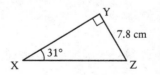

$$\text{Tan X} = \frac{\text{Opposite}}{\text{Adjacent}}$$

$$\text{Tan 31°} = \frac{7.8}{\text{XY}}$$

$$\text{Tan 31°} \times \text{XY} = 7.8 \ (\text{multiplying by XY})$$

$$\text{XY} = \frac{7.8}{\text{Tan 31°}}$$

$$= 13.0 \, \text{cm}$$

1 Find AB.

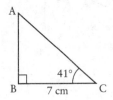

2 Find angle R.

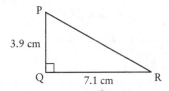

3 Find XY.

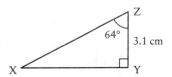

4 Find x and y.

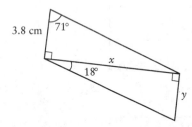

5 ABC is an isosceles triangle with AB = AC.

   a Find the altitude of triangle ABC.

   b Calculate the area of triangle ABC.

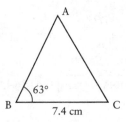

6 In the diagram, PT is a tangent to the circle, centre O, which has a radius of 6.5 cm. PT = 20 cm. Find angle RST.

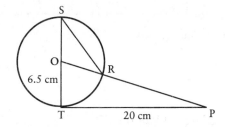

7 A chord of length 4 cm subtends an angle of 40° at O, the centre of the circle.
Find:

   a the distance of the chord from the centre

   b the radius of the circle.

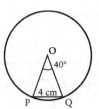

Sine and cosine ratios

The tangent ratio does not involve the hypotenuse, but the sides of a triangle can be paired in two more ways to give ratios which do involve the hypotenuse.

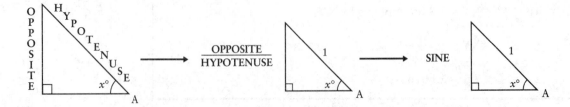

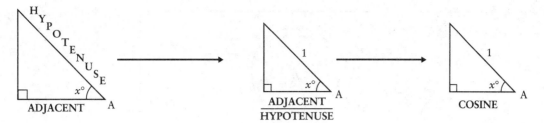

i.e.   Sine of angle A $= \dfrac{\text{Side opposite angle A}}{\text{Hypotenuse}}$

or   **Sin A** $= \dfrac{\textbf{Opposite}}{\textbf{Hypotenuse}}$

and   Cosine of angle A $= \dfrac{\text{Side adjacent to angle A}}{\text{Hypotenuse}}$

or   **Cos A** $= \dfrac{\textbf{Adjacent}}{\textbf{Hypotenuse}}$

Example 1

In the triangle shown, find angle A.

$$\text{Sin A} = \frac{\text{Opposite}}{\text{Hypotenuse}} = \frac{\text{BC}}{\text{AC}} = \frac{4}{10}$$

$$= 0.4$$

Angle A $= \boxed{\text{INV}}\ \boxed{\text{sin}}\ 0.4$

$$= 23.6°$$

Example 2

In the triangle shown, find BC.

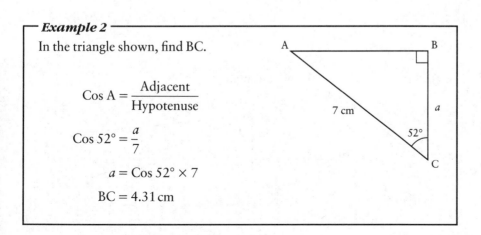

$$\text{Cos A} = \frac{\text{Adjacent}}{\text{Hypotenuse}}$$

$$\text{Cos }52° = \frac{a}{7}$$

$$a = \text{Cos }52° \times 7$$

$$\text{BC} = 4.31\,\text{cm}$$

Example 3

Find the length of the radius OQ.

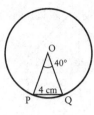

Draw in OR perpendicular to PQ, then ∠ ROQ = 20° and RQ = 2 cm.

$$\text{Sin Y} = \frac{\text{Opposite}}{\text{Hypotenuse}}$$

$$\text{Sin } 20° = \frac{2}{\text{OQ}}$$

$$\text{Sin } 20° \times \text{OQ} = 2$$

$$\text{OQ} = \frac{2}{\text{Sin } 20°} = 5.85 \text{ cm}$$

Exercise 151

1 a Find PQ.   **b** Find XY.

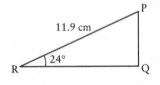
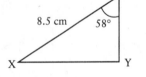

2 a Find AB.   **b** Find PR.

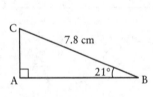
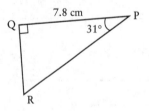

3 a Find angle C.   **b** Find angle R.

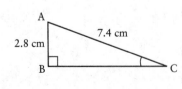
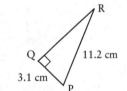

4 a Find angle A.   **b** Find angle P.

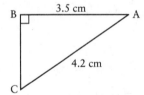
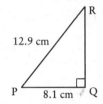

5 Find x and y.

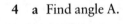

6 Find p and q.

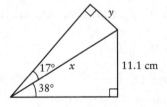
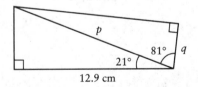

7 Find AC and angle CAD.

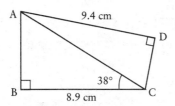

8 Draw in the altitude AD.
Hence find angle BAC.

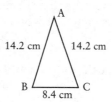

Choosing the ratio to solve a problem

When an angle is to be found in a right-angled triangle decide which two sides have been given. Then use the trigonometric ratio which includes these two sides.

When a side is to be found in a right-angled triangle and one angle (other than the right angle) is given, consider which side is given (opposite, adjacent or hypotenuse), and which is the side required. Then use the trigonometric ratio which includes these two sides.

Exercise 152 Miscellaneous questions

1 Find angle X.

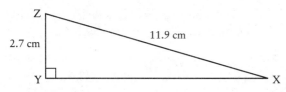

2 Find PQ.

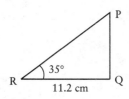

3 Sides AB and AC are equal.
Draw in the altitude AD.
Find AB.

4 Sides AB and DC are parallel.
Find p and q.

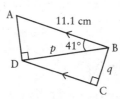

5 A ladder, 3.6 m long, rests against a window sill which is 3.3 m above the ground.
How far from the wall is the foot of the ladder?

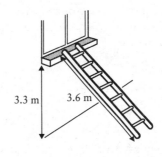

6 A ramp is to be constructed to allow wheelchair access to a building.

The height of the step into the building is 15 cm. The slope of the ramp is to have an angle of 10°.

a What is the length of the top of the ramp?

b How far from the doorstep will the ramp protrude?

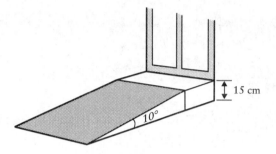

7 A dormer window is shown in the diagram below.

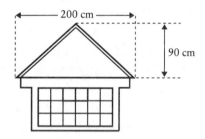

Calculate the pitch of the dormer roof (i.e. the angle the roof makes with the vertical).

8 The diagram shows the side view of the dormer window.
Calculate the pitch of the roof of the house.

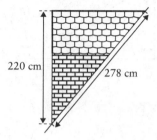

9 A step ladder is 125 cm long. When fully opened the distance between the foot of the ladder and its support is 82 cm.

a How high above the ground is the top of the ladder?

b What is the angle between the ladder and its support when it is fully opened?

29.3 Bearings

An instrument for measuring the direction, on the ground, of one location from another, is called a **compass**.

The compass direction is called a **bearing**.

A compass rose, often shown on maps, gives the bearings in terms of the four cardinal points, north, south, east and west.

A mariner's compass, or a surveyor's prismatic compass, gives a bearing as an angle measured from north in a clockwise direction.

This is the most usual method of giving a bearing, which is always written as a three-digit number. For example:

SW is 225°
E is 090°
N is 000°

To find the bearing of a point B from a point A,

1 Join A and B.
2 Draw in the **north** line at A.
3 Find the angle between north and AB, measured **clockwise**.
4 Record the angle, in degrees, as a three-digit figure.

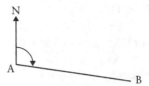

Example

In the diagram, the bearing of A from B is 250°.

Find the bearing of B from A.

Draw the north line at A.

The bearing of A from B is 250°.

$$\therefore \text{Angle } p = 360° - 250°$$
$$= 110°$$

$$\therefore \text{Angle } q = 180° - 110°$$
$$= 70°$$

$\therefore$ The bearing of B from A is 070°.

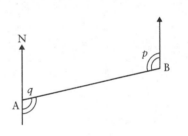

Exercise 153

1 Write the following directions as three-digit bearings:

a East, **b** SE, **c** NW, **d** N 30° W, **e** S 15° W.

2 In each of the following, find the bearing of B from A.

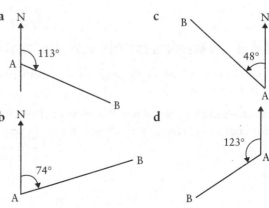

3 The bearing of Q from P is 192°.
What is the bearing of P from Q?

4 The bearing of R from S is 061°.
What is the bearing of S from R?

5 The bearing of D from C is 324°.
What is the bearing of C from D?

6 Paris is 200 km due north of Bourges and 145 km due west of Chalons.
What is the bearing of Bourges from Chalons?

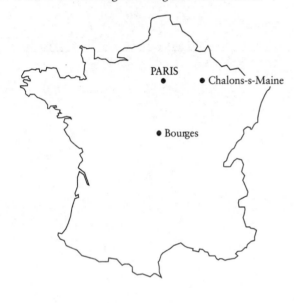

7 New Orleans is 360 miles due south of Memphis and Jacksonville is 500 miles due east of New Orleans.

a On what bearing will a plane fly when travelling from Memphis to Jacksonville?

b On what bearing will a plane fly when travelling from Jacksonville to Memphis?

8 A ship leaves harbour, H, on a bearing 310°. After sailing 3 miles, the ship is due north of a coastal town, T, which is due west of the harbour. How far from the harbour is the town?

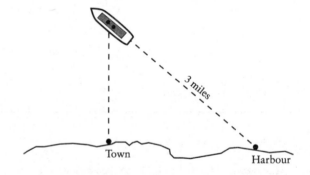

*29.4   Angles of elevation and depression

If you look at an object, the angle formed between the horizontal and your line of sight is called

the angle of elevation, if it is above the horizontal,
the angle of depression if it is below the horizontal.

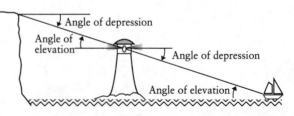

Example 1

Bill stands 200 m from a church and measures the angles of elevation of the top and the bottom of the steeple. These are 19° and 15° as shown below.

Calculate the height of the steeple.

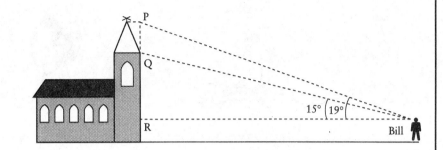

$$\frac{PR}{200} = \tan 19°$$

$$\therefore PR = 200 \tan 19°$$
$$= 68.86$$

$$\frac{QR}{200} = \tan 15°$$

$$\therefore QR = 200 \tan 15°$$
$$= 53.59$$

$$PQ = PR - PQ$$
$$= (68.86 - 53.59)$$
$$= 15.27$$
$$= 15.3 \text{ (to 3 s.f.)}$$

∴ The height of the steeple is 15.3 metres.

Example 2

A coastguard watching a yacht race sees a yacht at an angle of depression of 4°. He knows that he is 300ft above sea level.
How far from him is the yacht?

Draw a figure.

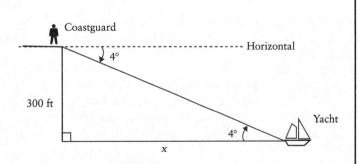

From the figure

$$\frac{300}{x} = \tan 4°$$

$$300 = x \tan 4°$$

$$x = \frac{300}{\tan 4°}$$

$$= 4290 \, \text{ft}$$

∴ The distance is 4290 feet.

Exercise 154

1 From a boat, the angle of elevation of the top of a cliff is 74°.
The height of the cliff is known to be 800ft.
How far is the cliff from the boat?

2 The height of a tree is 58 m. Jacky, who is 2 m tall, is 200 m away from the tree on horizontal ground.
What is the angle of elevation of the top of the tree from Jacky's eyes?

3 John sees a plane at an angle of elevation of 3.8°.
The plane is 2.1 miles horizontally from John.
What is its height?
Give your answer in feet.
(5280 ft = 1 mile.)

(*Note*. In this type of question, where the distances are large, the height of a person is negligible and can be ignored.)

4 A plane flies horizontally over Sue at a height of 2 miles. When she first sees it, the plane is at an angle of elevation of 74° and it takes 9.2 seconds to pass over her and once again to be at an elevation of 74°.
What is its speed? (Speed = distance/time.)

5 Ken is 30 m from a flagpole. The angle of elevation of its top is 12.4°, and the angle of depression of the bottom is 2.1°.

How high is the flagpole?

6 In a penalty shoot-out, the player is 36ft away from the goal mouth, and the goal has height 8 ft. What is the maximum angle of elevation at which the ball can be kicked to go into the goal if it travels in a straight line?

7 A space shuttle is rising at 2000 ft per second. Avril is 2 miles away from the launch pad.
What is the angle of elevation of the space shuttle after:

a 2 minutes,   **b** 5 minutes?

8 A balloon is positioned 420 metres above a showground.

Angus is 3 miles away from the showground. What is the angle of elevation of the balloon?
(1 mile = 1.6 kilometres.)

9 Joan is standing on a church tower 15 m high. She is looking at the top of a lamp post which has height 8 m. She finds that the angle of depression of the top of the lamp post is 8°.

The church tower and the lamp post stand on level ground.

a Draw a diagram to show this information.

b Calculate the distance from the church tower to the lamp post.

(SEG Summer (hH) 1989)

*29.5 Three-dimensional problems

Use of Pythagoras' theorem in three-dimensional problems

Problems set in three dimensions usually involve finding right-angled triangles within the three-dimensional structure.

It is advisable to show in separate clear diagrams every right-angled triangle to be used.

Example

The diagram shows a cuboid ABCDWXYZ.

$AB = 7\,m$, $BC = 5\,m$, $AW = 2.3\,m$.
Find the length of the diagonal AY.

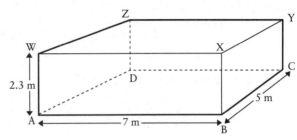

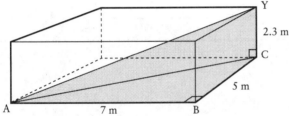

The right-angled triangles to be used are ABC, to find AC and ACY, to find AY.

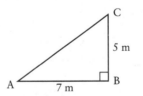
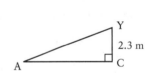

Using Pythagoras in triangle ABC gives:

$$AC^2 = AB^2 + BC^2$$
$$AC^2 = 49 + 25$$
$$= 74$$

Using Pythagoras in triangle ACZ gives:

$$AY^2 = AC^2 + CY^2$$
$$= 74 + 2.3^2$$
$$= 79.29$$
$$AY = 8.90\,m$$

The angle between a line and a plane

Find a right-angled triangle containing the line and a line in the plane (which is a projection of the line on to the plane).

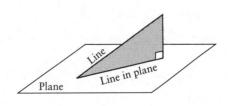
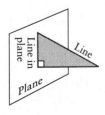

Example

Find the angle between:

a the line AZ and the plane ABCD

b the line AY and the plane ABCD.

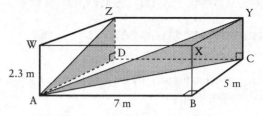

a Triangle ADZ contains the line AZ and the line AD which is in the plane ABCD. The angle ADZ is 90°.

Using ADZ gives:

$$\text{Tan ZAD} = \frac{2.3}{5.0} = 0.46$$

$$\angle\text{ ZAD} = 24.7°$$

b Using CAY gives:

$$\text{Tan CAY} = \frac{2.3}{8.602}$$

$$\angle\text{ CAY} = 15.0°.$$

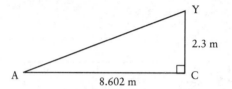

The angle between two planes

Two planes, which are not parallel, meet in a line. To find the angle between two planes which meet in a line:

1 identify this common line;

2 find two lines which are perpendicular to this common line, one in each plane.

The angle between these two lines is the angle between the two planes.

Example

ABCDV is a pyramid on a square base ABCD, with the apex V vertically above the centre of the square.

The side of the square is 12 cm and each slant edge is 10 cm, as shown in the diagram.
Find the angle between planes ABCD and ABV.

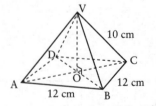

The common line of the two planes is AB.
Mark P, the mid-point of AB, then PV, in the plane ABV, is perpendicular to AB.

PO, in the plane ABCD, is perpendicular to AB.

The angle between the two planes is the angle between the two lines PO and PV, i.e. the angle VPO.

The right-angled triangles required are VAP and VPO.

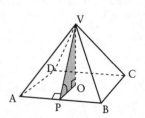

In triangle VAP, by Pythagoras
$$VP^2 = VB^2 - PB^2$$
$$= 100 - 36 = 64$$
$$\therefore VP = 8 \text{ cm}$$

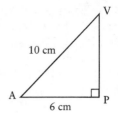

In triangle VPO
$$PO = \tfrac{1}{2} BC = 6 \text{ cm}$$

$$\text{Cos VPO} = \frac{\text{Adjacent}}{\text{Hypotenuse}} = \frac{6}{8}, \text{ so that } \angle VPO = 41.4°$$

$\therefore$ The angle between the two planes is 41.4°.

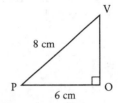

Exercise 155

1 A cardboard box, used for packing televisions, is 26 inches long, 12 inches wide and 19 inches high.
What is the length of the longest diagonal of the box?

2 A room is 6.2 m by 4.3 m. Its height is 2.4 m.
What is the length of the longest diagonal?

3 The volume of a cuboid container is 1254 cubic feet. Its length is 21 feet and its width is 8 feet.
What is its height?
What is the length of the longest diagonal?

4 The volume of a water butt, in the shape of a cuboid, is 2200 litres. The width is 0.8 m and the height is 1.4 m.
What is the length of the diagonal?

5 A brick has length 10 inches and width 6 inches. The diagonal through the centre of the brick is 12.9 inches.
What is its height?

6 OABCD is a pyramid on a rectangular base of dimensions 16 cm and 12 cm. O is 15 cm vertically above the centre of the base.
What is the angle which OA makes with the base?

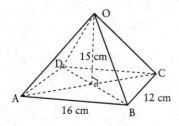

7 The diagram shows a pyramid on a square base, of side 7.8 cm, with VP = VQ = VR = VS = 8 cm.
Calculate:

a the height of the pyramid,

b the angle which VQ makes with the base.

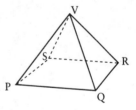

8 The barn shown in the diagram has a horizontal rectangular floor ABCD, with AB = 12 m and BC = 8 m. AP, BQ, CR and DS are all vertical and of length 5 m. X and Y are the highest points of semicircular arcs PXS and QYR. (Take π as 3.14 or use the π button on your calculator.)

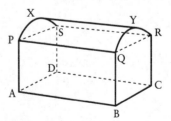

a Find the area of one end of the barn, for example BQYRC.

b Find the volume of the barn.

c Find the angle of depression of A from Y, giving your answer in degrees correct to the nearest 0.1 of a degree. (SEG 1990)

9 ABCDEFGH is a conservatory with face CDHG attached to a vertical wall. The horizontal base ABCD is a rectangle with AB 10 m, and BC 6 m. AE is 3.1 m, and DH is 3.8 m.
Find the angle which the plane EFGH makes with the horizontal.

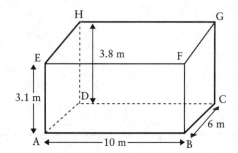

10 PQRSTUVW is a cuboid with PQ = 15 cm, QR = 11 cm and RV = 7.1 cm.
Find the angle which plane WRQ makes with plane PQRS.

11 An exhibition centre is in the shape of a pyramid with a hexagonal base of side 15 m. Each slant height of the pyramid is 18 m. Find:

a the vertical height of the pyramid,

b the angle which each face makes with the horizontal.

12 Part of a hillside slopes uniformly at an angle of 34° with the horizontal. A straight path makes an angle of 49° with the line of greatest slope. Find the angle which the path makes with the horizontal.

30 Graphs

30.1 Graphs and curves

There are many different types of graph, some of which you may meet in the section on statistics.

This unit deals with line graphs. The lines may be straight or curved, but they illustrate a relationship between two quantities. This relationship must be clear to anyone looking at the graph.

Graphs are used by many bodies to convey information quickly and with impact!

For example, these graphs were used by:

a holiday company

a Government department

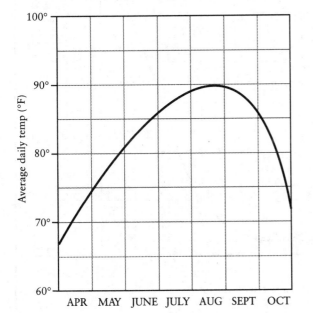

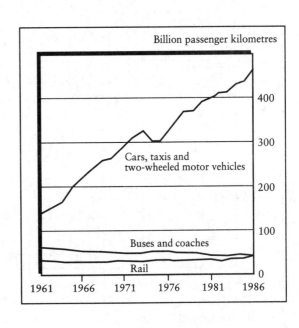

(Source: Department of Transport)

Many graphs involve time, which is always measured along the horizontal axis. The vertical axis then provides a measure of something that changes with time.

Whenever one measurable quantity changes as a result of another quantity changing it will be possible to draw a graph.

30.2   The interpretation of graphs

Once a graph has been drawn, information can be found from it very quickly.

Example

The graph shows a cross-section through a river bed. The distance from point A on one bank of the river to point B, directly opposite on the other bank is 20 m.

Soundings were taken of the depth of the water at various points and plotted against the distance from A.

Find, from the graph:

a the depth of the river at a distance of 12 m from A

b the distances from A at which the depth of the river is 2.8 m.

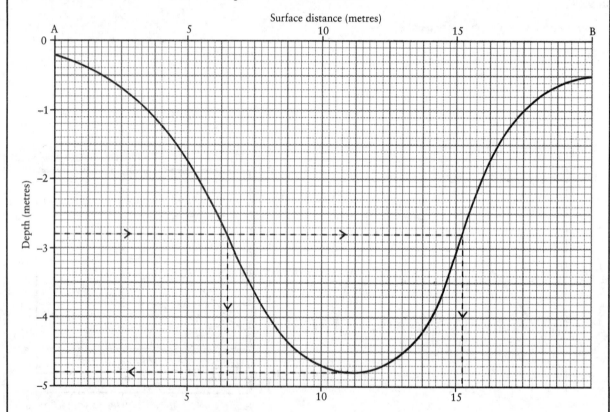

a The distance 12 m is located on the horizontal axis and a vertical line is drawn from this point to meet the graph. A horizontal line is then drawn across to the vertical axis and the distance is read off.

The distance is −4.8 m, i.e. the depth of the water is 4.8 m.

b The depth of 2.8 metres is located on the vertical axis at the point −2.8 and a horizontal line is drawn from this point which meets the curve at two points.
Vertical lines are drawn down to the horizontal axis and the two distances are read off.

The distances from A are 6.50 m and 15.25 m.

In the following exercise the graphs have been drawn for you.
Answer the questions relating to each graph by reading off the values from the graph.

In the first question guide lines have been drawn to make it easier for you.

1 This graph shows the number of births (measured along the vertical axis) in
   a given year (measured along the horizontal axis).
   (Based on figures from the *Annual Abstract of Statistics*, 1988)

   a Estimate the number of children born in 1915.

   b In which years were, approximately, 825 000 children born?

   c Which year had the lowest number of births?

   d In which year was there a 'baby boom'?

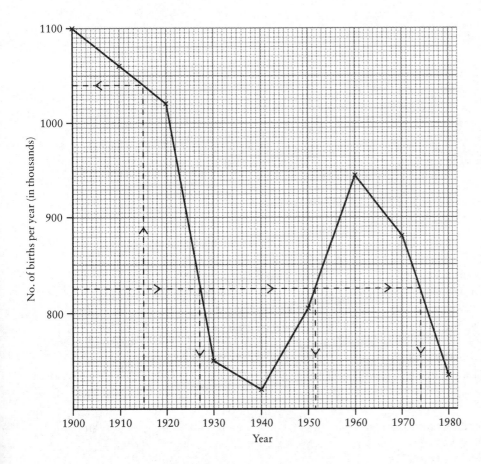

2 The next graph shows the average gross weekly earnings of men from 1970
   to 1986. (Based on figures from *Social Trends*, 1988.)

   a Estimate the average weekly earnings in 1974.

   b Estimate the average weekly earnings in 1984.

   c Why is the answer to **b** a better estimate than the answer to **a**?

   d In which year did the average weekly wage reach £100?

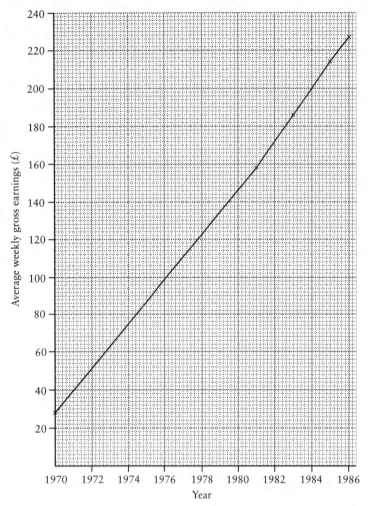

3 Two children, Ben and Sarah, were walking along the river bank when they spotted a bottle floating in the river. They each threw stones to see who could hit it first. The graph shows their first attempts.

a The bottle was 17 m from the bank. Who came nearest to hitting it?

b How far from the bottle did each stone hit the water?

c On their second attempt they threw their stones at the same time. The stones followed exactly the same paths as before, but this time they collided. How far above the river were they when the collision occurred?

d If the children were both 1.5 m tall, how high was the river bank? (Assume the children's hands were 1.5 m above the ground.)

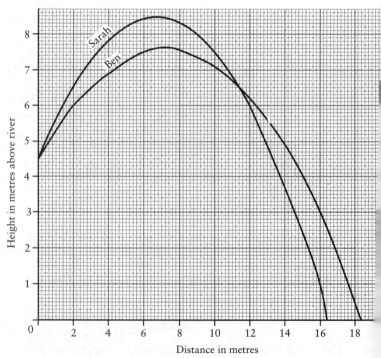

4 Nessa enjoys writing to her many pen-friends. One pen-friend lives in France and a favourite topic of conversation is the weather. The next graph shows the number of wet days each month in their two home cities.

The solid line shows the number of wet days where Nessa lives and the dotted line is for her friend Louis.

a Which is the wettest month for each city?

b Which is the driest month for each city?

c In which month does the smallest difference in the number of wet days occur?

d Nessa lives in London. Where do you think Louis lives?

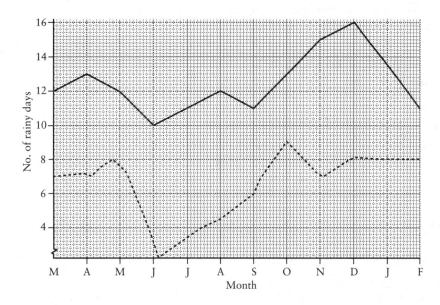

30.3   Graph plotting

When plotting graphs a certain procedure should be followed.

1 Choose the scales.

2 Draw the axes.

3 Scale the axes.

4 Plot the points.

5 Join the points.

The information from which the graph is to be drawn is usually given in a table.

Example

£1000 is invested in a savings account for 5 years. The amount in the account at the end of each year is shown on the table:

Number of years (x)	0	1	2	3	4	5
Amount in £ (y)	1000	1100	1210	1331	1464	1610

Plot the graph and find

a the amount in the account after $2\frac{1}{2}$ years

b the length of time for which the money must be invested to amount to £1500.

1 Choose the scales:

  (i) Find the range of values for each axis.
      On the horizontal axis the range is 5.
      On the vertical axis the values range from 1000 to 1610, i.e. the range is 610.

  (ii) Count the number of large squares in each direction on the graph paper.

      The graph paper provided here has 18 large squares in both directions and a grid with five small divisions in each large division.

      The aim is to draw as large a graph as possible, but the scale must be easy to read.

      Taking the 18 large division width for the horizontal axis, the range of 5 does not divide exactly into 18.

      A convenient scale would be 2 large divisions representing 1 year. On the vertical axis, a range of 610 does not divide into 18 large divisions, but 14 cm is divisible by 700. There will be little wastage of space if a scale of 2 large divisions representing £100 is chosen. The axis is scaled from £1000 to £1650.

2 Draw the axes.

  Once the scales have been chosen, there is no problem in placing the axes.

  The origin can be placed 2 large divisions from the left and 2 large divisions from the bottom edges of the graph paper.

  This allows space for labelling the axes. Label the origin O.

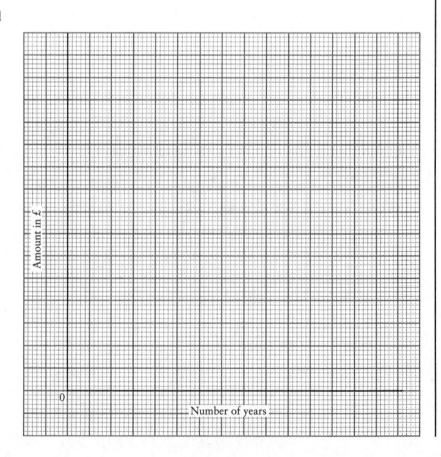

3   Scale the axes:

Mark the position of each year along the horizontal axis as shown on the following graph.

Mark the position of each £50 along the vertical axis as shown.

4   Plot the points.

Plot the six points given as explained on p. 298.

5   Join the points.

The points are joined with a **smooth** curve.

Trace through the points with a pencil, but without touching the paper, to find the shape of the curve. When you are satisfied that the path of the curve is smooth, draw the curve through the points in one movement.

If the values of the coordinates have been rounded, the line may not pass exactly through each point:

6   To find the required information draw appropriate lines on the graph (see p. 297) and read off the values:

   a £1270   **b** 4.3 years

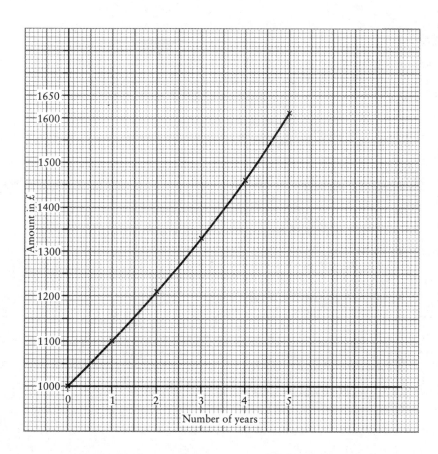

In the following exercises, you are given the information connecting the two variables in a table. In each case plot the points and join them with a smooth curve.

1   Mrs Lister weighed herself now and again through the year. After one Christmas she found her weight was over $10\frac{1}{2}$ stone.

She joined Weightwatchers and was determined to reach her target weight of $8\frac{1}{2}$ stone before the next Christmas.

The table shows her weight in stones and pounds at the beginning of each given month. (1 stone = 14 lb.)

Month/year	Jan/90	May/90	Sep/90	Jan/91	May/91	Sep/91
Weight in st–lb	9–2	9–12$\frac{1}{2}$	10–6	10–8	10–4	9–6

a   What weight was Mrs Lister in August 1990?

b   What was the heaviest weight she reached?

c   During which months did she weigh 9 stone 7 lb?

d   Do you think she would have reached her target weight before Christmas 1991?

2   While at a party, William spills a glass of red wine over the table cloth.

The stain spreads rapidly in a circular shape and the table shows how the area increases as the diameter of the stain increases.

Diameter (cm)	0	2	4	6	8	10
Area (cm²)	0	3	12	27	48	75

a   What area does the stain cover when the diameter is 9 cm?

b   What is the radius of the stain when its area is 50 sq cm?

c   The stain stops spreading when its diameter is 10.9 cm. What area does the stain eventually cover?

3   The table shows the average number of hours the sun shone during a day in the middle of each month, recorded at a Spanish resort.

	Jun	Jul	Aug	Sep	Oct	Nov	Dec	Jan	Feb	Mar	Apr	May	Jun
Hours of sunshine	16	15.5	14	12	10	8.5	8	8.5	10	12	14	15.5	16

a   During which months might you expect an average of $10\frac{1}{2}$ hours of sunshine in a day?

b   What were the average hours of sunshine at the beginning and end of May?

c   Which periods had the smallest increase in hours of sunshine?

4 One day when Heather is having a bath, she notices that the water level is gradually decreasing and realises that the plug is leaking. When she started her bath the depth of water was 300 mm. Plot the points given in the table below to show the depth of water remaining at a given time.

Time (in minutes)	0	1	2	3	4	5
Depth of water (in mm)	300	110	41	15	5.5	2.0

a What was the depth of the water after $2\frac{1}{2}$ min?

b After how long was the bath half empty?

c What was the drop in water level after $1\frac{1}{2}$ min?

5 Two young children have a swing in the garden. On one particular day it is Lucy's turn on the swing and Tony pulls the swing back through an arc of 100 cm and then releases it.

The table shows the position of the swing (distances are measured along the arc of the swing) at different times during its motion.

Time (in seconds)	0	0.2	0.4	0.6	0.8	1.0	1.2	1.4
Distance (in cm)	100	89	57	13	−34	−74	−97	−98
Time (in seconds)	1.6	1.8	2.0	2.2	2.4	2.6	2.8	3.0
Distance (in cm)	−77	−38	9	54	87	100	91	61

a How long does it take for one complete swing?

b In what position is the swing after 0.7 seconds?

c How long does it take to travel through 50 cm?

d How long does it take to travel through 250 cm?

30.4 Conversion graphs

A graph is a very useful means of converting quickly from one quantity to another.

Ranjit is planning a summer holiday in France. He sees in the newspaper that the exchange rate is 10.2 FF = £1, but he wants to be able to compare French prices with prices at home quickly and easily. Before he goes he draws a pocket-size conversion graph which he can keep handy while on holiday.

He knows that:

- A conversion graph is a straight line, so he only needs to plot two points (though a third point is useful as a check).
- Most conversion graphs go through the point (0,0).
- Either variable can be measured along the vertical or horizontal axis.

As Ranjit will mainly be changing from francs to pounds he measures francs along the horizontal axis.

He knows that £1 = 10.2 FF and that the graph goes through the origin. He plots these two points. They are very close together so he finds another point further from the origin.

£10 = 102 FF is an easy one to calculate.

He then draws a straight line through the three points.
(If the line does not pass through all the points there is a mistake and the calculations must be checked.)

This is what his graph looks like:

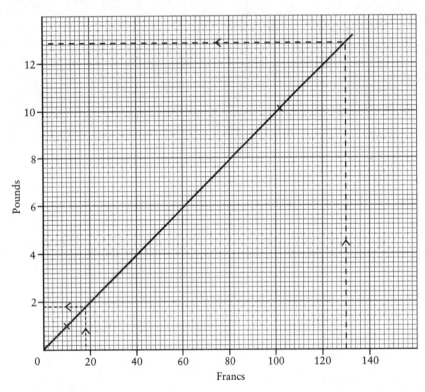

From the graph he can see that an 18-franc bottle of French wine is cheap at less than £1.80 and a meal for two costing 130 francs is about £12.80.

Exercise 158

1   Five miles is approximately equal to eight kilometres.
    Draw a graph to give a rough conversion from kilometres to miles.

    a How many miles are 35 km?

    b Convert the distances given on the signboard to miles.

    c A Frenchman on holiday travels 60 miles. How far has he travelled in km?

A10	
ORLEANS	106
TOURS	245
POITIERS	350
BORDEAUX	560

2   Weather forecasts now give all temperatures in degrees Celsius, but some people are more used to degrees Fahrenheit.
    Draw a graph to convert between degrees Celsius and degrees Fahrenheit.
    (10°C = 50°F, 20°C = 68°F, 30°C = 86°F.)

This graph does not go through the origin.

a The freezing point of water is 0°C.
What is this in degrees Fahrenheit?

b Convert a temperature of 25°C to °F.

c Convert a temperature of 83°F to °C.

d Convert a temperature of 14°F to °C.

3 Nessa's Canadian pen-friend sends her a recipe for apple pie.

Unfortunately, some of the ingredients are measured in millilitres (ml).
Nessa has a jug which measures fluid ounces (fl oz) and she finds in her
cookery book that 35 fl oz = 1000 ml.
Draw a graph which will change ml to fl oz.

a Use your graph to convert the quantities in Nessa's recipe to the nearest
fl oz:

175 ml milk	1.5 litres sliced apples
250 ml sugar	30 ml softened butter
125 ml flour	cinnamon and nutmeg

4 Mrs Kurtz works part-time in a fabric shop. The materials are sold by the
metre, but people still ask for them by the yard.
Draw a graph to convert yards to metres (1 yard = 0.9 m).

From the graph find (to the nearest metre) how many metres are equivalent
to:
a 3 yards, **b** $5\frac{1}{2}$ yards, **c** 12 yards.

30.5 Travel graphs

Example

One Sunday at 12 o'clock, a family set out for a
picnic at a well-known beauty spot, 25 miles from
their home.

They travelled at a steady speed and arrived at
their destination at 12.45 pm.

After $3\frac{1}{4}$ hours at the picnic spot they set off for
home.

Unfortunately the traffic was very heavy and they
only managed to average 20 mph.

They arrived home in time for their favourite
programme.

Find:

a the time at which they arrived home

b at what times they were 10 miles from home

c their average speed on the outward journey.

The family's day can be represented by a 'travel graph' which measures distance from home along the vertical axis and time along the horizontal axis.

The vertical axis is scaled from 0 to 25 miles and the horizontal axis from 12 noon to 6.00 pm.

At 12 noon they were at 0 miles (home) and at 12.45 pm they were at 25 miles (the picnic spot). These two points are plotted and the line joining them represents the family's outward journey.

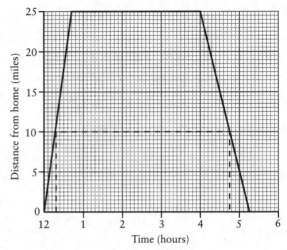

For $3\frac{1}{4}$ hours they are not travelling and this is represented by a horizontal line from 12.45 pm to 4.00 pm, 25 miles from home.

If their average speed on the return journey was 20 mph then after 1 hour they were 5 miles from home.

The point (5.00, 5) is plotted and a line is drawn through it to meet the horizontal axis.

The graph is shown opposite.

From the graph we can see that:

a they arrived home at 5.15 pm.

b on the outward journey they were 10 miles from home after 0.3 hours, i.e. at 12.18 pm, on the homeward journey they were 10 miles from home after 4.75 hours, i.e. at 4.45 pm.

c Their average speed on the outward journey is given by

$$\text{Average speed} = \frac{\text{Total distance}}{\text{Total time}} = \frac{25 \text{ miles}}{45 \text{ mins}}$$

$$= \frac{25 \text{ miles}}{0.75 \text{ hours}}$$

$$= 33.3 \text{ mph}$$

Exercise 159

1   The graph below represents Mr Phillip's journey to work one morning.

The first 10 miles of his journey is through a built-up area and then he joins the motorway. Unfortunately, on this particular morning, there has been an accident and he is held up for some time. He leaves the motorway at the next exit and completes his journey on the ordinary roads.

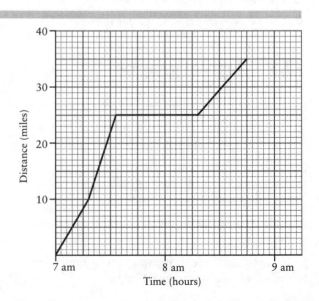

a How long did the journey through the built-up area take him?

b What was his average speed on the motorway section of the journey, until he was stopped by the accident?

c For how long was he held up?

d At what time did he arrive at work?

e How many miles did he travel from home to work?

f What was his average speed for the whole journey?

2   Draw a graph to represent the following journey:

Susan left home at 10.00 am and drove to her friend Valerie's house, 15 minutes away, for coffee. Her average speed was 32 mph.

The two friends drank coffee and chatted for 45 minutes and then Susan drove her friend to the local supermarket, 2 miles away, arriving at 11.12 am.

They spent 24 minutes shopping and then Susan drove home, dropping Valerie off on the way. Her average speed for the return journey was 25 mph.

a How far did she drive to her friend's house?

b At what time did they leave Valerie's house?

c At what time did Susan arrive home?

3   Barry has a friend who lives 5 miles away. They decide to spend a day together cycling and agree to meet at a convenient spot which is 2 miles from Barry's house.

His friend, Ali, leaves home at 9.00 am and cycles at a steady speed of 12 mph.

Barry sleeps in and does not leave until 10 minutes later. He arrives at the rendezvous 10 minutes after Ali.

Draw a graph to represent their journeys.

a At what time did each boy arrive at the rendezvous?

b At what speed did Barry cycle?

4   On another occasion, because Barry is unreliable, they decide that they will keep cycling towards each other until they meet.
Ali leaves home at 9.30 am and Barry leaves at 9.35 am, both cycling at 12 mph.
Draw a graph to find:

a at what time they meet,

b how far each boy will have travelled.

5   Nessa is invited to stay with her French pen-friend in Nice. In order to save money, she travels down by coach. At the beginning of August the whole of Paris seems to start their summer holiday and travel south on the A6. The traffic is very heavy during this first weekend and there are usually tailbacks around the main cities, particularly Lyon, which can cause long delays.

Below is Nessa's itinerary for travelling from Calais to Nice along the A6 and a map showing the distances involved.

Arrive Calais	1400	Continental time
Arrive Paris	1700	Dinner stop
Depart Paris	1800	
Arrive Valance	0400	Refreshment stop
Depart Valance	0430	
Arrive Nice	0800	

Use this information to draw a travel graph of Nessa's journey and use it to calculate the average speeds, to the nearest km/h, for the three sections of the journey:

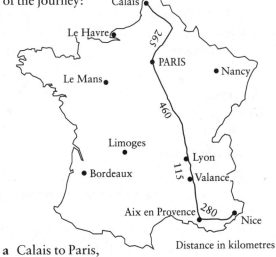

Distance in kilometres

a Calais to Paris,

b Paris to Valance,

c Valance to Nice.

30.6 Cartesian coordinates

In the early seventeenth century, the French mathematician René Descartes introduced the idea of a grid for locating and plotting points.

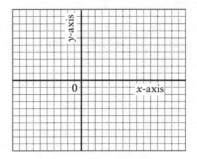

Onto this grid, called a **cartesian** graph, are drawn two straight lines at right angles to one another, called the **rectangular axes**.

Usually the axes cross at the **origin**, O, or starting point. The axes are scaled as number lines with positive numbers to the right and above the origin and negative numbers to the left and below the origin.

The position of any point (A, say) can then be described by two numbers: one referring to the horizontal axis x and one to the vertical axis y.

The two numbers are called the **cartesian coordinates**, after Descartes.

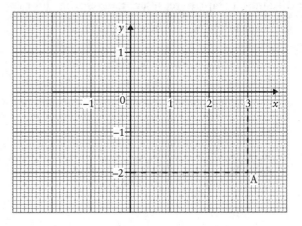

In general, the coordinates are called x- and y-coordinates, written as (x, y), and are plotted by first counting along the horizontal or x-axis and then along the vertical or y-axis.

In the diagram opposite A is at the point $(3, -2)$.

It is usual to plot a point on a graph using a cross (+ or ×) or a dot (·).

30.7 Straight-line graphs

Drawing straight-line graphs

An equation of the form $y = mx + c$ can be represented graphically by a straight line.

To draw a straight line requires **two** points, but it is better to find **three** points as a check.

> ### Example
>
> Draw a graph to represent the line $y = 3x - 2$ for values of x from -3 to 3.
>
> 1 Choose the scales.
>
> Choose three values for x between -3 and 3, e.g. $-3, 0, 3$. Calculate the corresponding y-values and write the three pairs of coordinates in a table:
>
> | x | -3 | 0 | 3 |
> |---|---|---|---|
> | y | -11 | -2 | 7 |
>
> The range for y is 18 and for x is 6.
> Suitable scales are 1 large division to 1 unit on the x-axis
> and 1 large division to 2 units on the y-axis.

2, 3 Draw and scale the axes.

As the negative values of x equal the positive values, set the y-axis near the centre of the graph paper.

The x-axis must be drawn above the halfway mark to allow for more negative values of y.

The axes are scaled in both positive and negative directions.

4, 5 Plot and join the points.

If several graphs are drawn using the same set of axes, each graph should be labelled with its equation.

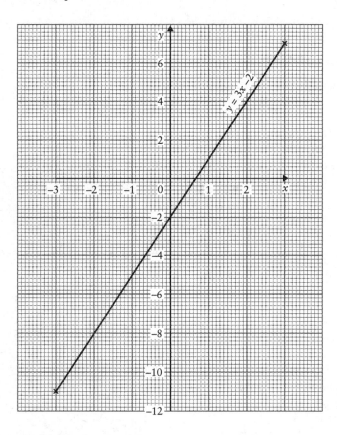

═══ **Exercise 160** ═══

Calculate and plot three points for each of the following lines.

Draw the graphs of each set of equations using the same axes.

1 $y = x$
$y = x + 3$
$y = x - 1$
$y = x - 2$

2 $y = 2x$
$y = 2x + 3$
$y = 2x - 1$
$y = 2x - 2$

3 $y = -x$
$y = -x + 3$
$y = -x - 1$
$y = -x - 2$

4 $y = -3x$
$y = -3x + 2$
$y = -3x - 1$
$y = -3x - 3$

5 $y = \frac{1}{2}x$
$y = \frac{1}{2}x + 2$
$y = \frac{1}{2}x - 1$
$y = \frac{1}{2}x - 3$

6 $y = -\frac{1}{2}x$
$y = -\frac{1}{2}x + 2$
$y = -\frac{1}{2}x - 1$
$y = -\frac{1}{2}x - 3$

7　a　What do the equations in each set have in common:

　　　(i)　in their algebraic form,

　　　(ii)　in their graphical form?

　　b　What conclusion can you draw from this?

8　a　What do the equations $y = x$, $y = 2x$, $y = -x$, $y = -3x$, $y = \frac{1}{2}x$, and $y = -\frac{1}{2}x$ have in common?

　　b　What conclusion can you draw from this?

9　What does the constant term on the RHS of an equation tell you about the graph of that line?

10　Compare the graphs in questions 3, 4 and 6. What does a negative coefficient of x tell you about the graph of that line? (See p. 220 if you need a reminder about what a 'coefficient' is.)

Gradients and intercepts

All linear (straight-line) equations are of the general form:

$$y = mx + c$$

where m is the gradient of the line and c is the intercept along the y-axis, providing the origin O is at the intersection of the axes.

The **gradient** of a line is the increase in the vertical value for every unit (i.e. 1) increase in the horizontal value.

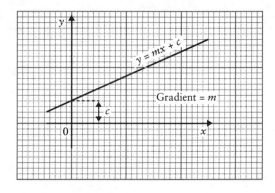

Example 1

Find the gradient of the straight line shown on the following graph:

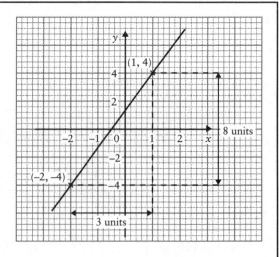

To calculate the gradient of a line:

1　Find two convenient points on the graph (i.e. the x-values should be integers). Convenient points on the graph given would be $(-2, -4)$ and $(1, 4)$.

2　Find the horizontal and vertical differences between the two points.

Horizontal difference $= 1 - (-2)$
$= 3$

Vertical difference $= 4 - (-4)$
$= 8$

3　Calculate the gradient by $m = \dfrac{\text{Increase in } y\text{-value}}{\text{Increase in } x\text{-value}}$

$= \dfrac{8}{3}$

Example 2

Find the equation of the line shown.

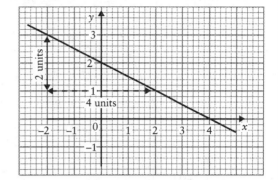

1 Gradient of line $= m = \dfrac{\text{Increase in } y\text{-value}}{\text{Increase in } x\text{-value}}$

$$= \dfrac{1 - 3}{2 - (-2)}$$

$$= -\dfrac{2}{4} = -\dfrac{1}{2}$$

Alternatively,

$\diagup$ is a positive gradient and $\diagdown$ is a negative gradient

From the graph:

$$\text{Gradient of line} = m = -\dfrac{2}{4} = -\dfrac{1}{2} \text{ or } -0.5$$

2 The line crosses the y-axis at the point $(0, 2)$, i.e. the intercept along the y-axis is $c = 2$

3 The equation of the line is $y = mx + c$

$$\therefore y = -\dfrac{1}{2}x + 2$$

$$2y = -x + 4$$

$$x + 2y = 4$$

Exercise 161

1 What are the gradients of the following lines?

  a $y = 3x + 2$ **c** $y = 4 - 2x$ **e** $3y = 2x + 6$ **g** $x + y = -2$ **i** $x + 3y = 6$

  b $y = 2x - 5$ **d** $2y = x - 2$ **f** $4y = 2 - 8x$ **h** $2x + y + 4 = 0$

2 For each of the equations in question 1, state the intercept along the y-axis.

3 The following pairs of points lie on a straight line. Calculate the gradient of each line.

  a $(3, 6) ; (1, 2)$ **c** $(4, 6) ; (2, 5)$ **e** $(2, -5) ; (1, -4)$ **g** $(-1, 6) ; (1, -2)$

  b $(3, 7) ; (1, 1)$ **d** $(0, 5) ; (2, 3)$ **f** $(4, 7) ; (-2, -2)$ **h** $(-4, -1) ; (0, -3)$

4   Find the equation of each of the following straight line graphs:

a

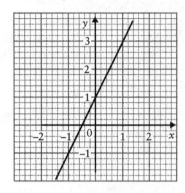

c

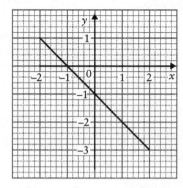

b

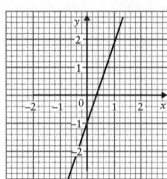

d

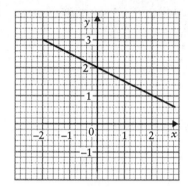

5   Each of the following sketch graphs represents one of the equations:

a $y = 2x + 1$ c $y = 3 - x$ e $x = -1$

b $2y = x$ d $y = 2$ f $x + y = -1$

Which is which?

(i)

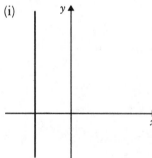

(iii)

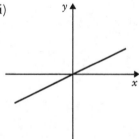

(v)

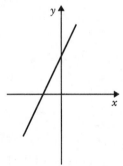

(ii)

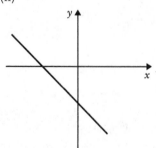

(iv)

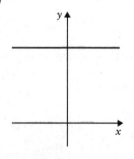

(vi)

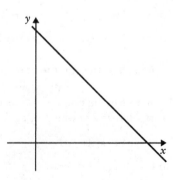

6 In an experiment, weights were hung on to a piece of elastic and the stretched length of the elastic was measured.

The results are shown in the table:

Weight (g)	1	2	3	4	5
Length (cm)	28	35	45	50	60

a Plot the points on to a graph and draw a straight line which best fits the points.

(Not all, and possibly only a few, points will lie on the line. Measurements from experiments can be inaccurate.)

b Find the equation of your line.

c Use your equation to calculate the length of the elastic if a weight of 2.8 g is hung on it. Does your graph give the same value?

d What was the original (i.e. unstretched) length of the elastic?

7 The table below shows the pulse rates for healthy people of different ages taking exercise.

Age	20	25	30	35	40	45	50
Pulse rate	120	117	115	110	108	105	100

a Plot the points and fit a straight line to the points.

b Find the equation of your line.

c What are the appropriate pulse rates for:

  (i) someone 55 years of age

  (ii) someone 18 years of age?

30.8   Graphs of simple curves

It is sufficient to plot three points if a straight-line graph is to be drawn, but more points are required if a **curve** is to be drawn accurately.

Choose suitable x-coordinates and then calculate the y-coordinates. The results are displayed on a table of x-values and y-values, as shown in the example below.

After plotting the points, they are joined with a **smooth curve**.

Avoid drawing straight lines between points, and be particularly careful when joining the lowest (or highest) points not to put too sharp a point on the curve or to flatten the curve between these points.

The solid line is correct, the dotted line is incorrect:

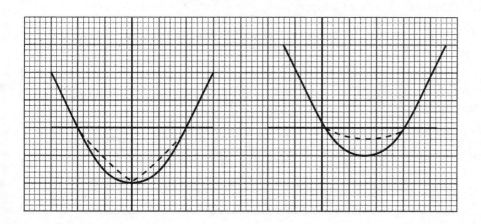

Example

The graph $y = x^2$ can be used to find squares and square roots of numbers.

a Construct a table of values for this graph for $-3 \leqslant x \leqslant 4$.

b Draw the graph of $y = x^2$ for this range of x-values.

c Use the graph to find: (i) the square of 2.5, (ii) the square root of 8.

a The x values are $-3, -2, -1, 0, 1, 2, 3, 4$. Calculating $y = x^2$ gives a table of values:

x	-3	-2	-1	0	1	2	3	4
y	9	4	1	0	1	4	9	16

A suitable scale is 2 large divisions to 1 unit on the x-axis
and 1 large division to 1 unit on the y-axis.

As there are no negative values for y, the x-axis is drawn near the bottom of the page. The y-axis is drawn slightly to the left of centre.

b The completed graph is shown below:

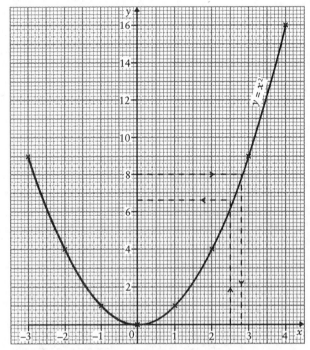

c (i) The square of 2.5 is given by y when $x = 2.5$
Vertical and horizontal lines are drawn, as shown on the graph, and the value of y is found to be 6.3.

That is, the square of $2.5 = 6.3$ (to 1 d.p.).

(ii) The square root of 8 is given by x when $y = 8$.
From the graph, $x = 2.8$ and $x = -2.8$.

That is, the square root of $8 = 2.8$ (to 1 d.p.).

Exercise 162

1 A car starts from rest and accelerates at a steady rate. The distance d metres travelled in t seconds is given by

$$d = \frac{t^2}{5}$$

  a Draw a graph to show the distances travelled during the first 30 seconds.

  b From the graph, find:

    (i) the distance travelled after 24 seconds

    (ii) the time taken to travel 50 m

    (iii) the distance travelled between the 8th and 18th seconds

    (iv) the average speed over this interval of time.

2 The area of an equilateral triangle of side x is given by

$$A = 0.433x^2$$

  a Draw up a table of values for the area of a triangle of side 1, 2, 3, 4, 5, 6 units giving A correct to 2 d.p.

  b Draw the graph of A against x.

  c Use the graph to find:

    (i) the area of a triangle of side 3.5 units

    (ii) the length of side of the triangle if its area is 10 square units.

3 The reciprocal of a number is given by the relationship

$$y = \frac{1}{x}$$

  a Copy and complete the table of values given below for

$$y = \frac{1}{x}$$

x	-5	-4	-3	-2	-1	1	2	3	4	5
y	-0.20		-0.33				0.50		0.25	

(There is no value for y for $x = 0$.)

  b Plot the points and join with two smooth curves.

  c From the graph, find the reciprocal of:
    (i) 1.5, (ii) -3.5, (iii) -0.4, (iv) 0.84

4 A market gardener requires a rectangular area of ground for growing potatoes.

  a Write down a formula for the length of the rectangle in terms of its breadth and area.

  b An area of 360 square metres is allowed for the potatoes.
    Draw up a table of values of length against breadth for breadths of 10 m to 20 m.

  c Draw the graph, joining the points with a smooth curve.

  d What length will the potato plot be if its width is:
    (i) 15.5 m, (ii) 19 m?

5 The time taken to complete a job depends on the number of workers employed. The time can be calculated from the formula.

$$T = \frac{k}{N}$$

where T is the time (in hours) and N is the number of employees.

  a If three workers take 12 hours to complete the job, find the value of k.

  b Using this value of k, draw up a table of values of T against $N = 4, 6, 8, 10$ and 12.

  c Draw the graph, joining the points with a smooth curve.

  d Use your graph to find how long the job will take if seven workers are employed.

*30.9   Graphs of quadratic functions

All quadratic functions of the form

$$y = ax^2 + bx + c$$

have the same general shape when represented by a graph.

This shape is a **smooth** curve and is called a **parabola**.

It is symmetrical about a vertical line through the lowest or highest point.

When drawing up a table of values for the quadratic function, it is easier if the values of y are calculated in stages.

Alternatively, the values for x may be substituted into the equation and the values for y obtained by using a calculator.

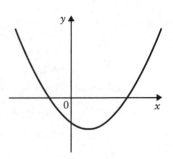

Example

Draw the graph of $y = 2x^2 - 3x - 6$ for values of x between -3 and 4.

The table of values is:

x	-3	-2	-1	0	1	2	3	4
$2x^2$	18	8	2	0	2	8	18	32
$-3x$	9	6	3	0	-3	-6	-9	-12
-6	-6	-6	-6	-6	-6	-6	-6	-6
y	21	8	-1	-6	-7	-4	3	14

Alternatively, when $x = -3$
$$y = 2 \times (-3)^2 - 3 \times (-3) - 6$$
From the calculator $y = 21$.
The other values for y are calculated in the same way.

The range for x is from -3 to 4, i.e. a range of 7.
The range for y is from -7 to 21, i.e. a range of 28.

The graph paper used in the examination is usually
18 cm wide and 24 cm high.
Therefore use a scale of 2 cm to 1 unit for the x-axis
and 1 cm to 2 units for the y-axis.

The y-axis is drawn to the left of centre, e.g. 8 cm from the LHS.
The x-axis is drawn 4 cm from the bottom.

The points plotted are joined with a smooth curve, taking care not to flatten the curve between $x = 0$ and $x = 1$.

The completed graph is shown opposite:

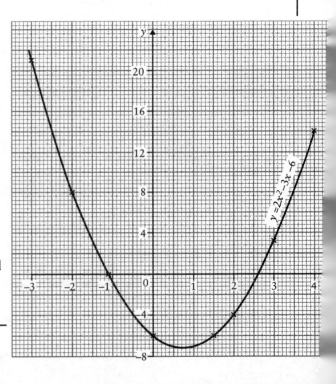

Exercise 163

1 a Copy and complete the table given below and hence draw the graph of $y = 4x - 3x^2$.

x	-2	-1	0	1	2	3
$4x$		-4			8	12
$-3x^2$	-12					-27
y	-20			1	-4	-15

b In what way is this parabola different from the one in the example on p. 306?
What is the reason for this difference?

c Find the value of y when $x = 1.5$.

d Find the value of x when $y = -10$.

2 a Draw up a table of values for x and y, taking values of x from -1 to 7, for the function
$$y = x^2 - 6x + 8$$

b Draw the graph of y against x.

c From the graph find:

  (i) the value of y when $x = 1.25$

  (ii) the values of x when $y = 9$

  (iii) the lowest value of y

  (iv) the equation of the line of symmetry.

3 Romeo throws a pebble at Juliet's window. The path followed by the pebble is given by the equation
$$y = 2x - \frac{x^2}{100}$$

where x is the horizontal distance and y is the vertical distance from Romeo.

a Draw up a table of values, taking x from 0 to 5.

b Draw the graph which shows the path the pebble takes.

c Juliet's window is 1 metre high and the window sill is 4 metres above the ground. Will the pebble hit the window if it is thrown from a distance of: (i) 3 m, (ii) 2.5 m from the house?

d If Romeo is to hit the window, between what distances from the house should he be when throwing the pebble?

4 Stopping distances, in metres, on wet roads can be calculated using the formula
$$d = \frac{v}{5} + \frac{v^2}{80}$$

where v is the speed of the car in km/h.

a Draw up a table of values of d against v for speeds from 0 to 100 km/h.

b Plot the graph of d against v.

c Use the graph to estimate the minimum distance the car should be from a junction when the brakes are applied if the car is travelling on wet roads at:

  (i) 24 km/h, (ii) 64 km/h.

d The driver sees a hazard 100 m ahead and applies the brakes immediately.
If she is travelling at 80 km/h on wet roads, can she stop in time?

5 The total surface area of a cylinder of height 9.5 cm is estimated using the formula
$$A = 6r^2 + 60r$$

a Complete the table below which gives the approximate areas for radii from 0 to 10 cm.

r	0	2	4	6	8	10
$6r^2$						
$60r$						
A						

b Plot the graph of A against r.

c Use the graph to estimate:

  (i) the surface area when the radius is 4.8 cm

  (ii) the radius of the cylinder if the surface area is 800 cm^2.

d The formula for calculating the surface area of a cylinder of height 9.5 cm is:
$$A = 2\pi(r^2 + 9.5r)$$

Calculate the surface area of the cylinder (correct to 3 s.f.) when $r = 6$ cm.

e Calculate the percentage error in the estimated value for the surface area given in the table, when $r = 6$ from the correct value calculated in **d**. Give your answer correct to 2 s.f.

*30.10   The gradient of a curve

A straight-line graph has a constant gradient, but if the graph is a curve, the gradient is continually changing.

For the function

$$s = 25t + 3t^2$$

the distance s is changing at different rates at different times.

The gradient, which is $\dfrac{\text{Distance travelled}}{\text{Time taken}}$, gives the **average** rate of change of

distance over the given time interval, i.e. the average speed or average velocity.

At a particular instant, the rate of change of distance, i.e. the **speed or velocity**, is given by the gradient of the curve at that point in time.

The **gradient of a curve** at a given point is the **gradient of the tangent** drawn to the curve at that point.

Example

A car is braking for traffic lights. The lights change to green when the car's speed is 4 m/s. The driver accelerates and, during the next 6 seconds, the distance travelled, s metres, after a time t seconds, is given by the formula:

$$s = 4t + t^2$$

a Draw up a table of values to show the distance travelled by the car in the first 6 seconds.

b Draw the graph of s against t.

c Find the average speed of the car during the third and fourth seconds.

d By drawing a tangent to the graph, find the speed of the car after 2 seconds.

a

t	0	1	2	3	4	5	6
s	0	5	12	21	32	45	60

b See opposite.

c   Average speed $= \dfrac{\text{Distance travelled}}{\text{Time taken}}$

$\qquad = \dfrac{32 - 12}{2}$ from the graph

$\qquad = 10\,\text{m/s}$

d   Speed after 2 s = Gradient of the tangent at $t = 2$

$\qquad = \dfrac{36 - 12}{5 - 2} = \dfrac{24}{3}$

$\qquad = 8\,\text{m/s}$

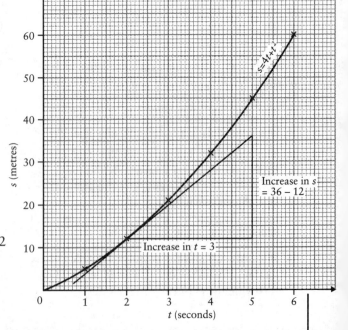

1 A cyclist rounds a corner and sees a tree across the road. He brakes and the distance travelled, s metres, in the next t seconds is given by

$$s = 6t - t^2$$

a Draw the graph of s against t for values of t from 0 to 3 seconds.

b What is the gradient of the curve at $t = 3$?

c How far does the cyclist travel after he starts braking?

d By drawing a tangent to the curve, find the speed of the cyclist immediately before he started to brake.

2 **a** Draw the graph of $y = 2x^2 + 3x - 1$ for $-3 \leqslant x \leqslant 2$.

b By drawing appropriate lines to the graph, find the gradient of the curve at:

(i) $x = -2.5$, (ii) $x = 1.25$.

3 A train entering a station reduces its speed so that the distance, s metres, covered in time t seconds, is given by

$$s = 36t - \tfrac{1}{4}t^2$$

a Draw the graph of s against t for $0 \leqslant t \leqslant 40$.

b By drawing an appropriate line on the graph, find the speed after 30 seconds.

4 **a** Draw the graph of y against x for the function

$$y = 3x^2 + 5x + 2$$

taking values of x from -4 to $+4$.

b Find the gradient of the curve at the points where:

(i) $x = -3$, (ii) $x = 2\tfrac{1}{2}$.

5 A stain is spreading in such a way that the area, $A\,\text{cm}^2$, of the stain after t seconds is given by

$$A = \tfrac{1}{2}t^2 + 3t + 12$$

a Draw up a table of values for A against t when $t = 0, 2, 4, 6, 8, 10$.

b Draw the graph of A against t.

c Find the amount by which the area increased during the third second.

d By drawing a suitable line on the graph, find the rate of increase of A when $t = 5$.

6 A stone is catapulted vertically upwards and its height, h metres, after time t seconds, is given by

$$h = 20t - 5t^2$$

a Plot the graph of h against t for values of t from 0 to 4 seconds.

b Use the graph to find:
(i) the maximum height to which the stone rises
(ii) the times when the stone is at a height of 12 m.

c By drawing an appropriate line on the graph, find the speed of the stone after:

(i) 1 second, (ii) 2.5 seconds.

*31 Further Algebra

31.1   Simultaneous equations

Eddie has just bought a new pen and his friend Pete would also like one, but Eddie can't remember how much it cost. He does remember that he bought the pen and a ruler for £2.20.

Writing this information in algebra gives:

$$p + r = 220$$

This does not help Pete to work out the cost of the pen as there are too many values of p and r which would fit the equation.

Fiona then remembers that she bought two of the same type of pen and a ruler for £3.50.

Pete can now work out the cost of the pen:

$$p + r = 220$$
and $$2p + r = 350$$

The extra pen which Fiona bought must have cost $(350 - 220)$ pence, i.e. 130 pence.

Therefore the pen cost £1.30 and the ruler 90 pence.

If an equation has two unknown values (variables), it cannot be solved on its own.

Two variables require **two** equations which are solved at the same time, i.e. **simultaneously**.

Three variables would require three simultaneous equations, etc.

To solve a pair of simultaneous equations the method of **elimination** is used.

Solving simultaneous equations

Example 1

Solve the equations: $4x + 3y = 24$ (1)

$\qquad\qquad\qquad\quad 2x + 3y = 18$ (2)

The only difference between the left-hand sides of the two equations is that equation (1) has $2x$ more, which must be balanced by the extra 6 on the right-hand side.

Subtracting equation (2) from equation (1) gives:

$$\begin{array}{r} 4x + 3y = 24 \\ -\underline{\quad 2x + 3y = 18 \quad} \\ 2x \qquad\quad = 6 \\ \therefore x = 3 \end{array}$$

The value of y can be found by substituting the value of x into one of the equations.

Substitute $x = 3$ into (2):
$$2 \times 3 + 3y = 18$$
$$6 + 3y = 18$$
$$3y = 12$$
$$y = 4$$

The solution is $x = 3, y = 4$.

(This solution can be checked using the other equation:

Substitute $x = 3, y = 4$ into equation (1)
$$\text{LHS} = 4 \times 3 + 3 \times 4 = 12 + 12 = 24 = \text{RHS}$$
So the solution is correct.)

Example 2

Solve the equations
$$3x - y = 11 \qquad (1)$$
$$2x + y = 9 \qquad (2)$$

The number of ys in each equation is the same, except for the sign, so, in this case y is eliminated by *adding* the two equations.

Adding equation (1) to equation (2) gives:

$$
\begin{aligned}
& 3x - y = 11 \\
+\ & \underline{2x + y = 9} \\
& 5x = 20 \\
\therefore\ & x = 4
\end{aligned}
$$

Substituting $x = 4$ into (2):
$$2 \times 4 + y = 9$$
$$8 + y = 9$$
$$y = 1$$

The solution is $x = 4, y = 1$.

(To check, substitute $x = 4, y = 1$ into (1).)

Example 3

Solve the equations
$$3x + 4y = 18 \qquad (1)$$
$$4x - 3y = -1 \qquad (2)$$

Neither x nor y has the same coefficient in each equation, so this needs to be remedied first.

Method	Working
(i) Decide which variable is to be eliminated.	Eliminate y.
(ii) Multiply one or both equations so that this variable has the same coefficient in each equation (not counting the signs).	Multiply equation (1) by 3 and equation (2) by 4 $$9x + 12y = 54$$ $$16x - 12y = -4$$

(iii) Add or subtract the equations, depending on the signs of the variable to be eliminated.	As the coefficients of y have opposite signs, add the equations: $+\quad\begin{array}{r}9x + 12y = 54\\ 16x - 12y = -4\end{array}$ $\overline{\quad 25x \qquad = 50}$
(iv) Solve for the remaining variable.	$x = 2$
(v) Substitute into one of the original equations to find the eliminated variable and hence the complete solution.	Substitute $x = 2$ into (1) $3 \times 2 + 4y = 18$ $4y = 12$ $y = 3$ Solution is $x = 2, y = 3$
(vi) Substitute into the other original equation to check the solution.	Substitute into (2) LHS $= 4 \times 2 - 3 \times 3$ $= 8 - 9$ $= -1 = $ RHS

Exercise 165

Solve the following pairs of simultaneous equations and check your solutions:

1. $2x + y = 8$
$x + y = 6$

2. $x + 3y = 12$
$x + y = 10$

3. $x + y = 12$
$x - y = 2$

4. $3x - 2y = 9$
$x + 2y = 15$

5. $x + 3y = 6$
$2x + y = 7$

6. $4x + 2y = 10$
$3x + 5y = 11$

7. $3x + 2y = 18$
$4x - y = 2$

8. $2x - 3y = 10$
$3x + 4y = 15$

9. $x - y = -1$
$2x + 3y = 28$

10. $7x + y = 9$
$3x + y = -3$

11. $2x + 5y = 15\frac{1}{2}$
$3x - 4y = -5\frac{1}{2}$

12. $6x + 5y = -17$
$3x + 2y = -8$

Solving problems

Example

An ice cream seller charges a customer £3.20 for three ice cream cornets and two choc-ices.
The next customer is charged £4.50 for four ice cream cornets and three choc-ices.
What are the prices of the ice cream cornet and the choc-ice?

Let x pence be the cost of an ice cream cornet
and y pence be the cost of a choc-ice.

Then $\qquad 3x + 2y = 320 \qquad$ (1)
and $\qquad 4x + 3y = 450 \qquad$ (2)

Eliminate y:
Multiply (1) by 3: $\qquad 9x + 6y = 960 \qquad$ (3)
Multiply (2) by 2: $\qquad 8x + 6y = 900 \qquad$ (4)

Subtract (4) from (3): $x = 60$

Substituting $x = 60$ in (1): $3 \times 60 + 2y = 320$
$$2y = 140$$
$$y = 70$$

An ice cream cornet costs 60p and a choc-ice costs 70p.

(Check by substituting $x = 60$ and $y = 70$ in (2).)

1 Five pounds of apples and three pounds of bananas cost £4.52

   Three pounds of apples and two pounds of bananas cost £2.80.
   What is the price per pound of the apples and the bananas?

2 Two numbers are chosen such that twice the first plus the second is 17 and four times the first minus the second is 25.
   What are the two numbers?

3 The perimeter of a rectangle is 38 metres. The difference between the length and the breadth is 3 metres.
   What is the area of the rectangle?

4 A holiday for 2 adults and 3 children costs £748. The same holiday for 3 adults and 4 children costs £1056.
   What would be the cost of the holiday for 2 adults and 2 children?

5 A potter makes two types of ware. Type A takes $\frac{1}{2}$ hour to make and sells for £3 each. Type B takes 1 hour to make and sells for £5 each.

   One day he worked for 8 hours and sold the pottery he made for £45.
   How many of each type did he make?

6 The ages of a man and his daughter total 27 years.

   In 4 years, time the father will be four times as old as his daughter.
   How old are they both now?

7 At the beginning of a mathematics lesson a teacher says to her class
   'If you subtract twice the number of sweets in my left hand from the number of sweets in my right hand the answer is 3 sweets, but if you add four times the number in my left hand to the number in my right hand the answer is seven times as many.
   The first person to tell me, correctly, how many sweets I have in my hands wins the sweets.'

   How many sweets will the winner receive?

Graphical solution

Simultaneous equations can also be solved by a graphical method.

The graphs of the two equations are drawn and the solution to the equations is the coordinates of the point of intersection of the graphs since these values of x and y will satisfy both equations.

(For the method of plotting straight-line graphs see p. 298.)

Example

Solve, graphically, the equations $x + 2y = 8$
$$3x + 4y = 18$$

To draw the lines, use the points where $x = 0$ and $y = 0$.

For $x + 2y = 8$ the points are $(0, 4)$ and $(8, 0)$
For $3x + 4y = 18$ the points are $(0, 4\frac{1}{2})$ and $(6, 0)$

The axes are now drawn and scaled and the points plotted and joined.

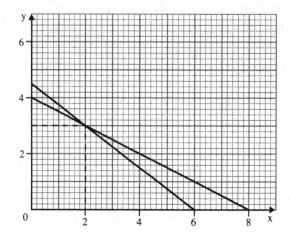

The graphs intersect at the point $(2, 3)$.

Substitute $x = 2$, $y = 3$ into equation (1): LHS $= 2 + 2 \times 3 = 8 =$ RHS
Substitute $x = 2$, $y = 3$ into equation (2): LHS $= 3 \times 2 + 4 \times 3 = 6 + 12$
$$= 18 = \text{RHS}$$

The solution to the equations is therefore $x = 2$, $y = 3$.

Exercise 167

Solve the following equations by the graphical method:

1. $x + y = 5$
    $3x + 2y = 12$

4. $3x - 2y = 0$
    $3x + 4y = 18$

7. $y = 3x + 5$
    $2y = 5x + 9$

2. $x - 2y = 1$
    $x + y = 4$

5. $2x + 5y = 15$
    $2x - y = 3$

8. $y = 7x$
    $2y = 5x - 9$

3. $4x + 5y = 40$
    $x - y = 1$

6. $x + 3y = 11$
    $2x + y = 7$

9. $y = 6 - 2x$
    $3y + 4x = 10$

31.2 Removing brackets

To multiply out a bracket in an expression of the form $2x(x + 3)$

each term in the bracket must be multiplied by $2x$, i.e. $2x(x + 3) = 2x^2 + 6x$

To multiply out a bracket in an expression of the form $(2x + 4)(x + 3)$

each term in the second bracket is multiplied by $2x$ and by 4, i.e.

$$(2x + 4)(x + 3) = 2x(x + 3) + 4(x + 3)$$
$$= 2x^2 + 6x + 4x + 12$$

Collecting like terms gives $(2x + 4)(x + 3) = 2x^2 + 10x + 12$

Example 1

Multiply $(3x - 2)$ by $(2x + 1)$

$$(3x - 2)(2x + 1) = 3x(2x + 1) - 2(2x + 1)$$
$$= 6x^2 + 3x - 4x - 2$$
$$= 6x^2 - x - 2$$

Brackets of this type can be removed without the intermediate stages being shown if the following stages of working are carried out mentally and only the answer is written down.

To expand $(2x + 4)(x + 3)$:

The final answer will have three terms, i.e. an x^2, an x and a constant term.

The x^2 term is calculated by multiplying the x term in the first bracket by the x term in the second bracket:

$$(2x + 4)\ (x + 3) \rightarrow 2x^2$$

The x term is calculated by multiplying the x term in one bracket by the constant term in the other bracket and adding the two resultant terms.

$$(2x + 4)\ (x + 3) \rightarrow 4x + 6x = 10x$$
$$\text{add}$$

The constant term is calculated by multiplying the constant term in the first bracket by the constant term in the second bracket.

$$(2x + 4)\ (x + 3) \rightarrow 12$$
$$\therefore (2x + 4)(x + 3) = 2x^2 + 10x + 12$$

— *Example 2* —

Expand $(3x - 4)(2x - 3)$

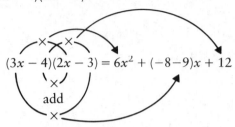

$$(3x - 4)(2x - 3) = 6x^2 + (-8-9)x + 12$$

add

$$\therefore (3x - 4)(2x - 3) = 6x^2 - 17x + 12$$

Exercise 168

Remove the brackets from the following:

1 $(x + 3)(x + 4)$	6 $(x - 3)(x - 7)$	11 $(2x + 5)(2x - 5)$	16 $(2x - 3)(2x - 3)$
2 $(x + 2)(x - 1)$	7 $(x + 1)(x - 2)$	12 $(3x - 4)(3x + 4)$	17 $(2x + 1)(x + 4)$
3 $(x - 5)(x - 3)$	8 $(x - 6)(x - 2)$	13 $(x + 2)(x + 2)$	18 $(3x + 2)(x - 2)$
4 $(x + 1)(x + 4)$	9 $(x + 3)(x - 3)$	14 $(x - 1)(x - 1)$	19 $(2x + 1)(3x + 2)$
5 $(x + 4)(x - 2)$	10 $(x + 1)(x - 1)$	15 $(3x + 1)(3x + 1)$	20 $(3x - 2)(4x + 3)$

31.3 Quadratic factors

All the expressions in Exercise 168 were **quadratic expressions**.

In each case, the brackets multiplied out to give **four** terms (which then simplified to three terms) and the final expression always contained an x^2 term, but no higher power of x.

$$\left. \begin{array}{l} 2x^2 + 7x - 3 \\ x^2 - 6x \\ 4y^2 - 9 \end{array} \right\} \quad \begin{array}{l} \text{are all examples} \\ \text{of quadratic} \\ \text{expressions.} \end{array}$$

We saw that $(2x + 4)(x + 3) = 2x^2 + 10x + 12$.
Therefore, $(2x + 4)$ and $(x + 3)$ are **factors** of the quadratic $2x^2 + 10x + 12$.

Quadratic expressions and quadratic equations occur in many branches of mathematics and it is necessary to be able to factorise them.

When the coefficient of x^2 is unity

Example 1

Factorise $x^2 + 5x + 6$.

Reversing the process of multiplying out brackets, we can see that
$$x^2 + 5x + 6 = (x + ?)(x + ?)$$
and 6 must factorise as 6×1 or 3×2.

The key to factorising the quadratic expression is the x term.

$5x$ is the sum of two x terms and the coefficients of these two terms must be the factors of 6.
$$3 \times 2 = 6 \text{ and } 3x + 2x = 5x$$
$$\therefore x^2 + 5x + 6 = (x + 3)(x + 2)$$

Example 2

Factorise $x^2 - x - 6$.

The factors of 6, when added together, must equal the coefficient of x, which is -1.

The factors of 6 are -6×1, 6×-1, -3×2, 3×-2.
Only $-3 + 2 = -1$
$$\therefore x^2 - x - 6 = (x - 3)(x + 2)$$

Example 3

Factorise $x^2 + 6x$.

This is a much simpler quadratic expression to factorise than the standard type because it has a **common factor** which is x.
$$x^2 + 6x = x(x + 6)$$

Exercise 169

Factorise the following:

1 $x^2 + 8x + 7$ 4 $x^2 - 5x + 6$ 7 $x^2 - 6x + 9$ 10 $x^2 + 11x - 12$

2 $x^2 + 7x + 10$ 5 $x^2 + 2x - 8$ 8 $x^2 - 4x - 12$ 11 $x^2 - 5x - 6$

3 $x^2 + x - 6$ 6 $x^2 - 2x - 15$ 9 $x^2 - 7x + 12$ 12 $x^2 - 6x - 16$

13 $x^2 - 7x$	15 $6x^2 - 3x$	17 $x^2 + 12x + 36$
14 $2x^2 + 6x$	16 $x - 2x^2$	18 $x^2 - 8x + 16$

When the coefficient of x^2 is not unity

The coefficient of x^2 will not necessarily be 1.
In this section the method for factorising quadratic expressions is extended to include those where the coefficient of x^2 is 2 or 3.

Example

Factorise $2x^2 + 11x + 12$.

The method previously used must be adapted to take into account the coefficient 2.

As 2 will only factorise as 2×1, it can be seen that

$$2x^2 + 11x + 12 = (2x + a)(x + b)$$

As before, $a \times b = 12$, but now we require

$$a + 2b = 11 \text{ (not } a + b \text{ as before)}$$

Factors of 12 are $a \times b =$ 1×12 and $a + 2b = 1 + 24 = 25$
$\qquad\qquad\qquad\qquad\quad$ $2 \times \ 6 \qquad\qquad\qquad = 2 + 12 = 14$
$\qquad\qquad\qquad\qquad\quad$ $3 \times \ 4 \qquad\qquad\qquad = 3 + \ 8 = 11$
$\qquad\qquad\qquad\qquad\quad$ $4 \times \ 3 \qquad\qquad\qquad = 4 + \ 6 = 10$
$\qquad\qquad\qquad\qquad\qquad\quad \cdot \qquad\qquad\qquad\qquad\quad \cdot$
$\qquad\qquad\qquad\qquad\qquad\quad \cdot \qquad\qquad\qquad\qquad\quad \cdot$
$\qquad\qquad\qquad\qquad\qquad\quad \cdot \qquad\qquad\qquad\qquad\quad \cdot$

$$\therefore\ 2x^2 + 11x + 12 = (2x + 3)(x + 4)$$

Alternatively list all possible solutions and select the one which 'works':

$$(2x + 1)(x + 12) = 2x^2 + 25x + 12$$
$$(2x + 2)(x + 6) \ = 2x^2 + 14x + 12$$
$$(2x + 3)(x + 4) \ = 2x^2 + 11x + 12$$

Exercise 170

Factorise:

1 $2x^2 + 5x + 2$	5 $3x^2 + 8x + 5$	9 $2x^2 - 11x + 12$
2 $2x^2 + 7x + 6$	6 $3x^2 - 6x - 9$	10 $3x^2 + 20x + 12$
3 $2x^2 + 9x - 5$	7 $3x^2 - 8x + 4$	11 $3x^2 - 22x - 16$
4 $2x^2 - 13x - 7$	8 $3x^2 - 13x - 10$	12 $2x^2 - 3x - 14$

Difference of two squares

If you look back to Exercise 168 questions 9–12 on p. 316, you will see that these brackets always multiplied out to give a quadratic expression with two terms and not the more usual three terms.

Moreover, the two terms are both perfect squares separated by a minus sign. This type of quadratic expression is known as **the difference of two squares**.

Once a quadratic has been recognised as the difference of squares, it is easily factorised:

$$a^2 - b^2 = (a - b)(a + b)$$

Example 1

Factorise $x^2 - 9$.

$$x^2 - 9 = (x - 3)(x + 3)$$

Example 2

Factorise $9x^2 - 16$.

$$9x^2 - 16 = (3x)^2 - (4)^2$$
$$= (3x - 4)(3x + 4)$$

Example 3

Factorise $8x^2 - 2$.

$8x^2$ and 2 are not perfect squares, but 2 is a **common factor** of the expression.

$$8x^2 - 2 = 2(4x^2 - 1)$$

$4x^2 - 1 = (2x)^2 - 1^2$ which is the difference of two squares

$$\therefore 8x^2 - 2 = 2(2x - 1)(2x + 1)$$

(*Note.* If a quadratic has a common factor, always take out the common factor before factorising the quadratic into two brackets.)

Exercise 171

Factorise:

1 $x^2 - 1$	4 $4x^2 - 9$	7 $49 - x^2$	10 $2x^2 - 8$
2 $x^2 - 16$	5 $9x^2 - 1$	8 $36 - 25x^2$	11 $9x^2 - 36$
3 $x^2 - 25$	6 $16x^2 - 25$	9 $25 - 49x^2$	12 $12x^2 - 75$

31.4   Quadratic equations

An equation which can be arranged so that a quadratic expression is equal to zero is called a **quadratic equation**.

A quadratic equation is of the form

$$ax^2 + bx + c = 0$$

where b and c can be any integer and a can be any integer except 0.

Of course, the equation does not have to be in terms of x. It could be in terms of y or z or x^2, provided the powers of the variable are 2, 1 and 0.

Solution of quadratic equations by factorising

If the product of two numbers is zero, we know that one of the numbers must itself be zero:

$xy = 0$ means that either $x = 0$ or $y = 0$ (or both $x = 0$ and $y = 0$)

This concept is very important for the solution of quadratic equations by factors.

Example 1

Solve the equation $x^2 - 5x - 14 = 0$.

$$x^2 - 5x - 14 = 0$$

Factorising gives: $(x - 7)(x + 2) = 0$

$\therefore$ either $(x - 7) = 0$ or $(x + 2) = 0$

$\therefore$ either $x = 7$ or $x = -2$

$\therefore$ The solution is $x = 7$ or -2.

Note. A quadratic equation always has *two* possible solutions. If the quadratic is a perfect square, the two solutions will be the same and they are called repeated roots: see example 4 below.

Example 2

Solve the equation $2x^2 - 5x + 3 = 0$.

$$2x^2 - 5x + 3 = 0$$

Factorising gives: $(2x - 3)(x - 1) = 0$

so either $(2x - 3) = 0$ and $x = 1\frac{1}{2}$

or $(x - 1) = 0$ and $x = 1$

$\therefore x = 1\frac{1}{2}$ or 1

Example 3

Solve $6x^2 + 8x - 8 = 0$.

The LHS has a common factor of 2. Dividing the equation by 2 gives:

$$3x^2 + 4x - 4 = 0$$

Factorising gives: $\qquad (3x - 2)(x + 2) = 0$

so either $\qquad\qquad\qquad (3x - 2) = 0$ and $x = \frac{2}{3}$

or $\qquad\qquad\qquad\qquad (x + 2) = 0$ and $x = -2$

$$\therefore x = \tfrac{2}{3} \text{ or } -2$$

Example 4

Solve $x^2 - 6x + 9 = 0$.

$$x^2 - 6x + 9 = 0$$

Factorising gives: $\qquad (x - 3)(x - 3) = 0$

so either $\qquad\qquad\qquad x - 3 = 0$ or $x - 3 = 0$

either $\qquad\qquad\qquad\qquad x = 3$ or $x = 3$

$$\therefore x = 3$$

(In an example like this we say that the equation has **repeated roots**.)

Exercise 172

Solve the following equations:

1 $x^2 - 5x + 4 = 0$	**5** $2x^2 + 7x + 3 = 0$	**9** $3x^2 - 20x + 12 = 0$
2 $x^2 + 11x + 28 = 0$	**6** $x^2 + 3x = 0$	**10** $2x^2 + 3x - 14 = 0$
3 $x^2 - 49 = 0$	**7** $3x^2 - 2x - 8 = 0$	**11** $2x^2 - 2x - 40 = 0$
4 $x^2 + 6x + 9 = 0$	**8** $5x^2 - 2x = 0$	**12** $6x^2 + 3x - 3 = 0$

Solving quadratic equations by the formula

The equation $2x^2 - 3x - 1 = 0$ does not have simple factors.

To solve this type of equation the **quadratic formula** must be used.

(The formula will be given to you in an examination, but you must know how to use it.)

For the equation of $ax^2 + bx + c = 0$

The solution is

$$x = \frac{-b \pm \sqrt{b^2 - 4ac}}{2a}$$

Example 1

Solve $x^2 + 3x + 1 = 0$.

Compare $\qquad x^2 + 3x + 1 = 0$
with $\qquad\quad ax^2 + bx + c = 0$

Then $a = +1, b = +3, c = +1$
and substituting for a, b and c in the formula gives:

$$x = \frac{-3 \pm \sqrt{9 - 4 \times 1 \times 1}}{2 \times 1}$$

$$= \frac{-3 \pm \sqrt{5}}{2}$$

$$= \frac{-3 + 2.236}{2} \text{ or } \frac{-3 - 2.236}{2}$$

$$= -0.38 \text{ or } -2.62 \text{ (to 2 d.p.)}$$

Example 2

Solve $2x^2 - 3x - 1 = 0$.

In this equation, $a = 2, \quad b = -3, \quad c = -1$ and substituting in the formula gives

$$x = \frac{+3 \pm \sqrt{9 - 4 \times 2 \times (-1)}}{2 \times 2}$$

$$= \frac{3 \pm \sqrt{17}}{4}$$

$$= \frac{3 + 4.123}{4} \text{ or } \frac{3 - 4.123}{4}$$

$$= 1.78 \text{ or } -0.28 \text{ (to 2 d.p.)}$$

Exercise 173

Solve the following equations using the quadratic formula and give your answers correct to 2 d.p.

1   $2x^2 + 2x - 3 = 0$ 4 $3x^2 - x - 1 = 0$ 7 $4x^2 - 8x - 16 = 0$ 10 $3x^2 - 10x = -5$

2   $2x^2 + 4x + 1 = 0$ 5 $x^2 + 6x - 10 = 0$ 8 $3x^2 - 6x + 2 = 0$ 11 $x(x + 2) = 5$

3   $x^2 + 2x - 2 = 0$ 6 $x^2 - 7x + 9 = 0$ 9 $2x^2 + 5x = 6$ 12 $x(x - 3) = -1$

Problems involving quadratic equations

Example

One side of a rectangle is 3 centimetres longer than the other. The area
(in cm^2) of the rectangle is twice its perimeter (in cm). Find the
dimensions of the rectangle.

$$\text{Let the length of one side} = x\,\text{cm}$$

$$\text{The length of the other side} = (x + 3)\,\text{cm}$$

$$\text{The area} = x(x + 3)\,\text{cm}^2$$

$$\text{The perimeter} = (4x + 6)\,\text{cm}$$

$$\text{Since Area} = 2 \times \text{Perimeter}$$

$$x(x + 3) = 2(4x + 6)$$

$$x^2 + 3x = 8x + 12$$

$$x^2 - 5x - 12 = 0$$

$$x = \frac{+5 \pm \sqrt{25 + 48}}{2}$$

$$= \frac{5 \pm \sqrt{73}}{2}$$

As x cannot be negative, $\qquad x = \dfrac{5 + 8.544}{2} = 6.77\,\text{cm (to 2 d.p.)}$

$$\therefore \text{The dimensions are } 6.77\,\text{cm and } 9.77\,\text{cm.}$$

Exercise 174

1 A carpet is 3 metres too long for a square floor in
one direction and 1 metre too short in the other.
The area of the carpet is 21 square metres.

   a If the floor is x metres square, write down, in
   terms of x, the dimensions of the carpet.

   b Write down an equation in x for the area of the
   carpet.

   c Solve the equation and find the dimensions of
   the floor.

2 The number of diagonals D of a polygon with n
sides is given by the formula:

$$D = \tfrac{1}{2}n(n - 3)$$

   a Find D if $n = 7$.

   b Find n if $D = 20$.

3 In a tennis tournament each team plays every
other team.

   a If there are x number of teams, write down an
   expression, in terms of x, for the total number
   of games played.

   (*Remember.* Team A plays team B is the same
   game as team B plays team A.)

   b If a total of 28 games is played, write down an
   equation in x.

   c Solve the equation to find how many teams
   played in the tournament.

4 A farmer has 40 metres of fencing with which he intends to make a rectangular enclosure for a few sheep and their lambs.

   a If one side of the rectangle is x metres, what is the length of the other side, in terms of x?

   b Write down an expression for the area in terms of x.

   c The farmer calculates that he requires $96\,m^2$ of grass in the sheep pen.
   What are the dimensions of the pen?

   d If the farmer had required an enclosure of area $80\,m^2$, but not less, what would the dimensions of the rectangle have been?
   (Use the quadratic formula to solve the equation and give your answer to the nearest $0.1\,m$.)

   e State the dimensions of the pen which would enclose the largest area of grass, using the 40 m of fencing.

5 A rectangle has sides of 7 cm and 4 cm.
Two strips, each x cm wide, are cut from the rectangle.
One strip is cut parallel to the shorter side and the other is cut parallel to the longer side.

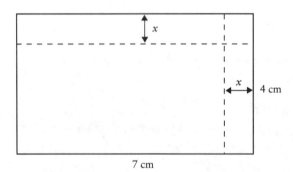

7 cm

The resulting rectangle is half the area of the original rectangle.

Form an equation in x and use the quadratic formula to find the value of x correct to 2 decimal places.

6 A photograph is 5 cm square. It is mounted on a piece of card so that there is a border which is x cm wide at the top and bottom and $\frac{1}{2}x$ cm wide at each side.

The area of the card is $80\,cm^2$.

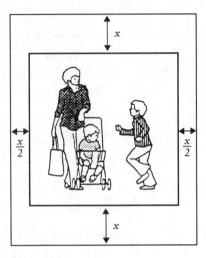

Write down an equation for the area of the card, in terms of x, and solve it to find the value of x correct to the nearest millimetre.

What are the dimensions of the card?

7 **a** A cube with dimensions $4 \times 4 \times 4$ is made of 64 small blocks, as shown in the diagram. The outside of the cube is painted. How many of the blocks will have just one side painted?

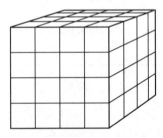

   b A cube with dimensions $5 \times 5 \times 5$ would be made from 125 blocks. How many of these blocks would have one side painted?

   c Find a formula for the number of blocks with one side painted (S) if the cube has dimensions $n \times n \times n$.

   d Calculate the value of n if: (i) $S = 24$, (ii) $S = 96$.

31.5   The transformation of formulae

> **Example**
>
> The profit (£P) made when an article is sold for a price £S with a percentage profit of $x\%$, can be calculated from the formula:
>
> $$P = \frac{Sx}{100 + x}$$
>
> **a** Find P when $x = 12$ and $S = 392$.
>
> **b** Rearrange the formula to make x the subject.
>
> **c** Find the percentage profit if a profit of £33 is made when an article is sold for £253.
>
>
>
>
> **a**
> $$P = \frac{392 \times 12}{100 + 12} = \frac{4704}{112} = £42$$
>
>
> **b**
> $$P = \frac{Sx}{100 + x}$$
>
> (i) Deal with the fraction by multiplying each side by $(100 + x)$:
> $$P(100 + x) = Sx$$
>
> (ii) Multiply out the bracket:
> $$100P + Px = Sx$$
>
> (iii) Collect x terms on one side, other terms on the opposite side (in this case, it is easier to collect x terms on RHS to avoid a negative sign for $100P$):
> $$100P = Sx - Px$$
>
> (iv) Take out a common factor of x:
> $$100P = x(S - P)$$
>
> (v) Divide each side by $(S - P)$ to find the value of x:
> $$\frac{100P}{S - P} = x$$
>
>
> **c** $P = £33, \quad S = £253$
>
> $$\text{Profit \%} = x = \frac{100P}{S - P} = \frac{100 \times 33}{253 - 33} = 15\%$$

Exercise 175

1 The formula $s = \frac{1}{2}(u + v)t$ gives the distance travelled in time t by a vehicle, where u is its initial speed and v is the speed after travelling a distance s.

    a Rearrange the formula to give v in terms of s, u and t.

    b Calculate v if $s = 46$, $u = 5$, $t = 4$.

    c Calculate v if $s = 120$ m, $u = 24$ m/s, $t = 10$ s. What does this value of v tell you about the vehicle?

2 The focal length of a lens is given by the formula

$$\frac{1}{f} = \frac{1}{u} + \frac{1}{v}$$

    a Calculate v if $f = 3.0$ cm and $u = 4.8$ cm.

    b Rearrange the formula to give u in terms of f and v.

    c Find u if $f = 2.6$ cm and $v = 6.5$ cm.

3 The time, in seconds, for a full swing, or oscillation, of a pendulum, length l metres, is given by

$$T = 2\pi\sqrt{\frac{l}{10}}$$

    a Rearrange the formula to make l the subject.

    b The pendulum of a clock is to make a complete oscillation every one second. What length, to the nearest millimetre, must the pendulum be?

4 If £p is invested for 2 years at a compound interest rate of $r\%$, the amount in the account at the end of the 2 years is given by

$$A = P\left(1 + \frac{r}{100}\right)^2$$

    a Rearrange the formula to make r the subject.

    b £2500 was invested and at the end of 2 years had amounted to £3775. At what rate of compound interest was the money invested?

    The formula can be simplified if $\left(1 + \dfrac{r}{100}\right)$ is

    replaced by R.

    c Rewrite the formula for A in terms of P and R.

    d Rearrange the formula to make R the subject.

    e (i) Find R if $A = £4143.75$ and $P = £3750$.
        (ii) At what rate was the money invested?

5 A boy, standing near the edge of a cliff, throws a stone up into the air. The velocity, v metres per second, of the stone, when it is a distance s metres from the boy's hand, is given by

$$v^2 = 100 - 20s$$

    a Rearrange the formula to give s in terms of v.

    b Find s when (i) $v = 5$ m/s
                    (ii) $v = 0$ m/s
                    (iii) $v = 10$ m/s
                    (iv) $v = 15$ m/s
    and describe the position of the stone in each case.

6 A window is made in the shape of a rectangle of width $2x$ and height $2h$ surmounted by a semicircle.

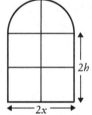

    a Write down an equation for the perimeter of the window, P, in terms of x and h.

    b Rearrange the above equation to give x in terms of P and h.

    c Find x if $P = 15.7$ and $h = 2$.

7 A rectangle has sides of length x and y.

    a Write down an expression for the perimeter of the rectangle in terms of x and y.

    b Write down an expression for the area of the rectangle in terms of x and y.

    c The area of the rectangle is three times the length of the perimeter. Write down an equation which describes this relationship.

    d Rearrange the equation to give a formula for x in terms of y.

    e Find a value for x if $y = 10$ cm.

▧ *Exercise 176 (Miscellaneous questions)* ▧

1 **a** Solve the equations:

  (i) $2x - y = 8$ (ii) $2x - 4y = 1$
    $3x + y = 7$ $3x - 5y = 2\frac{1}{2}$

  b Solve, by drawing a graph, the equations

  $$x - y = -2$$
  $$2y = 3x + 1$$

2 A school buys two types of calculator. One type costs £9 each, the other costs £30 each.

  The school ordered 74 calculators and paid £1170.
  How many of each type were bought?

3 Expand:

  a $(x + 5)(x + 2)$

  b $(x - 7)(x + 7)$

  c $(2x - 1)(x - 8)$

4 Factorise:

  a $x^2 - 8x - 9$ **d** $3x^2 + 12x + 9$

  b $3x^2 + 12x$ **e** $2x^2 + 11x - 6$

  c $3x^2 - 12$

5 Solve:

  a $x^2 - 4x - 5 = 0$ **c** $9x^2 - 16 = 0$

  b $4x^2 - 8x = 0$ **d** $2x^2 + 7x + 6 = 0$

6 Solve, giving your answers correct to 2 d.p.

  a $x^2 + 2x - 7 = 0$ **b** $2x^2 - 5x + 1 = 0$

7 A rectangle has a side of length x centimetres. The other side of the rectangle is 2 centimetres longer.

  a Write down an expression, in terms of x, for the area of the rectangle.

  b Write down an expression, in terms of x, for the perimeter of the rectangle.

  c The area, in square centimetres, is equal to the perimeter, in centimetres.
  Write down an equation in x.

  d Simplify your equation and solve it to find x, using the formula:

  $$x = \frac{-b \pm \sqrt{b^2 - 4ac}}{2a}$$

  (Give your answer correct to 3 s.f.)

8 A firm makes novelties for Christmas crackers. Each novelty costs x pence to make and they are sold for £y per 1000.

  a Write down an expression for the cost of manufacturing N novelties.

  b Write down an expression for the selling price (in pence) of N novelties.

  c Write down a formula for the profit P pence in terms of N, x and y.

  d Calculate the profit if $N = 3000$, $x = 3$ and $y = 50$.

  e Rearrange the formula to make N the subject.

  f If the cost of manufacture and the selling price are unchanged, how many novelties must be sold in order to make a profit of £100?

*32 Sets

32.1 Introduction

A set is a collection of numbers or objects which can be clearly described in words or symbols or by a list.

For example:

The set of square integers can be represented as:

$$S = \{\text{the square integers}\}$$
$$S = \{1, 4, 9, 16, \ldots\}$$
$$S = \{x^2 : x \epsilon Z\}$$

It is important to be able to translate the symbols into words, and vice versa.

Curly brackets, { }, are used to denote a set.

S = {square integers}, translates as S is 'the set of' square integers.

The symbol ϵ means **member** or **element** of a set.

If $S = \{1, 4, 9, 16, \text{------}\}$
$4 \epsilon S$ means 4 is a member of, or belongs to, the set of square integers, i.e. 4 is a square number.
$3 \notin S$ means 3 is *not* a member of S, i.e. 3 is not a square number.

$S = \{x^2 : x \epsilon Z\}$ translates as: S is the set of square numbers, x^2, such that x is an integer, i.e. S is the set of square integers.

The elements of the set of colours of the rainbow can all be listed and this is a **finite set**.

The elements of the set of square integers cannot all be listed, only sufficient to establish a pattern are given, and this is an **infinite set**.

A set which contains no elements is called an **empty set** or **null set**.

For example, the set of square roots of a negative number has no members and $S = \{ \ \}$ or $S = \emptyset$.

A set which contains *all* the elements under consideration is called the **universal set** and is given the symbol $\mathscr{E}$.

A set which contains some, but not all, of the elements of a set S is called a **subset** of S.

If $S = \{3, 6, 9, 12, 15\}$ and $A = \{6, 12\}$,
then A is a subset of S and this is written $A \subset S$.

If $B = \{9, 18\}$, B is not a subset of S, i.e. $B \notin S$.

Example 1

List the elements of the following sets, in full if the set is finite, or the first five elements if the set is infinite.

a $A = \{\text{vowels}\}$ **b** $B = \{2^n: n \text{ is a positive integer}\}$

a A is finite and $A = \{a, e, i, o, u\}$

b B is the set of powers of 2 where the power is $1, 2, 3, \ldots$.
$B = \{2, 4, 8, 16, 32, \ldots\}$

Example 2

Describe, in words, the following sets:

a $A = \{+, -, \times, -\}$ **b** $B = \{6, 8, 10, 12\}$

a A is the set of symbols used in the four rules of arithmetic.

b B is the set of even numbers between 5 and 13.

Example 3

$E = \{\text{European countries}\}$ $C = \{\text{Common Market countries}\}$

Translate each of the following statements in set language into a sentence:

a $C \subset E$ **b** $E \not\subset C$ **c** France $\in E$ **d** USSR $\notin C$

a All Common Market countries are European.

b All European countries are not in the Common Market.

c France is a European country or France is in Europe.

d The USSR is not a member of the Common Market.

Exercise 177

1 State whether the following sets are finite, infinite or empty:

a numbers greater than 1000

b odd numbers divisible by 2

c odd numbers less than 200

d factors of 36

e cubic numbers.

2 List the elements of the following sets, in full, if the set is finite, or the first five elements if the set is infinite:

a $A = \{\text{the last four letters of the alphabet}\}$

b $B = \{x: x \text{ is a multiple of 5}\}$

c $C = \{\text{prime numbers greater than 3}\}$

d $D = \{\text{the days of the week}\}$

e $E = \{x: x \text{ is an even number less than 16}\}$

f $F = \{x: x \text{ is an integer, } 4 \leq x < 12\}$

3 Describe, in words, the following sets:

   a A = {Europe, Asia, Africa, N America,
            S America, Australasia, Antarctica}

   b B = {1, 2, 3, 4, 6, 12}

   c C = {isosceles, equilateral, scalene}

   d D = {−2, +2}

   e E = {1, 3, 6, 10, 15}

4 A = {letters of the alphabet}
   V = {vowels}
   C = {consonants}

   For each of the following statements, state whether it is true or false:

   a $V \subset A$ **b** $C \not\subset A$ **c** $y \in V$ **d** $e \notin C$

   e {a, e, i, o, w} $\subset V$ **f** {vowels after v} = $\emptyset$

5 A = {1, 2, 3}

   List all the subsets of A (not including A and $\emptyset$).

6 P = {prime numbers except 2}
   Q = {odd numbers}
   E = {even numbers}

   Translate each of the following statements in set language into a sentence:

   a $P \not\subset E$ **b** $4 \notin Q$ **c** $5 \in P$ and Q
   d $P \subset Q$ **e** $5, 9 \in Q$

32.2   Union and intersection

The union of two sets is the set which contains all the elements of the original two sets.

The symbol for **union** is ∪. An example can show how the symbol is used.

Example 1

There are 9 students in class 6B.
Ann, Carl, Damian, Fahzid, and Harry swim.
Carol, Eddie, Harry, Inez, Kathy and Ossie play tennis.
Write down the complete list of people who play tennis or swim.

If S = {the members of 6B who swim}
        = {Ann, Carol, Damian, Fahzid, Harry}

and T = {the members of 6B who play tennis}
        = {Carol, Eddie, Harry, Inez, Kathy, Ossie}

then
$S \cup T$ = {Ann, Carol, Damian, Eddie, Fahzid, Harry, Inez, Kathy, Ossie}

(Note that, although Carol and Harry both swim and play tennis, they are not listed twice in the union set.)

Members who belong to both sets are said to be in the **intersection** of the sets.

The symbol for **intersection** is ∩. Again we can see how this is used in an example.

Example 2

Look at the previous example. Make a list of the people who both swim *and* play tennis (the intersection of the two sets).

$$S \cap T = \{\text{Carol, Harry}\}$$

S' is the complement of the set S and means any members of the universal set which are not in S.

Example 3

Referring to class 6B again: who are the students who do not swim?

$$S' = \{\text{Eddie, Inez, Kathy, Ossie}\}$$

Number of members in a set

In Example 1 on p. 330 the set S contains 5 members or elements. This is written as

$$n(S) = 5$$

i.e. the number of elements in S is 5 or 5 people in 6B swim.

Also $n(T) = 6$, $n(S \cup T) = 9$ and $n(S \cap T) = 2$.

(6 people play tennis, 9 people swim or play tennis, 2 people swim and play tennis.)

Example

$\mathscr{E} = \{2, 4, 6, 8, 10, 12, 14, 16, 18, 20\}$
$A = \{4, 8, 12, 16, 20\}$
$B = \{6, 12, 18\}$

Find:

a $A \cup B$ **b** $A \cap B$ **c** A' **d** $(A \cup B)'$ **e** $n(\mathscr{E})$ **f** $n(A \cap B)'$

a $A \cup B = \{4, 6, 8, 12, 16, 18, 20\}$

b $A \cap B = \{12\}$

c $A' = \{2, 6, 10, 14, 18\}$

d $(A \cup B)' = \{2, 10, 14\}$

e $n(\mathscr{E}) = 10$

f $n(A \cap B)' = 9$

Exercise 178

1 $\mathscr{E}$ = {students in LVI}

B = {boys in LVI}

S = {students in LVI who smoke}

G = {girls in LVI}

Describe the following sets in words:

a $S \cup G$

b $S \cap B$

c B'

d $(G \cap S)'$

e $G \subset S$

f $G' \cup B'$

g $n(\mathscr{E}) = 10$

h $n(S' \cap B) = 5$

2 $\mathscr{E} = \{1, 2, 3, 4, 5, 6, 10, 12, 15, 20, 30, 60\}$

A = {1, 2, 3, 4, 5, 6}

B = {3, 6, 12, 15, 30, 60}

List the elements of the following sets:

a $A \cap B$

b $A \cup B$

c A'

d $(A \cup B)'$

e $A' \cap B'$

f $(A \cap B)'$

g $A' \cup B'$

h $A' \cup B$

3 P = {all parallelograms}   R = {all rectangles}

Q = {all quadrilaterals}   S = {all squares}

Write the following statements in set notation:

a A rhombus is a parallelogram.

b Parallelograms are quadrilaterals.

c All quadrilaterals are not parallelograms.

d All squares are rectangles and all rectangles are parallelograms.

e No trapeziums are parallelograms.

f A kite is a quadrilateral, but not a parallelogram.

4 The sketch below shows the villages of Lower and Upper Amble. The boundary line of the parish of Browsing cuts through Lower Amble.

B = {dwellings in the Parish of Browsing}
L = {dwellings in Lower Amble}
U = {dwellings in Upper Amble}

Each dwelling is represented by a solid block (■).

Find:

a $n(U)$ c $n(L \cap U)$ e $n(L' \cup B)$

b $n(L \cap B)$ d $n(L \cap B')$ f $n(L \cap B)'$

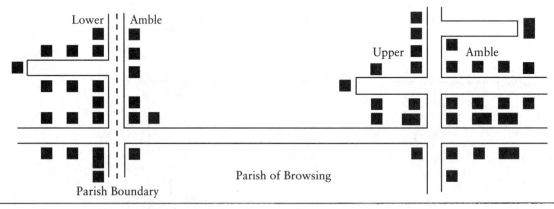

5 Find the single set which is equivalent to:

a $A \cap \mathscr{E}$ c $A \cap A$ e $A \cup A$

b $A \cup \mathscr{E}$ d $A \cap A'$ f $A \cup A'$

6 T = {students in a tutorial group}

B = {boys in the tutorial group}

M = {students in the tutorial group studying mathematics}

From the statements below, draw a conclusion. Write your conclusion in words and in set notation.

a $n(B \cap M) = 8$ and $B \subset M$

b $n(M) = 13$

c $n(B' \cap M') = 4$

Summary of set notation

$S = \{\ \}$ S is the set of . . .

$a \in S$ a is an element of the set S (a belongs to S).

$a \notin S$ a is not an element of set S

$A \subset S$ A is a subset of set S

$\mathscr{E}$ the universal set

$\emptyset$ the empty set

$A \cap B$ A intersection B
             (the set of elements in both A and B)

$A \cup B$ A union B
             (the set of elements in either A or B)

A' the complement of A
             (the set of elements which are not in A)

$(A \cap B)'$ the complement of (A intersection B)
             (the set of elements which are in neither A nor B)

$A' \cap B$ (the set of elements which are in B but not in A)

$A' \cup B$ (the set of elements which are in B, including those in A and B,
             but not in A only)

$n(S)$ the number of elements in set S

32.3   Venn diagrams

A Venn diagram is a method of using diagrams to represent the relationship between sets.

This system was devised by the mathematician John Venn.

The universal set, $\mathscr{E}$, is represented by a rectangle:

Other sets can be represented by any closed shape, but circles or ovals are the most common shapes.

The relationships between two sets A and B are represented as follows:

The intersection of A and B   **The union of A and B**

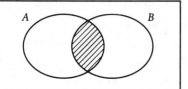   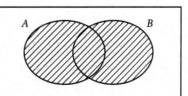

The complement of *A*.

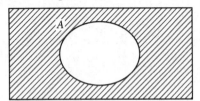

A and B do not intersect.

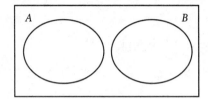

A is a subset of B.

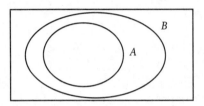

Example 1

Show, by drawing and shading a Venn diagram, that $(A \cup B)'$ is the same as $A' \cap B'$.

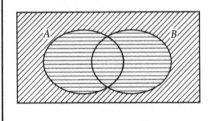

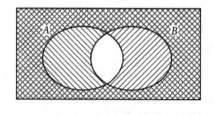

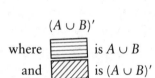

$(A \cup B)'$

where ⬜ is $A \cup B$

and ⬜ is $(A \cup B)'$

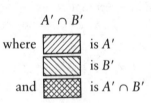

$A' \cap B'$

where ⬜ is A'

⬜ is B'

and ⬜ is $A' \cap B'$

Example 2

$\mathscr{E} = \{0, 1, 2, 3, 4, 5, 6, 7, 8, 9, 10\}$
$A = \{1, 2, 3, 4, 5, 6, 7, 8, 9\}$
$B = \{2, 4, 6, 8, 10\}$
$C = \{1, 4, 9\}$

Show, on a Venn diagram, how the elements of sets $\mathscr{E}$, A, B and C are related.

The diagram must show three overlapping sets inside the universal set, $\mathscr{E}$.

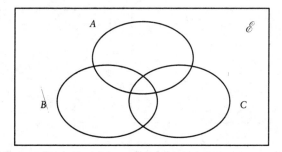

Place the numbers in their correct positions, working from the centre outwards.

The set of numbers which belong to A, B and C	$= A \cap B \cap C$	$= \{4\}$
The set of numbers which belong to A and B only	$= A \cap B \cap C'$	$= \{2, 6, 8\}$
The set of numbers which belong to B and C only	$= B \cap C \cap A'$	$= \emptyset$
The set of numbers which belong to A and C only	$= A \cap C \cap B'$	$= \{1, 9\}$
The set of numbers which belong to A only	$= A \cap B' \cap C'$	$= \{3, 5, 7\}$
The set of numbers which belong to B only	$= B \cap C' \cap A'$	$= \{10\}$
The set of numbers which belong to C only	$= C \cap A' \cap B'$	$= \emptyset$

The set of numbers which belong to neither

$$A, B \text{ nor } C = (A \cup B \cup C)' = \{0\}$$

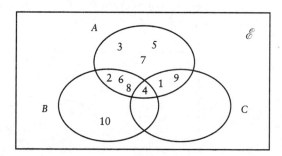

Note that $C \subset A$.

1   Illustrate the following sets by drawing and shading a Venn diagram:

   a $A \cap B$ **c** A' **e** $A \cap B'$

   b $A \cup B$ **d** $A' \cup B$ **f** $(A \cap B') \cup (B \cap A')$

2   Show, by drawing and shading Venn diagrams, that
   $(A \cap B)' = A' \cup B'$

3 Write down, using set notation, the sets illustrated (by shading) in the
following diagrams:

a

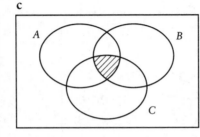

b

c

d

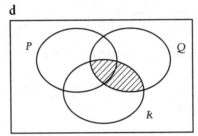

e

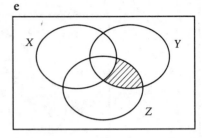

f

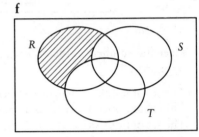

4 List the following sets and then illustrate them on
a Venn diagram:

$\mathscr{E} = \{$integers from 1 to 10$\}$
$A = \{$powers of 2$\}$
$B = \{$even numbers$\}$

List the elements of A', $A \cap B$, $A \cap B'$, $(A \cup B)'$.

5 Illustrate the following sets on a Venn diagram:

$\mathscr{E} = \{10, 11, 12, 13, 14, 15, 16, 17, 18, 19, 20\}$
$A = \{11, 13, 17, 19\}$
$B = \{14, 15, 16, 17, 18\}$
$C = \{10, 12, 14, 16, 18\}$

List the elements of $B \cap C$, $A \cup B$, $A \cap B \cap C$,
$(A \cup B \cup C)'$, $A \cap B' \cap C'$.

6 From the Venn diagram on the right find:

a $A \cap B \cap C$ **c** $B \cup C$ **e** $(A \cup B \cup C)'$

b $n(A \cap B \cap C)$ **d** $n(B \cup C)$ **f** $n(A \cup B \cup C)'$

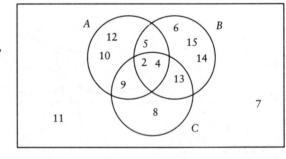

7 In a group of 12 teenagers, eight have a driving
licence. Six of these drive a car and three ride a
motor bike.

a If $D = \{$car drivers$\}$ and $M = \{$motorbike
riders$\}$, draw a Venn diagram to illustrate the
sets D, M and $\mathscr{E}$.

b How many of the teenagers can only drive a
car?

8 In a family of eight, four drink tea and six drink
coffee.

a If $T = \{$tea drinkers$\}$ and $C = \{$coffee
drinkers$\}$, draw a Venn diagram to illustrate
the family's drinking preferences.

b From your diagram, find:
(i) $n(T \cap C)$, (ii) $n(T \cap C')$, (iii) $n(C \cap T')$.

c Translate the results of **b** into a sentence.

Problems

Example

In the summer of 1990, a group of 50 people were asked which sports they had watched on television, the World Cup soccer in Italy, the golf Open Championship at St Andrew's or the tennis at Wimbledon.

The results were:

31 people watched the soccer   10 watched tennis and golf
20 watched the golf   12 watched soccer and golf
24 watched the tennis   5 watched none of the three sports
13 watched tennis and soccer x watched all three.

a Draw a Venn diagram to illustrate this information, giving the numbers in the subsets in terms of x.

b Form an equation in x and solve it to find the value of x.

a No. who watched all three games $= x$
No. who watched soccer and tennis but not golf $= 13 - x$
No. who watched soccer and golf but not tennis $= 12 - x$
No. who watched tennis and golf but not soccer $= 10 - x$
No. who watched soccer only $= 31 - (12 + 13 - x)$ $= 6 + x$
No. who watched golf   only $= 20 - (12 + 10 - x)$ $= x - 2$
No. who watched tennis only $= 24 - (10 + 13 - x)$ $= 1 + x$

The Venn diagram is

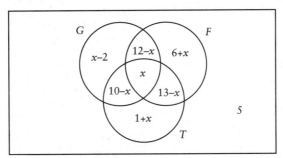

b Adding the amounts in all the regions gives an equation in x which is:

$$x + (13 - x) + (12 - x) + (10 - x) + (6 + x) + (1 + x) + (x - 2) + 5 = 50$$
$$x + 45 = 50$$
$$x = 5$$

Exercise 180

1 In order to discover what customers look for when shopping, an electrical store asked 40 customers to state which of the following three requirements would be their first preference:

    A, helpful sales staff
    B, good after sales service
    C, well-displayed goods with clearly marked prices.

x people said all three were equally important.

The Venn diagram on the next page shows the numbers in each of the subsets, in terms of x, where

    A = {people who put A first}
    B = {people who put B first}
    C = {people who put C first}.

a Write down an equation in x, and hence find its value.

b Redraw the Venn diagram putting the correct numbers in each subset.

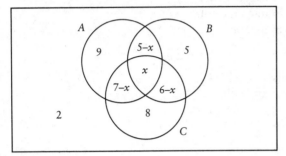

2 Twenty-eight students were asked what they usually eat for breakfast. If

$\mathscr{E}$ = {students questioned}
B = {students who eat bacon and/or eggs}
C = {students who eat cereal}
T = {students who eat toast}

the Venn diagram below shows the number of students in each set.

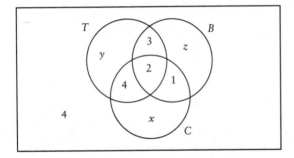

13 students eat cereal and 14 students eat toast.

a How many students eat cereal and toast?

b Find the values of x, y and z.

3 As part of an investigation before a sales drive, 100 householders were asked whether they owned a dishwasher, a video recorder, or a compact disc player, and if so which.

49 owned none of the three items.

5 owned only a dishwasher and a video recorder.
2 owned only a dishwasher and a compact disc player.
6 owned only a video recorder and a compact disc player.

20 owned a dishwasher.
31 owned a video recorder.
19 owned a compact disc player.

x is the number that owned all three.

a Draw a Venn diagram to illustrate this information, marking on it the numbers in each of the subsets, using x where necessary.

b Write down an equation in x, and hence find its value.

c How many householders own only a dishwasher?

(SEG, Summer 1989)

4 A garage, which does MOT testing, tested 22 cars in one day.

9 cars passed the test

Of the cars that failed the test:

10 were found to have faulty brakes
 6 were found to have faulty lights
 8 were found to have faulty tyres

2 failed on brakes and lights only
4 failed on brakes and tyres only
1 failed on lights and tyres only

x failed on all three.

a By drawing a Venn diagram and forming an equation in x, find the value of x.

b Redraw the Venn diagram showing the numbers of cars in all the subsets.

5 In the LVI of a sixth-form college, it was found that 44% of the students were retaking exams in one or more of Mathematics, English and French.

27% were retaking Mathematics
25% were retaking English
 8% were retaking French.

 4% were retaking at least French and English
12% were retaking at least English and Mathematics
 3% were retaking at least French and Mathematics.

x% were retaking all three subjects.

a By drawing a Venn diagram and forming an equation in x, find the value of x.

b What percentage of students were retaking only one of the three subjects?

*33 Vectors

33.1   Vector representation

Some quantities in mathematics need to be specified by both their magnitude and direction. For example:

- The wind speed is 50 mph from the NE.
- Everyone take two paces forward.
- The lift accelerated downwards at $2 \, \text{m/s}^2$.

These quantities are called **vectors**.

Velocity, displacement and acceleration are vectors. Force is also a vector.

Many quantities are completely specified by their magnitude. For example:

- I have been waiting for 40 minutes.
- The temperature today is 24°C.
- The car was travelling at 70 mph.

These quantities are called **scalars**.

Time, temperature, speed, mass, area are all scalar quantities.

A vector can be represented in several ways:

 (i) Geometrically

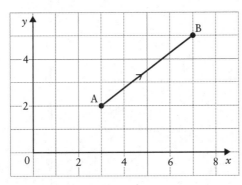

The vector is represented by the line from point A (3, 2) to point B (7, 5).

Its magnitude is the length of the line AB and its direction is the angle it makes with the x-axis.

(ii) As a column vector

The displacement from A at (3, 2) to B at (7, 5) is 4 along and 3 up.

To distinguish the vector from the coordinates of a point, it is written as $\begin{pmatrix} 4 \\ 3 \end{pmatrix}$ with the x-value on top of the y-value.

Because the vector goes from A to B, we write the vector as $\overrightarrow{AB} = \begin{pmatrix} 4 \\ 3 \end{pmatrix}$

The vector from B to A is $\overrightarrow{BA} = \begin{pmatrix} -4 \\ -3 \end{pmatrix}$.

(iii) **In unit base vectors**

A **unit vector** is a vector of magnitude 1.

The unit vectors $\begin{pmatrix} 1 \\ 0 \end{pmatrix}$ and $\begin{pmatrix} 0 \\ 1 \end{pmatrix}$ are important because their directions are along the x-axis and y-axis respectively.

$\begin{pmatrix} 1 \\ 0 \end{pmatrix}$ is given the letter **i** and $\begin{pmatrix} 0 \\ 1 \end{pmatrix}$ is given the letter **j**.

i and **j** are always printed in bold type.
When they are written, they should be underlined: $\underline{i}$, $\underline{j}$.

All two-dimensional vectors can be written in terms of **i** and **j**.

For example, the vector $\begin{pmatrix} 4 \\ 3 \end{pmatrix}$, which is a displacement of 4 in the x-direction and 3 in the y-direction, is written as:

$$4\mathbf{i} + 3\mathbf{j}$$

Example

Find the displacement vector $\overrightarrow{PQ}$ which joins P(3, -2) and Q(7, -4) and write each vector (i) as a column vector, (ii) in terms of the base vectors **i** and **j**.

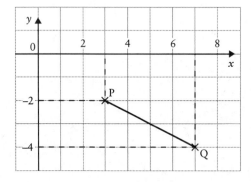

The displacement along the x-axis is $7 - 3$ $= 4$
The displacement along the y-axis is $-4 - (-2) = -4 + 2 = -2$

(i) As a column vector $\overrightarrow{PQ} = \begin{pmatrix} 4 \\ -2 \end{pmatrix}$

(ii) In terms of **i** and **j** $\overrightarrow{PQ} = 4\mathbf{i} - 2\mathbf{j}$

Exercise 181

1   Find the displacement vectors in the following diagrams and write each
    vector (i) as a column vector, (ii) in terms of the base vectors **i** and **j**.

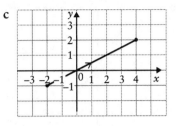

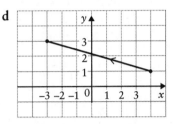

2   Plot the points A(1, 1),   B(1.5, 3),   C(6, 5) and D(8, 1).
    Write, as column vectors:

    a $\overrightarrow{AB}$ **b** $\overrightarrow{AD}$ **c** $\overrightarrow{CB}$ **d** $\overrightarrow{AC}$ **e** $\overrightarrow{DB}$

3   Find the displacement vector $\overrightarrow{PQ}$ which joins each of the following points,
    giving your answer (i) as a column vector, (ii) in terms of the base vectors
    i and **j**.

    a P(3, 5) Q(7, 6)   **c** P(−5, 2) Q(4, 1)   **e** P(6, −4) Q(−1, 3)

    b P(0, 6) Q(4, 2)   **d** P(6, 3) Q(−4, −2)   **f** P(−2, −7) Q(−5, 0)

33.2   The modulus of a vector

The **modulus** of a vector is another term for the magnitude.

The modulus of $\overrightarrow{AB}$ is written as AB or $|\overrightarrow{AB}|$.

To calculate the modulus, we use Pythagoras' theorem:

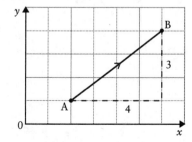

$$\overrightarrow{AB} = \begin{pmatrix} 4 \\ 3 \end{pmatrix}$$

To find the length of AB: $AB^2 = 4^2 + 3^2$
                                $= 25$
                         $AB = 5$ or $|\overrightarrow{AB}| = 5$

If $\overrightarrow{AB} = \begin{pmatrix} x \\ y \end{pmatrix} = x\mathbf{i} + y\mathbf{j}$, then $|\overrightarrow{AB}| = \sqrt{(x^2 + y^2)}$

1 Find the modulus of $\overrightarrow{AB}$ in each of the following:

a
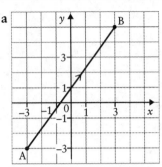

b
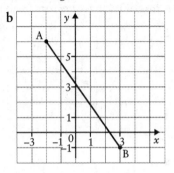

2 Calculate the modulus of the vectors:

a $\begin{pmatrix} 12 \\ 5 \end{pmatrix}$ b $\begin{pmatrix} -8 \\ 15 \end{pmatrix}$ c $3\mathbf{i} - 4\mathbf{j}$

3 Calculate the modulus of:

a the vector joining the points $(1, 3)$ and $(5, 5)$

b the vector joining the points $(-1, 3)$ and $(5, -5)$.

33.3 Combining vectors

Multiplication by a scalar

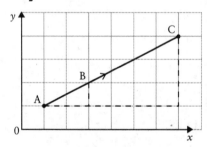

In the diagram above, the vector $\overrightarrow{AC}$ is three times longer than vector $\overrightarrow{AB}$.

$$\overrightarrow{AC} = 3\,\overrightarrow{AB} = 3\begin{pmatrix} 2 \\ 1 \end{pmatrix} = \begin{pmatrix} 6 \\ 3 \end{pmatrix}$$

Multiplying a vector $\begin{pmatrix} x \\ y \end{pmatrix}$ by a scalar k gives

$$k\begin{pmatrix} x \\ y \end{pmatrix} = \begin{pmatrix} kx \\ ky \end{pmatrix}$$

In the example above, both $\overrightarrow{AB}$ and $\overrightarrow{AC}$ can be written in terms of

the vector $\begin{pmatrix} 2 \\ 1 \end{pmatrix}$.

This is because, although the **magnitudes** of $\overrightarrow{AB}$ and $\overrightarrow{AC}$ are different, their **directions** are the same.

Example

Points P, Q, R and S have coordinates $(-1, 1)$, $(2, 3)$, $(4, 0)$ and $(1, -2)$ respectively.

a Find $\overrightarrow{QR}$, $\overrightarrow{QS}$, $\overrightarrow{PR}$ and $\overrightarrow{SP}$.

b State which pairs of vectors are equal in magnitude, parallel or both.

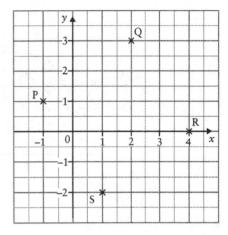

a $\overrightarrow{QR} = \begin{pmatrix} 4-2 \\ 0-3 \end{pmatrix} = \begin{pmatrix} 2 \\ -3 \end{pmatrix}$

$\overrightarrow{QS} = \begin{pmatrix} 1-2 \\ -2-3 \end{pmatrix} = \begin{pmatrix} -1 \\ -5 \end{pmatrix}$

$\overrightarrow{PR} = \begin{pmatrix} 4-^-1 \\ 0-1 \end{pmatrix} = \begin{pmatrix} 5 \\ -1 \end{pmatrix}$

$\overrightarrow{SP} = \begin{pmatrix} -1-1 \\ 1-^-2 \end{pmatrix} = \begin{pmatrix} -2 \\ 3 \end{pmatrix}$

b $\overrightarrow{QR} = -\overrightarrow{SP}$

∴ QR and PS are both equal in magnitude and parallel.

$|\overrightarrow{QS}| = |\overrightarrow{PR}|$

∴ QS and PR are equal in magnitude.

Exercise 183

1 If $\mathbf{a} = \begin{pmatrix} 5 \\ -12 \end{pmatrix}$ calculate

(i) $4\mathbf{a}$ (ii) $\frac{1}{2}\mathbf{a}$ (iii) $-2\mathbf{a}$ (iv) $-\mathbf{a}$ (v) $|\mathbf{a}|$ (vi) $|3\mathbf{a}|$

2 State which of the following pairs of vectors are parallel:

a $\begin{pmatrix} 3 \\ 6 \end{pmatrix}$ and $\begin{pmatrix} 1 \\ 3 \end{pmatrix}$ **b** $\begin{pmatrix} -2 \\ 8 \end{pmatrix}$ and $\begin{pmatrix} 1 \\ -4 \end{pmatrix}$

c $5\mathbf{i} + 10\mathbf{j}$ and $\frac{1}{4}\mathbf{i} + \frac{1}{2}\mathbf{j}$ **d** $4\mathbf{i} + ^-7\mathbf{j}$ and $2\mathbf{i} - 3\mathbf{j}$

e $\overrightarrow{PQ}$ and $\overrightarrow{RS}$ if P, Q, R and S have coordinates $(2, 5)$, $(4, 9)$, $(-1, 3)$ and $(-2, 1)$ respectively.

3 Calculate the displacement vectors $\overrightarrow{AB}$ and $\overrightarrow{CD}$ in each of the following and state whether AB and CD are equal, parallel or both.

a A(1, 3)   B(3, 7)   C(1, -1)   D(4, 5)

b A(2, 0)   B(6, 2)   C(4, 1)   D(8, -1)

c A($\frac{1}{2}$, 2)   B(2, 5)   C(-2, -7)   D(-$\frac{1}{2}$, -4)

d A(-1, 3$\frac{1}{2}$)   B(-2, 4)   C(0, -1)   D(3, -2$\frac{1}{2}$)

e A(3, -2)   B(0, 4)   C(3, 7)   D(6, 13)

Addition and subtraction

If we move from point A to point B and then from point B to point C, $\overrightarrow{AB}$ followed by $\overrightarrow{BC}$ produces the vector $\overrightarrow{AC}$, i.e.

$$\overrightarrow{AC} = \overrightarrow{AB} + \overrightarrow{BC}$$

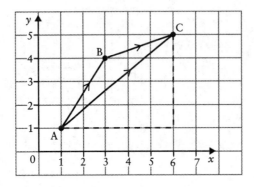

Addition

In the diagram above, $\overrightarrow{AB} = \begin{pmatrix} 2 \\ 3 \end{pmatrix}$ and $\overrightarrow{BC} = \begin{pmatrix} 3 \\ 1 \end{pmatrix}$

The displacement from A to C is 5 along and 4 up, i.e.

$$\overrightarrow{AC} = \begin{pmatrix} 2 \\ 3 \end{pmatrix} + \begin{pmatrix} 3 \\ 1 \end{pmatrix} = \begin{pmatrix} 5 \\ 4 \end{pmatrix}$$

Vectors are added by adding the corresponding x- and y-coordinates.

$$\begin{pmatrix} x_1 \\ y_1 \end{pmatrix} + \begin{pmatrix} x_2 \\ y_2 \end{pmatrix} = \begin{pmatrix} x_1 + x_2 \\ y_1 + y_2 \end{pmatrix}$$

Subtraction

$$\overrightarrow{BC} = \begin{pmatrix} 3 \\ 1 \end{pmatrix}$$

$$\therefore \overrightarrow{CB} = \begin{pmatrix} -3 \\ -1 \end{pmatrix}$$

$$\overrightarrow{AB} - \overrightarrow{BC} = \overrightarrow{AB} + (-\overrightarrow{BC})$$

$$= \overrightarrow{AB} + \overrightarrow{CB}$$

$$= \begin{pmatrix} 2 \\ 3 \end{pmatrix} + \begin{pmatrix} -3 \\ -1 \end{pmatrix}$$

$$= \begin{pmatrix} 2 - 3 \\ 3 - 1 \end{pmatrix}$$

$$= \begin{pmatrix} -1 \\ 2 \end{pmatrix}$$

Vectors are subtracted by subtracting the corresponding x- and y-coordinates.

$$\begin{pmatrix} x_1 \\ y_1 \end{pmatrix} - \begin{pmatrix} x_2 \\ y_2 \end{pmatrix} = \begin{pmatrix} x_1 - x_2 \\ y_1 - y_2 \end{pmatrix}$$

Example

If $\mathbf{a} = \begin{pmatrix} 7 \\ -4 \end{pmatrix}$ and $\mathbf{b} = \begin{pmatrix} -3 \\ 6 \end{pmatrix}$, calculate

(i) $\mathbf{a} - \mathbf{b}$ (ii) $4\mathbf{b} - 3\mathbf{a}$ (iii) $|4\mathbf{b} - 3\mathbf{a}|$

(i) $\mathbf{a} - \mathbf{b} = \begin{pmatrix} 7 \\ -4 \end{pmatrix} - \begin{pmatrix} -3 \\ 6 \end{pmatrix} = \begin{pmatrix} 7 - (-3) \\ -4 - 6 \end{pmatrix} = \begin{pmatrix} 10 \\ -10 \end{pmatrix}$

(ii) $4\mathbf{b} - 3\mathbf{a} = 4\begin{pmatrix} -3 \\ 6 \end{pmatrix} - 3\begin{pmatrix} 7 \\ -4 \end{pmatrix} = \begin{pmatrix} -12 \\ 24 \end{pmatrix} - \begin{pmatrix} 21 \\ -12 \end{pmatrix} = \begin{pmatrix} -33 \\ 36 \end{pmatrix}$

(iii) $|4\mathbf{b} - 3\mathbf{a}| = \sqrt{(33^2 + 36^2)} = \sqrt{(1089 + 1296)} = \sqrt{2385} = 48.8$

Exercise 184

1   If $\mathbf{a} = \begin{pmatrix} 4 \\ 5 \end{pmatrix}$ $\mathbf{b} = \begin{pmatrix} 2 \\ -1 \end{pmatrix}$ and $\mathbf{c} = \begin{pmatrix} 0 \\ -\frac{1}{2} \end{pmatrix}$ calculate:

   (i) $\mathbf{a} + \mathbf{b}$ (iv) $3\mathbf{a} - 2\mathbf{b}$ (vii) $\mathbf{c} + \frac{1}{4}(\mathbf{b} - \mathbf{a})$

   (ii) $\mathbf{b} - \mathbf{a}$ (v) $\frac{1}{2}\mathbf{b} - \mathbf{c}$ (viii) $\mathbf{a} - 2(\mathbf{b} - \mathbf{c})$

   (iii) $2\mathbf{a} + \mathbf{c}$ (vi) $\frac{1}{2}(\mathbf{a} - \mathbf{b})$

2   If $\mathbf{a} = 2\mathbf{i} + \mathbf{j}$ and $\mathbf{b} = 7\mathbf{i} - 2\mathbf{j}$, calculate:

   (i) $\mathbf{a} - \mathbf{b}$ (iii) $\mathbf{b} + 2\mathbf{a}$ (v) $\frac{1}{2}(\mathbf{a} + \mathbf{b})$

   (ii) $|\mathbf{a} - \mathbf{b}|$ (iv) $|\mathbf{b} + 2\mathbf{a}|$ (vi) $|\frac{1}{2}(\mathbf{a} + \mathbf{b})|$

3   $\mathbf{a} = \begin{pmatrix} 4 \\ 2 \end{pmatrix}$ $\mathbf{b} = \begin{pmatrix} 3 \\ 5 \end{pmatrix}$ $\mathbf{c} = \begin{pmatrix} -4 \\ 2 \end{pmatrix}$ $\mathbf{d} = \begin{pmatrix} -1 \\ -5 \end{pmatrix}$

   (i) Calculate $\mathbf{p} = 2\mathbf{a} + 3\mathbf{b}$.

   (ii) Calculate $\mathbf{q} = -3\mathbf{c} - 5\mathbf{d}$.

   (iii) What conclusion can you draw about $\mathbf{p}$ and $\mathbf{q}$?

33.4   Vector geometry

In geometrical figures, the vectors are often represented by single letters and the questions are solved algebraically.

Example 1

In the triangle ABC, $\overrightarrow{AB} = \mathbf{a}$ and $\overrightarrow{BC} = \mathbf{b}$.
BD divides AC in the ratio 3 : 1.

Find, in terms of $\mathbf{a}$ and $\mathbf{b}$:

(i) $\overrightarrow{AC}$ (ii) $\overrightarrow{AD}$ (iii) $\overrightarrow{BD}$

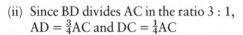

(i) $\overrightarrow{AC} = \overrightarrow{AB} + \overrightarrow{BC} = \mathbf{a} + \mathbf{b}$

(ii) Since BD divides AC in the ratio 3 : 1,
AD = $\frac{3}{4}$AC and DC = $\frac{1}{4}$AC

i.e. $\overrightarrow{AD} = \frac{3}{4}\overrightarrow{AC} = \frac{3}{4}(\mathbf{a} + \mathbf{b})$

(iii) $\overrightarrow{BA} = -\overrightarrow{AB}$

$\overrightarrow{BD} = \overrightarrow{BA} + \overrightarrow{AD} = -\mathbf{a} + \frac{3}{4}(\mathbf{a} + \mathbf{b}) = -\frac{1}{4}\mathbf{a} + \frac{3}{4}\mathbf{b} = \frac{1}{4}(3\mathbf{b} - \mathbf{a})$

Example 2

In quadrilateral OPQR, $\overrightarrow{OP} = \mathbf{p}$, $\overrightarrow{OR} = \mathbf{r}$, $\overrightarrow{PQ} = \mathbf{p} + \mathbf{r}$. RS = 2OR.

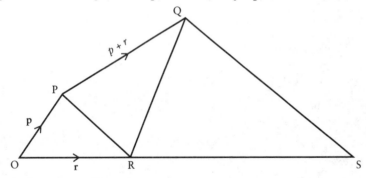

(i) Find, in terms of $\mathbf{p}$ and $\mathbf{r}$,
$\overrightarrow{PR}$, $\overrightarrow{RQ}$, $\overrightarrow{RS}$ and $\overrightarrow{QS}$

(ii) What conclusion can you make about PR and QS?

(i) $\overrightarrow{PR} = \overrightarrow{PO} + \overrightarrow{OR} = -\mathbf{p} + \mathbf{r} = \mathbf{r} - \mathbf{p}$
$\overrightarrow{RQ} = \overrightarrow{RP} + \overrightarrow{PQ} = (\mathbf{p} - \mathbf{r}) + (\mathbf{p} + \mathbf{r}) = 2\mathbf{p}$
$\overrightarrow{RS} = 2\overrightarrow{OR} = 2\mathbf{r}$
$\overrightarrow{QS} = \overrightarrow{QR} + \overrightarrow{RS} = -2\mathbf{p} + 2\mathbf{r} = 2(\mathbf{r} - \mathbf{p})$

(ii) $\overrightarrow{QS} = 2\overrightarrow{PR}$ i.e. QS $\parallel$ PR and QS = 2PR

Exercise 185

1 In triangle OPQ, $\overrightarrow{OP} = \mathbf{p}$ and $\overrightarrow{OQ} = \mathbf{q}$. R and S divide OP and OQ respectively in the ratio of 3 : 1.

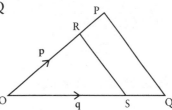

(i) Find, in terms of **p** and **q**: $\overrightarrow{PQ}$, $\overrightarrow{OR}$, $\overrightarrow{OS}$ and $\overrightarrow{RS}$.

(ii) What can you deduce about RS and PQ?

2 ABCD is a quadrilateral. P, Q, R and S are the mid-points of AB, BC, CD and DA respectively.

$\overrightarrow{AD} = \mathbf{a}$, $\overrightarrow{AB} = \mathbf{b}$, $\overrightarrow{BC} = \mathbf{c}$.

(i) Find, in terms of **a**, **b** and **c**,

$\overrightarrow{AP}$, $\overrightarrow{CD}$, $\overrightarrow{CR}$, $\overrightarrow{PS}$, $\overrightarrow{QR}$, $\overrightarrow{PQ}$, $\overrightarrow{SR}$.

(ii) What can you deduce about PQRS?

3 OPQRST is a regular hexagon. $\overrightarrow{OP} = \mathbf{p}$ and $\overrightarrow{PQ} = \mathbf{q}$.

Find, in terms of **p** and **q**:

$\overrightarrow{RS}$, $\overrightarrow{ST}$, $\overrightarrow{QT}$, $\overrightarrow{QR}$, $\overrightarrow{PR}$ and $\overrightarrow{OR}$.

4 The diagram, not drawn to scale, shows a trapezium OABC.
   X and Y are the mid-points of diagonals OB and AC respectively.

$\overrightarrow{OA} = \mathbf{a}$, $\overrightarrow{OC} = \mathbf{c}$,   and $\overrightarrow{CB} = \frac{1}{2}\mathbf{a}$

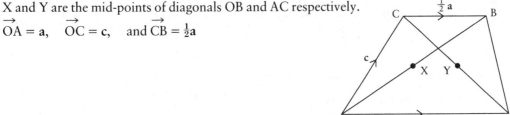

(i) Find each of the following vectors in terms of **a** and **c**, simplifying your answer where possible:

$\overrightarrow{OB}$, $\overrightarrow{OX}$, $\overrightarrow{AO}$, $\overrightarrow{AC}$, $\overrightarrow{AY}$, $\overrightarrow{OY}$, $\overrightarrow{XY}$.

(ii) What do you deduce about XY and OA?

5 In the diagram given below, $\overrightarrow{CA} = \mathbf{a}$ and $\overrightarrow{CB} = \mathbf{b}$.
   BE is parallel to CA, $BE = \frac{1}{2}CA$ and $AD = \frac{2}{3}AB$.

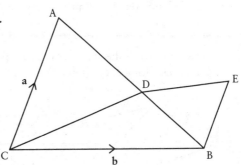

(i) Find, in terms of **a** and **b**:

$\overrightarrow{AB}$, $\overrightarrow{AD}$, $\overrightarrow{CD}$, $\overrightarrow{BE}$, $\overrightarrow{DB}$, $\overrightarrow{DE}$.

(ii) What can you deduce about CD and DE?

*34 Matrices

34.1 Matrix representation

In mathematics, a matrix stores information in a concise way. The information is written down in a rectangular array of rows and columns and each number has its precise position in the array.

A milkman delivers several different types of milk, e.g. gold top (GT), silver top (ST), and skimmed (SK). In addition he delivers eggs and bread.

A section from his daily order book might look like this:

	GT	ST	SK	Eggs	Bread
10 Cromarty St	0	2	2	2	1
12 Cromarty St	0	2	1	1	3
14 Cromarty St	1	2	1	0	0
16 Cromarty St	4	0	0	1	3

The milkman knows the order in which the types of milk are listed and the order in which he delivers to the houses, so he only needs to look at the numbers:

$$\begin{pmatrix} 0 & 2 & 2 & 2 & 1 \\ 0 & 2 & 1 & 1 & 3 \\ 1 & 2 & 1 & 0 & 0 \\ 4 & 0 & 0 & 1 & 3 \end{pmatrix}$$

This is a 4 × 5 matrix because it has 4 rows and 5 columns.

A matrix is described by its order which is always

number of rows × number of columns

If the number of rows = number of columns, the matrix is a **square matrix**.

A matrix such as $\begin{pmatrix} 1 \\ 2 \\ 3 \end{pmatrix}$ has order 3 × 1 and is a **column matrix**.

A matrix such as (1 2 3) has order 1 × 3 and is a **row matrix**.

Example

The diagram, or network, below shows the routes connecting four towns and the distances, in miles, along each route.

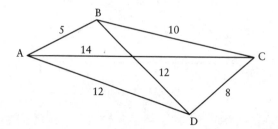

Write this information in matrix form.

What order is the matrix?

$$
\begin{array}{c}
\text{TO} \\
\begin{array}{cccc}
A & B & C & D
\end{array} \\
\text{FROM} \;
\begin{array}{c}
A \\ B \\ C \\ D
\end{array}
\begin{pmatrix}
0 & 5 & 14 & 12 \\
5 & 0 & 10 & 12 \\
14 & 10 & 0 & 8 \\
12 & 12 & 8 & 0
\end{pmatrix}
\end{array}
$$

It is a square matrix of order 4×4.

■ Exercise 186 ■

1 A small bakery makes three types of bread: white, wholemeal and granary; in two sizes: large and small.

Each day 20 large white, 15 small white, 30 large wholemeal, 25 small wholemeal, 24 large granary and 20 small granary are baked.
Arrange this information in a 3×2 matrix.

2 Jon has a collection of records, cassettes and CDs, some of which are jazz, some classical and some 'pop'.

One day he takes stock of his collection and finds that he has:
42 jazz, 21 classical and 19 'pop' records;
 8 jazz, 12 classical and 30 'pop' cassettes;
15 jazz, 35 classical and 10 'pop' CDs.

Complete the matrix below showing his collection:

$$
\begin{array}{c}
\;\;\text{Rec}\quad\text{Cass}\quad\text{CD} \\
\begin{array}{c}
\text{Jazz} \\ \text{Class} \\ \text{Pop}
\end{array}
\begin{pmatrix}
 & & \\
 & & \\
 & &
\end{pmatrix}
\end{array}
$$

3 State the order of each of the following matrices:

a $\begin{pmatrix} 1 & 0 \\ 2 & 3 \\ 0 & 4 \end{pmatrix}$ **c** $\begin{pmatrix} 7 & 4 \\ 9 & -2 \end{pmatrix}$ **e** $(2 \quad 0)$

b $\begin{pmatrix} 1 \\ 4 \\ 6 \end{pmatrix}$ **d** $\begin{pmatrix} 1 & 0 & 4 \\ 3 & 7 & 2 \end{pmatrix}$ **f** $\begin{pmatrix} 2 & 2 & 4 \\ 5 & -1 & 7 \\ 6 & 2 & 1 \end{pmatrix}$

4 The diagram below shows distances between 5 villages.
Represent these distances on a route matrix.

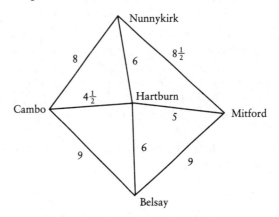

5 A 'relation' matrix for the set of numbers 3, 6, 9, 12 has been partially completed.
If the relation 'is a factor of' is true for numbers a and b, a '1' is recorded, otherwise a '0' is recorded.

$$
\begin{array}{c}
\quad\; 3 \quad 6 \quad 9 \quad 12 \\
\begin{array}{c}
3 \\ 6 \\ 9 \\ 12
\end{array}
\begin{pmatrix}
1 & & & \\
 & & 0 & 1 \\
 & & & \\
 & & &
\end{pmatrix}
\end{array}
$$

Copy and complete the matrix.

34.2 Addition and subtraction of matrices

In question 2 of exercise 186, p. 349, Jon's collection of records, cassettes and CDs was:

$$\begin{array}{c} \\ \text{Jazz} \\ \text{Class} \\ \text{Pop} \end{array} \begin{array}{ccc} \text{Rec} & \text{Cass} & \text{CD} \\ \left(\begin{array}{ccc} 42 & 8 & 15 \\ 21 & 12 & 35 \\ 19 & 30 & 10 \end{array} \right) \end{array}$$

Since counting his collection, Jon has bought 2 more classical CDs, 3 jazz CDs, 2 'pop' cassettes and 1 jazz cassette.
Writing the additions to his collection in matrix form, we have:

$$\begin{pmatrix} 0 & 0 & 3 \\ 0 & 1 & 2 \\ 0 & 2 & 0 \end{pmatrix}$$

Although he did not buy any records, the column for records must be included in the matrix so that each item occupies the same position in this matrix as in the first matrix (e.g. the number of classical CDs is still written in the second row, third column).

The total collection can be found by *adding* the two matrices:

$$\begin{pmatrix} 42 & 8 & 15 \\ 21 & 12 & 35 \\ 19 & 30 & 10 \end{pmatrix} + \begin{pmatrix} 0 & 0 & 3 \\ 0 & 1 & 2 \\ 0 & 2 & 0 \end{pmatrix} = \begin{pmatrix} 42 & 8 & 15+3 \\ 21 & 12+1 & 35+2 \\ 19 & 30+2 & 10 \end{pmatrix} = \begin{pmatrix} 42 & 8 & 18 \\ 21 & 13 & 37 \\ 19 & 32 & 10 \end{pmatrix}$$

Matrices can be added by adding the elements in the corresponding positions, but this can only be done if the matrices are of the *same order*.

Jon's recent purchases could be recorded as $\begin{pmatrix} 0 & 3 \\ 1 & 2 \\ 2 & 0 \end{pmatrix}$ as we know he did not

buy any records, but

$$\begin{pmatrix} 42 & 8 & 15 \\ 21 & 12 & 35 \\ 19 & 30 & 10 \end{pmatrix} + \begin{pmatrix} 0 & 3 \\ 1 & 2 \\ 2 & 0 \end{pmatrix}$$

does not make any sense, since the matrices are not of the same order.

The rules for subtraction are the same as those for addition.

Example

$$A = \begin{pmatrix} 2 & 3 & 1 \\ 5 & 2 & 4 \end{pmatrix} \qquad B = \begin{pmatrix} 0 & 1 \\ 4 & 6 \end{pmatrix} \qquad C = \begin{pmatrix} 3 & 6 & 7 \\ 1 & 2 & 3 \end{pmatrix}$$

Find, where possible: a $A + B$ b $A + C$ c $C - A$.

a The addition is not possible as the matrices are of different orders.

b $A + C = \begin{pmatrix} 2 & 3 & 1 \\ 5 & 2 & 4 \end{pmatrix} + \begin{pmatrix} 3 & 6 & 7 \\ 1 & 2 & 3 \end{pmatrix} = \begin{pmatrix} 2+3 & 3+6 & 1+7 \\ 5+1 & 2+2 & 4+3 \end{pmatrix}$

$\quad = \begin{pmatrix} 5 & 9 & 8 \\ 6 & 4 & 7 \end{pmatrix}$

c $C - A = \begin{pmatrix} 3 & 6 & 7 \\ 1 & 2 & 3 \end{pmatrix} - \begin{pmatrix} 2 & 3 & 1 \\ 5 & 2 & 4 \end{pmatrix} = \begin{pmatrix} 3-2 & 6-3 & 7-1 \\ 1-5 & 2-2 & 3-4 \end{pmatrix}$

$\quad = \begin{pmatrix} 1 & 3 & 6 \\ -4 & 0 & -1 \end{pmatrix}$

Exercise 187

1 $A = \begin{pmatrix} 1 & 2 \\ 3 & 4 \end{pmatrix}$ $B = \begin{pmatrix} 0 & 1 & 0 \\ 3 & 0 & 4 \end{pmatrix}$ $C = \begin{pmatrix} 2 & 3 & 4 \\ 1 & 4 & 2 \end{pmatrix}$

$D = \begin{pmatrix} 3 & 7 \\ 1 & 5 \end{pmatrix}$ $E = \begin{pmatrix} 1 & 2 & 0 \\ 3 & -1 & 0 \end{pmatrix}$

Find, where possible,

a $A + D$ **c** $C - D$ **e** $C - E$
b $B + E$ **d** $C - B$ **f** $B + C + E$.

2 Fiona takes the orders for lunch each day. The choices are:
a double or single portion of cold meat, chips, baked potato or salad.

Her own choice is a single portion of meat and of salad.
Ian orders double meat, double chips and single salad.
Moira orders single baked potato and double salad.
Neil orders double meat, single baked potato and single salad.

Write each order as a 4 × 2 matrix and find the total order as a single 4 × 2 matrix.

3 The matrix below represents Mrs Cotton's savings (in £) in various accounts:

	Current	Savings	Investment
Bank	520	1000	2500
Building Society	1120	4100	3000

One month later she has withdrawn the following amounts:
£200 from the bank current account, £100 from the building society current account, and £420 from the building society savings account. Form a new 2 × 3 matrix representing Mrs Cotton's withdrawals and hence find the new balance in each account in matrix form.

4   The following extract shows the numbers of games won, drawn or lost in home and away games for five football teams.

		Home			Away		
	P	W	D	L	W	D	L
Albion	26	6	7	1	6	0	6
Brunswick	26	7	2	3	8	3	3
Croydon	26	8	3	2	8	4	1
Melbourne	25	5	5	3	2	3	7
Ringwood	26	1	3	9	5	2	6

Write the results of the games as two 5×3 matrices and, by adding the matrices, find the total number of games won, drawn and lost by each team.

5   Find p, q and r if

a $\begin{pmatrix} 2 & 3 \\ p & 5 \end{pmatrix} - \begin{pmatrix} 1 & q \\ 2 & 2 \end{pmatrix} = \begin{pmatrix} 1 & 0 \\ 2 & r \end{pmatrix}$

b $\begin{pmatrix} p & 2 & 3 \\ 4 & -1 & 0 \end{pmatrix} + \begin{pmatrix} 5 & 6 & 2 \\ q & 3 & 1 \end{pmatrix} = \begin{pmatrix} 12 & 8 & 5 \\ 3 & r & 1 \end{pmatrix}$

c $\begin{pmatrix} 3 & p \\ 4 & -1 \\ 6 & -2 \end{pmatrix} - \begin{pmatrix} 1 & -2 \\ q & 0 \\ 4 & -4 \end{pmatrix} = \begin{pmatrix} 2 & 0 \\ 7 & -1 \\ 2 & r \end{pmatrix}$

34.3   Multiplication by a scalar

The bakery's daily production of loaves of bread was given earlier as

	Large	Small
White	20	15
Wholemeal	30	25
Granary	24	20

In a six-day week, the production will be six times as much, i.e.

$$6 \times \begin{pmatrix} 20 & 15 \\ 30 & 25 \\ 24 & 20 \end{pmatrix} = \begin{pmatrix} 6 \times 20 & 6 \times 15 \\ 6 \times 30 & 6 \times 25 \\ 6 \times 24 & 6 \times 20 \end{pmatrix} = \begin{pmatrix} 120 & 90 \\ 180 & 150 \\ 144 & 120 \end{pmatrix}$$

To multiply by a scalar, n, multiply each element in the matrix by n.

Example

$$A = \begin{pmatrix} 2 & 4 & 3 \\ 1 & 0 & 5 \end{pmatrix} \qquad B = \begin{pmatrix} 1 & 6 & -1 \\ -2 & 3 & 0 \end{pmatrix}$$

Find $2A + 3B$.

$$
\begin{aligned}
2A + 3B &= 2\begin{pmatrix} 2 & 4 & 3 \\ 1 & 0 & 5 \end{pmatrix} + 3\begin{pmatrix} 1 & 6 & -1 \\ -2 & 3 & 0 \end{pmatrix} \\
&= \begin{pmatrix} 4 & 8 & 6 \\ 2 & 0 & 10 \end{pmatrix} + \begin{pmatrix} 3 & 18 & -3 \\ -6 & 9 & 0 \end{pmatrix} \\
&= \begin{pmatrix} 7 & 26 & 3 \\ -4 & 9 & 10 \end{pmatrix}
\end{aligned}
$$

Exercise 188

1   Given that

$$A = \begin{pmatrix} 1 & 2 \\ 2 & -1 \end{pmatrix} \text{ and } B = \begin{pmatrix} -3 & 0 \\ 4 & 1 \end{pmatrix}$$

find

a $3A$　　　　**c** $3A + 2B$　　**e** $A - 3B$

b $2B$　　　　**d** $2A - B$　　　**f** $\frac{1}{2}(A + B)$

2   The cost of developing and printing a roll of film is:

Print size

No. of exposures	6″ × 4″	7″ × 5″
up to 24	£2.14	£4.04
up to 36	£2.84	£5.44

a Write the information in a 2 × 2 matrix and find the costs of developing and printing 4 films.

b The cost of a double set of prints is 1.6 times the cost of a single set. Find the costs for a double set of prints from one film, to the nearest 1p.

3   The calorie and fibre content of a selection of fruits is given below:

	Calories	Fibre
Apple	60	2.3
Banana	80	3.0
Orange	60	2.0
Peach	40	1.4
Pear	40	2.5

a Rita eats an apple, a banana, and a pear every day.
Write down a 3 × 2 matrix showing her daily intake of calories and fibre for each of these fruits.

b Calculate her calorie and fibre intake for one week, giving your answer in matrix form.

c Saskia eats five of the same fruits as Rita in one week. Find, in matrix form, the difference in the girls' calorie and fibre intake for each fruit.

4 Given that

$$A = \begin{pmatrix} 3 & 6 & -4 \\ 2 & -5 & 0 \end{pmatrix} \quad B = \begin{pmatrix} 1 & -2 & 6 \\ 6 & -2 & 1 \end{pmatrix} \quad C = \begin{pmatrix} -3 & 4 & -2 \\ 7 & -1 & -3 \end{pmatrix}$$

find:

a A + B + C   c 2B + C   e 2A + B − 3C

b 3A − 2C   d A − 4B   f 4C − 3(A + B)

34.4 Matrix multiplication

Combining rows and columns

King Edward's School played a rugby match against Tanleigh College. During the game, King Edward's scored five tries, three conversions, two drop goals and no penalties, which can be represented in matrix form as (5 3 2 0).

The points awarded were, 4 for a try, 2 for a conversion, 3 for a drop goal and 3 for a penalty.

King Edward's final score = $5 \times 4 + 3 \times 2 + 2 \times 3 + 0 \times 3 = 32$.

To combine as matrices we could write:
(5 3 2 0)(4 2 3 3)

or $$\begin{pmatrix} 5 \\ 3 \\ 2 \\ 0 \end{pmatrix} \begin{pmatrix} 4 \\ 2 \\ 3 \\ 3 \end{pmatrix}$$

In fact the conventional method is a combination of the two methods above and is written as

$$(5 \quad 3 \quad 2 \quad 0) \begin{pmatrix} 4 \\ 2 \\ 3 \\ 3 \end{pmatrix} = (5 \times 4 + 3 \times 2 + 2 \times 3 + 0 \times 3) = (32)$$

The answer is a 1 × 1 matrix.

Suppose that the opposing rugby team, Tanleigh College, scored four tries, three conversions, one drop goal and one penalty. By the above convention, their score was

$$(4 \quad 3 \quad 1 \quad 1)\begin{pmatrix}4\\2\\3\\3\end{pmatrix} = (4 \times 4 + 3 \times 2 + 1 \times 3 + 1 \times 3) = (28)$$

The results for both teams can be combined in matrix form:

	Tries	Con	Goals	Pen
King Edward's	5	3	2	0
Tanleigh	4	3	1	1

$$\begin{pmatrix}5 & 3 & 2 & 0\\4 & 3 & 1 & 1\end{pmatrix}\begin{pmatrix}4\\2\\3\\3\end{pmatrix} = \begin{pmatrix}32\\28\end{pmatrix}$$

Score — Total score

The total scores are written as a 2 × 1 matrix.

Example

If a pint of milk costs 29p for gold top, 27p for silver top and 26p for skimmed milk, calculate the daily milk bill for numbers 10, 12, 14 and 16 Cromarty Street.

The milk orders for these four houses, in matrix form, are

$$\begin{pmatrix}0 & 2 & 2\\0 & 2 & 1\\1 & 2 & 1\\4 & 0 & 0\end{pmatrix} \quad \text{and the cost matrix is} \quad \begin{pmatrix}29\\27\\26\end{pmatrix}$$

To multiply, the first row 'dives' into the column:

$$\begin{pmatrix}0 & 2 & 2\\0 & 2 & 1\\1 & 2 & 1\\4 & 0 & 0\end{pmatrix}\begin{pmatrix}0 \times 29\\+\\2 \times 27\\+\\2 \times 26\end{pmatrix} = \begin{pmatrix}0 + 54 + 52\end{pmatrix} = \begin{pmatrix}106\end{pmatrix}$$

The result is written in the *first row* of the answer matrix.

The process is repeated with the second row 'diving' into the column matrix:

$$\begin{pmatrix}0 & 2 & 2\\0 & 2 & 1\\1 & 2 & 1\\4 & 0 & 0\end{pmatrix}\begin{pmatrix}0 \times 29\\+\\2 \times 27\\+\\1 \times 26\end{pmatrix} = \begin{pmatrix}0 + 54 + 26\end{pmatrix} = \begin{pmatrix}106\\80\end{pmatrix}$$

and then with the third and fourth rows:

$$\begin{pmatrix} 0 & 2 & 2 \\ 0 & 2 & 1 \\ 1 & 2 & 1 \\ 4 & 0 & 0 \end{pmatrix} \begin{pmatrix} 29 \\ 27 \\ 26 \end{pmatrix} = \begin{pmatrix} & & \\ & & \\ 29 + 54 + 26 \\ 116 + 0 + 0 \end{pmatrix} = \begin{pmatrix} 106 \\ 80 \\ 109 \\ 116 \end{pmatrix}$$

The final answer is a 4×1 matrix which lists in order, the milk bills of 10, 12, 14 and 16 Cromarty Street:

$$\begin{pmatrix} 0 & 2 & 2 \\ 0 & 2 & 1 \\ 1 & 2 & 1 \\ 4 & 0 & 0 \end{pmatrix} \begin{pmatrix} 29 \\ 27 \\ 26 \end{pmatrix} = \begin{pmatrix} 106 \\ 80 \\ 109 \\ 116 \end{pmatrix}$$

Exercise 189

1 Combine the following row and column matrices, giving your answer as a 1×1 matrix:

a $(1 \quad 3 \quad 2) \begin{pmatrix} 4 \\ 2 \\ 1 \end{pmatrix}$ **c** $(20 \quad 10) \begin{pmatrix} 4 \\ 6 \end{pmatrix}$

b $(1 \quad 0 \quad 3 \quad 2) \begin{pmatrix} 2 \\ 4 \\ 3 \\ 1 \end{pmatrix}$ **d** $(\tfrac{1}{4} \quad \tfrac{1}{2} \quad \tfrac{1}{2}) \begin{pmatrix} 8 \\ 3 \\ 5 \end{pmatrix}$

2 Combine the following matrices, giving your answer in the appropriate matrix form:

a $\begin{pmatrix} 4 & 3 & 1 \\ 6 & 2 & 0 \end{pmatrix} \begin{pmatrix} 2 \\ 2 \\ 2 \end{pmatrix}$ **c** $\begin{pmatrix} 5 & 7 & 9 \\ 6 & 8 & 12 \end{pmatrix} \begin{pmatrix} 1 \\ 0 \\ 3 \end{pmatrix}$

b $\begin{pmatrix} 8 & 6 \\ 4 & 3 \\ 5 & 7 \end{pmatrix} \begin{pmatrix} 2 \\ 4 \end{pmatrix}$ **d** $\begin{pmatrix} 5 & 7 & 9 \\ 4 & 2 & 3 \\ 1 & 0 & 4 \end{pmatrix} \begin{pmatrix} 2 \\ 4 \\ 3 \end{pmatrix}$

3 a In 1990, the Post Office issued three booklets of stamps. One contained 10 first-class stamps, another contained 10 second-class stamps and the third contained 1 first-class and 2 second-class stamps.
Write this information as a 3×2 matrix.

b The cost of the stamps was 20p for first-class and 15p for second-class.
Write down the cost matrix and, by combining the two matrices, find the cost of each booklet of stamps.

4 A florist makes up three bouquets of flowers.
One contains 12 carnations, 3 gladioli and four lilies.
The second contains 8 carnations, 4 gladioli and 10 freesias.
The third contains 8 carnations, 3 gladioli, 3 lilies and 5 freesias.
The costs of the flowers are:

Carnations	36p each
Gladioli	49p each
Lilies	40p each
Freesias	42p each

Write the above information as two matrices and hence find the matrix of the cost of each bouquet.

5 The point count for a hand in bridge is calculated by counting 4 points for an Ace, 3 for a King, 2 for a Queen, 1 for a Jack, and 0 for any other card.

Four players, North, East, South and West receive the following hands during a game of bridge:

♠ 4 3
♥ A J 9 8 4 3
♦ 6 3
♣ K J 7

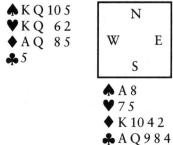

♠ K Q 10 5   ♠ J 9 7 6 2
♥ K Q 6 2   ♥ 10
♦ A Q 8 5   ♦ J 9 7
♣ 5   ♣ 10 6 3 2

♠ A 8
♥ 7 5
♦ K 10 4 2
♣ A Q 9 8 4

Complete the matrix below, which shows the composition of each hand, and, by forming a column matrix for the points, calculate the point score for each hand.

$$\begin{array}{c} \\ N \\ E \\ S \\ W \end{array}\begin{pmatrix} A & K & Q & J & O \\ 1 & 1 & 0 & 2 & 9 \\ & & & & \\ & & & & \\ & & & & \\ & & & & \end{pmatrix}$$

6 The matrix for the football results in Exercise 187 question 4 was:

	W	D	L
Albion	12	7	7
Brunswick	15	5	6
Croydon	16	7	3
Melbourne	7	8	10
Ringwood	6	5	15

The points awarded are, 2 for a win and 1 for a draw.

Calculate, in matrix form, the points scored by each team and hence place the teams in order of merit.

Multiplying matrices

The milk, eggs and bread order for 16 Cromarty Street was (4   0   0   1   3) and the associated cost matrix was

$$\begin{pmatrix} 29 \\ 27 \\ 26 \\ 63 \\ 54 \end{pmatrix}$$

Because of inflation there has been a rise in prices and the new cost matrix is

$$\begin{pmatrix} 31 \\ 29 \\ 28 \\ 66 \\ 56 \end{pmatrix}$$

The two bills, before and after the price rises, can be compared using matrix multiplication.

Using the rule for combining row and column twice gives:

$$
\begin{array}{c}
\text{Order} \\
(4 \quad 0 \quad 0 \quad 1 \quad 3)
\end{array}
\begin{array}{cc}
\text{Old} & \text{New} \\
\text{Cost} & \text{Cost} \\
\begin{pmatrix}
29 & 31 \\
27 & 29 \\
26 & 28 \\
63 & 66 \\
54 & 56
\end{pmatrix}
\end{array}
= (341 \quad 358)
$$

Note that this time the answer is a 2×1 matrix.

The matrix for the milk orders can be extended to include the orders of the other houses in the street and the old and new costs can be found for each house:

$$
\begin{pmatrix}
0 & 2 & 2 & 2 & 1 \\
0 & 2 & 1 & 1 & 3 \\
1 & 2 & 1 & 0 & 0 \\
4 & 0 & 0 & 1 & 3
\end{pmatrix}
\begin{pmatrix}
29 & 31 \\
27 & 29 \\
26 & 28 \\
63 & 66 \\
54 & 56
\end{pmatrix}
=
\begin{pmatrix}
286 & 302 \\
305 & 320 \\
109 & 117 \\
341 & 358
\end{pmatrix}
$$

Matrices which are to be added or subtracted must be of the same *order* (i.e. they must have a corresponding number of rows and columns).

This is not so for matrices which are to be multiplied, but there is a rule which determines whether two matrices can be combined by the multiplication process.

Before reading further, try to write down the rule for multiplication.

In order to combine two matrices by multiplication, the number of *columns* in the first matrix must equal the number of *rows* in the second matrix.

An $l \times m$ matrix can be combined with an $m \times n$ matrix and the answer will be an $l \times n$ matrix.

In this case the matrices are said to be **compatible**.

The result of combining, for example, the *second row* of the first matrix with the *third column* of the second matrix will be written in the *second row, third column* of the answer matrix:

$$
\begin{pmatrix}
1 & 3 & 2 \\
4 & 7 & 1
\end{pmatrix}
\begin{pmatrix}
1 & 3 & 1 \\
2 & 1 & 2 \\
1 & 2 & 3
\end{pmatrix}
=
\begin{pmatrix}
9 & 10 & 13 \\
19 & 21 & 21
\end{pmatrix}
$$

(2×3) matrix (3×3) matrix = (2×3) matrix

Example

$$
A = \begin{pmatrix} 1 & 2 \\ 3 & 1 \\ 1 & -3 \end{pmatrix}
\qquad
B = \begin{pmatrix} 4 & 1 & 2 \\ 3 & -7 & -1 \end{pmatrix}
$$

a In which order can **A** and **B** be combined?

b What order will the answer matrix be?

c Multiply the matrices **A** and **B**.

a **A** is a 3×2 matrix and **B** is a 2×3 matrix. Therefore they can be combined as **AB** or **BA**.

b The product **AB** will be a $(3 \times 2)(2 \times 3)$ matrix, i.e. (3×3)
The product **BA** will be a $(2 \times 3)(3 \times 2)$ matrix, i.e. (2×2)

c $\mathbf{AB} = \begin{pmatrix} 1 & 2 \\ 3 & 1 \\ 1 & -3 \end{pmatrix} \begin{pmatrix} 4 & 1 & 2 \\ 3 & -7 & -1 \end{pmatrix}$ and $\mathbf{BA} = \begin{pmatrix} 4 & 1 & 2 \\ 3 & -7 & -1 \end{pmatrix} \begin{pmatrix} 1 & 2 \\ 3 & 1 \\ 1 & -3 \end{pmatrix}$

$\mathbf{AB} = \begin{pmatrix} 4+6 & 1-14 & 2-2 \\ 12+3 & 3-7 & 6-1 \\ 4-9 & 1+21 & 2+3 \end{pmatrix}$ $\qquad \mathbf{BA} = \begin{pmatrix} 4+3+2 & 8+1-6 \\ 3-21-1 & 6-7+3 \end{pmatrix}$

$\mathbf{AB} = \begin{pmatrix} 10 & -13 & 0 \\ 15 & -4 & 5 \\ -5 & 22 & 5 \end{pmatrix}$ $\qquad \mathbf{BA} = \begin{pmatrix} 9 & 3 \\ -19 & 2 \end{pmatrix}$

Exercise 190

1 $\mathbf{A} = \begin{pmatrix} 4 & 1 \\ 2 & 3 \end{pmatrix}$ $\qquad \mathbf{B} = \begin{pmatrix} 2 & 2 \\ 1 & 0 \end{pmatrix}$

a Calculate: (i) **AB**, (ii) **BA**.

b What does this suggest about matrix multiplication?

2 Calculate

a $\begin{pmatrix} 1 & 0 \\ 0 & 1 \end{pmatrix} \begin{pmatrix} 2 & 3 \\ 1 & 1 \end{pmatrix}$ d $\begin{pmatrix} -1 & 0 \\ 0 & -1 \end{pmatrix} \begin{pmatrix} 0 & -1 \\ -1 & 0 \end{pmatrix}$ g $\begin{pmatrix} 3 & 4 \\ 2 & 3 \end{pmatrix} \begin{pmatrix} 3 & -4 \\ -2 & 3 \end{pmatrix}$

b $\begin{pmatrix} 4 & 1 \\ 2 & 3 \end{pmatrix} \begin{pmatrix} 1 & 0 \\ 0 & 1 \end{pmatrix}$ e $\begin{pmatrix} 0 & -1 \\ 1 & 0 \end{pmatrix} \begin{pmatrix} 0 & -1 \\ 1 & 0 \end{pmatrix}$ h $\begin{pmatrix} 5 & 3 \\ 2 & 1 \end{pmatrix} \begin{pmatrix} 1 & -3 \\ -2 & 5 \end{pmatrix}$

c $\begin{pmatrix} 0 & 1 \\ 1 & 0 \end{pmatrix} \begin{pmatrix} 0 & 1 \\ 1 & 0 \end{pmatrix}$ f $\begin{pmatrix} 4 & 3 \\ 1 & 1 \end{pmatrix} \begin{pmatrix} 1 & -3 \\ -1 & 4 \end{pmatrix}$ i $\begin{pmatrix} -3 & 1 \\ 5 & -2 \end{pmatrix} \begin{pmatrix} -2 & -1 \\ -5 & -3 \end{pmatrix}$

3 a Using the capital letter to represent the matrix, list all the products which could be calculated from:

$\mathbf{A} = \begin{pmatrix} 1 & 4 & 2 \\ 3 & 0 & 5 \end{pmatrix}$ $\qquad \mathbf{B} = \begin{pmatrix} 1 & 3 & 2 \\ 2 & 1 & 1 \\ 3 & 0 & 2 \end{pmatrix}$ $\qquad \mathbf{C} = \begin{pmatrix} 4 & 1 \\ 2 & 0 \\ 1 & 3 \end{pmatrix}$ $\qquad \mathbf{D} = \begin{pmatrix} 6 & 2 \\ 3 & 5 \end{pmatrix}$

b For each pair of matrices which can be multiplied, state the order of the answer matrix.

4 Using the matrices defined in question 2, calculate:

a DA **c** BC **e** BC − CD

b AB **d** DA + 2AB

34.5 Identity and inverse matrices

In this section the matrices are restricted to 2 × 2 matrices.

In arithmetic, adding zero to any number leaves the number unchanged. For example:

$$7 + 0 = 7$$

Zero is called the **identity element** for addition.

In matrix algebra, the same is true for the **zero matrix**. For example:

$$\begin{pmatrix} 2 & 1 \\ 3 & 7 \end{pmatrix} + \begin{pmatrix} 0 & 0 \\ 0 & 0 \end{pmatrix} = \begin{pmatrix} 2 & 1 \\ 3 & 7 \end{pmatrix}$$

and the zero matrix $\begin{pmatrix} 0 & 0 \\ 0 & 0 \end{pmatrix}$ is called the **identity matrix** for all 2 × 2 matrices.

The **inverse** of a number, for addition, is the number which when added to a given number will produce the **identity element**. For example:

$7 + (-7) = 0$ and -7 is the additive inverse of 7.

It is easy to see that as $\begin{pmatrix} a & b \\ c & d \end{pmatrix} + \begin{pmatrix} -a & -b \\ -c & -d \end{pmatrix} = \begin{pmatrix} 0 & 0 \\ 0 & 0 \end{pmatrix}$

the **additive inverse** of a 2 × 2 matrix $\begin{pmatrix} a & b \\ c & d \end{pmatrix}$ is $\begin{pmatrix} -a & -b \\ -c & -d \end{pmatrix}$

For arithmetic multiplication, the **identity element** is 1. That is:

$$a \times 1 = 1 \times a = a$$

and the **inverse** of any number a is $\frac{1}{a}$ (and vice versa). That is:

$$a \times \frac{1}{a} = \frac{1}{a} \times a = 1$$

In matrix multiplication, the **identity matrix** for 2 × 2 matrices is the

unit matrix $\begin{pmatrix} 1 & 0 \\ 0 & 1 \end{pmatrix}$. That is:

$$\begin{pmatrix} a & b \\ c & d \end{pmatrix}\begin{pmatrix} 1 & 0 \\ 0 & 1 \end{pmatrix} = \begin{pmatrix} 1 & 0 \\ 0 & 1 \end{pmatrix}\begin{pmatrix} a & b \\ c & d \end{pmatrix} = \begin{pmatrix} a & b \\ c & d \end{pmatrix}$$

For the inverse matrix we require a matrix which will multiply $\begin{pmatrix} a & b \\ c & d \end{pmatrix}$ to give $\begin{pmatrix} 1 & 0 \\ 0 & 1 \end{pmatrix}$.

Before reading on, see if you can write down the inverses of $\begin{pmatrix} 3 & 1 \\ 2 & 1 \end{pmatrix}$ and $\begin{pmatrix} 3 & 2 \\ 2 & 4 \end{pmatrix}$

The inverse of $\begin{pmatrix} 3 & 1 \\ 2 & 1 \end{pmatrix}$ is $\begin{pmatrix} 1 & -1 \\ -2 & 3 \end{pmatrix}$

Check: $\begin{pmatrix} 3 & 1 \\ 2 & 1 \end{pmatrix} \begin{pmatrix} 1 & -1 \\ -2 & 3 \end{pmatrix} = \begin{pmatrix} 1 & 0 \\ 0 & 1 \end{pmatrix}$

This suggests interchanging the elements on the first diagonal and changing the sign of the elements on the other diagonal.

However, if we do that for $\begin{pmatrix} 3 & 2 \\ 2 & 4 \end{pmatrix}$ we obtain $\begin{pmatrix} 3 & 2 \\ 2 & 4 \end{pmatrix} \begin{pmatrix} 4 & -2 \\ -2 & 3 \end{pmatrix} = \begin{pmatrix} 8 & 0 \\ 0 & 8 \end{pmatrix}$

which has the correct form for the inverse matrix, but is eight times too large.

The inverse of $\begin{pmatrix} 3 & 2 \\ 2 & 4 \end{pmatrix}$ is therefore $\frac{1}{8}\begin{pmatrix} 4 & -2 \\ -2 & 3 \end{pmatrix}$.

Carrying out the same operation for $\begin{pmatrix} a & b \\ c & d \end{pmatrix}$

we have $\begin{pmatrix} a & b \\ c & d \end{pmatrix} \begin{pmatrix} d & -b \\ -c & a \end{pmatrix} = \begin{pmatrix} ad - bc & 0 \\ 0 & ad - bc \end{pmatrix}$

and the **multiplicative inverse** of $\begin{pmatrix} a & b \\ c & d \end{pmatrix}$ is $\dfrac{1}{ad - bc}\begin{pmatrix} d & -b \\ -c & a \end{pmatrix}$

The expression $ad - bc$ is important and is called the **determinant** of the matrix.

Example

Find the inverse of **a** $\begin{pmatrix} 2 & 1 \\ 4 & -3 \end{pmatrix}$ **b** $\begin{pmatrix} 3 & 6 \\ 2 & 4 \end{pmatrix}$

a Determinant $= ad - bc = -6 - 4 = -10$

$\therefore$ Inverse $= \dfrac{-1}{10}\begin{pmatrix} -3 & -1 \\ -4 & 2 \end{pmatrix} = \begin{pmatrix} 0.3 & 0.1 \\ 0.4 & -0.2 \end{pmatrix}$

Check. $\begin{pmatrix} 2 & 1 \\ 4 & -3 \end{pmatrix} \begin{pmatrix} 0.3 & 0.1 \\ 0.4 & -0.2 \end{pmatrix} = \begin{pmatrix} 0.6+0.4 & 0.2-0.2 \\ 1.2-1.2 & 0.4+0.6 \end{pmatrix} = \begin{pmatrix} 1 & 0 \\ 0 & 1 \end{pmatrix}$

b Determinant $= ad - bc = 12 - 12 = 0$

The inverse should be $\dfrac{1}{0}\begin{pmatrix} 4 & -6 \\ -2 & 3 \end{pmatrix}$, but it is not possible to divide by

zero. Hence, if the determinant is zero (i.e. $ad - bc = 0$), the matrix has no inverse and is called a **singular matrix**.

1   Find the additive inverse of each of the following matrices:

a $\begin{pmatrix} 6 & 2 \\ 7 & 3 \end{pmatrix}$ b $\begin{pmatrix} 1 & 3 \\ -2 & 5 \end{pmatrix}$ c $\begin{pmatrix} 1 & -3 \\ -4 & 6 \end{pmatrix}$

2   Find the determinant of each of the following matrices:

a $A = \begin{pmatrix} 3 & 2 \\ 4 & 3 \end{pmatrix}$ d $D = \begin{pmatrix} 5 & 4 \\ 4 & 3 \end{pmatrix}$ g $G = \begin{pmatrix} 4 & 4 \\ -2 & -2 \end{pmatrix}$

b $B = \begin{pmatrix} 4 & 3 \\ 2 & 2 \end{pmatrix}$ e $E = \begin{pmatrix} -3 & -6 \\ 1 & 2 \end{pmatrix}$ h $H = \begin{pmatrix} -5 & 3 \\ -2 & 2 \end{pmatrix}$

c $C = \begin{pmatrix} 6 & 4 \\ 3 & 2 \end{pmatrix}$ f $F = \begin{pmatrix} 2 & 4 \\ 2 & 3 \end{pmatrix}$ i $I = \begin{pmatrix} 1 & 0 \\ 0 & 1 \end{pmatrix}$

3   State which of the matrices in question 2 is a singular matrix.

4   Find the inverse matrices, where possible, of the matrices defined in question 2.

5   $A = \begin{pmatrix} 2 & -3 \\ 4 & 1 \end{pmatrix}$ $B = \begin{pmatrix} 1 & 2 \\ -4 & 3 \end{pmatrix}$

   a   Find the additive inverse of $A + B$.

   b   Find the multiplicative inverse of $A + B$.

6   For the matrix $A = \begin{pmatrix} a & b \\ c & d \end{pmatrix}$, show that $AA^{-1} = A^{-1}A = I$,

where A^{-1} is the multiplicative inverse of A and I is the identity matrix.

34.6   Matrix transformations

An important use of matrices is in the geometry of transformations.

A triangle ABC has coordinates A(1, 2), B(1, 4), C(2, 4).

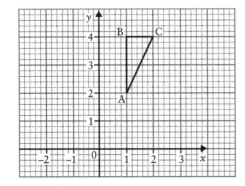

If the positions of the vertices of the triangle are written in vector form,

$$\begin{pmatrix} 1 \\ 2 \end{pmatrix}; \begin{pmatrix} 1 \\ 4 \end{pmatrix}; \begin{pmatrix} 2 \\ 4 \end{pmatrix}$$

the triangle can be represented by a matrix:

$$\begin{matrix} A \ B \ C \\ \begin{pmatrix} 1 & 1 & 2 \\ 2 & 4 & 4 \end{pmatrix} \end{matrix}$$

Premultiplying by a 2 × 2 matrix, $\mathbf{T} = \begin{pmatrix} -1 & 0 \\ 0 & 1 \end{pmatrix}$, will transform the triangle:

$$\begin{matrix} \mathbf{T} \\ \begin{pmatrix} -1 & 0 \\ 0 & 1 \end{pmatrix} \end{matrix} \begin{matrix} A & B & C \\ \begin{pmatrix} 1 & 1 & 2 \\ 2 & 4 & 4 \end{pmatrix} \end{matrix} = \begin{matrix} A' & B' & C' \\ \begin{pmatrix} -1 & -1 & -2 \\ 2 & 4 & 4 \end{pmatrix} \end{matrix}$$

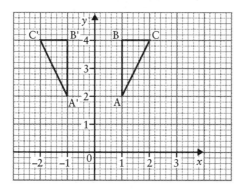

The transformation is a reflection in the y-axis.

Example

The vertices of a square have coordinates:
O(0, 0); A(0, 2); B(2, 2); C(2, 0).

A transformation is defined as

$$\begin{pmatrix} x' \\ y' \end{pmatrix} = \begin{pmatrix} 2 & 0 \\ 0 & 2 \end{pmatrix} \begin{pmatrix} x \\ y \end{pmatrix} \text{ where } \begin{pmatrix} 2 & 0 \\ 0 & 2 \end{pmatrix} \text{ is the matrix of the transformation.}$$

a Find the coordinates of the transformed figure O'A'B'C'.

b Plot the figures OABC and O'A'B'C' on to graph paper.

c Describe the transformation geometrically.

d Find the matrix which will transform O'A'B'C' to OABC.

a Writing OABC in matrix form and multiplying by the transformation matrix gives:

$$\begin{matrix} \mathbf{T} \\ \begin{pmatrix} 2 & 0 \\ 0 & 2 \end{pmatrix} \end{matrix} \begin{matrix} O & A & B & C \\ \begin{pmatrix} 0 & 0 & 2 & 2 \\ 0 & 2 & 2 & 0 \end{pmatrix} \end{matrix} = \begin{matrix} O' & A' & B' & C' \\ \begin{pmatrix} 0 & 0 & 4 & 4 \\ 0 & 4 & 4 & 0 \end{pmatrix} \end{matrix}$$

b The coordinates of O'A'B'C' are (0, 0); (0, 4); (4, 4); (4,0).

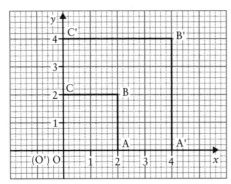

c The transformation is an **enlargement** of scale factor 2 with the centre of the enlargement at the origin.

d The matrix which will transform O'A'B'C' to OABC is the inverse

matrix of the transformation $\begin{pmatrix} 2 & 0 \\ 0 & 2 \end{pmatrix}$, i.e.

$$\frac{1}{4}\begin{pmatrix} 2 & 0 \\ 0 & 2 \end{pmatrix} = \begin{pmatrix} \frac{1}{2} & 0 \\ 0 & \frac{1}{2} \end{pmatrix}$$

Check.

$$\begin{array}{ccc} \mathbf{T}^{-1} & \text{O' A' B' C'} & \text{O A B C} \\ \begin{pmatrix} \frac{1}{2} & 0 \\ 0 & \frac{1}{2} \end{pmatrix} & \begin{pmatrix} 0 & 0 & 4 & 4 \\ 0 & 4 & 4 & 0 \end{pmatrix} = & \begin{pmatrix} 0 & 0 & 2 & 2 \\ 0 & 2 & 2 & 0 \end{pmatrix} \end{array}$$

Exercise 192

1 Plot the unit square OABC with vertices (0, 0) (0, 1) (1, 1) and (1, 0) on to graph paper and transform the square using the following matrices:

a $\begin{pmatrix} 1 & 0 \\ 0 & -1 \end{pmatrix}$ **d** $\begin{pmatrix} -1 & 0 \\ 0 & 1 \end{pmatrix}$ **g** $\begin{pmatrix} 0 & -1 \\ 1 & 0 \end{pmatrix}$

b $\begin{pmatrix} 3 & 0 \\ 0 & 3 \end{pmatrix}$ **e** $\begin{pmatrix} 0 & 1 \\ 1 & 0 \end{pmatrix}$ **h** $\begin{pmatrix} 0 & -1 \\ -1 & 0 \end{pmatrix}$

c $\begin{pmatrix} -1 & 0 \\ 0 & -1 \end{pmatrix}$ **f** $\begin{pmatrix} 0 & 1 \\ -1 & 0 \end{pmatrix}$

Describe each transformation geometrically.

2 **a** Draw the parallelogram with vertices (0, 0) (1, 0) (2, 1) and (1, 1) on to graph paper.

     b Find the images of the parallelogram under the transformations defined by the matrices:

$$A = \begin{pmatrix} 0 & 1 \\ 1 & 0 \end{pmatrix} \quad B = \begin{pmatrix} 0 & -1 \\ 1 & 0 \end{pmatrix} \quad C = \begin{pmatrix} 1 & 0 \\ 0 & -1 \end{pmatrix} \quad D = \begin{pmatrix} -1 & 0 \\ 0 & -1 \end{pmatrix}$$

$$E = \begin{pmatrix} 0 & -1 \\ -1 & 0 \end{pmatrix} \quad F = \begin{pmatrix} 0 & 1 \\ -1 & 0 \end{pmatrix} \quad G = \begin{pmatrix} -1 & 0 \\ 0 & 1 \end{pmatrix}$$

   c Describe each of the above transformations geometrically.

3   a Draw on graph paper the triangle ABC with coordinates A(1, 1), B(4, 6), C(2, 5).

   b Apply the transformation $\begin{pmatrix} x' \\ y' \end{pmatrix} = \begin{pmatrix} 0 & -1 \\ 1 & 0 \end{pmatrix}\begin{pmatrix} x \\ y \end{pmatrix}$ to ABC

     and plot triangle A′B′C′ on to the graph paper.

   c Apply the transformation $\begin{pmatrix} x'' \\ y'' \end{pmatrix} = \begin{pmatrix} 0 & -1 \\ 1 & 0 \end{pmatrix}\begin{pmatrix} x' \\ y' \end{pmatrix}$ to A′B′C′

     and plot triangle A″B″C″ on to the graph paper.

   d $T = \begin{pmatrix} 0 & -1 \\ 1 & 0 \end{pmatrix}$. Calculate T^2.

   e What is the connection between the transformation T^2 and triangle ABC?

4   The transformation of a point P with coordinates (x, y) on to a point P′ with coordinates $(x'\ y')$ is defined as

$$\begin{pmatrix} x' \\ y' \end{pmatrix} = \begin{pmatrix} 2 & 2 \\ 1 & 1 \end{pmatrix}\begin{pmatrix} x \\ y \end{pmatrix} \text{ where } \begin{pmatrix} 2 & 2 \\ 1 & 1 \end{pmatrix} \text{ is the transformation matrix.}$$

   A triangle has coordinates A(0, 0), B(2, 1), C(1, 3).

   a Plot triangle ABC on to graph paper.

   b Find the coordinates of the points A′, B′, C′ under the transformation and plot them on to the graph.

   c Describe the transformation geometrically.

   d A′B′C′ cannot be transformed back to ABC. Why not?

5   a The trapezium ABCD with coordinates A(0, 0) B(0, 1) C(2, 1) and D(3, 0) is transformed to A′B′C′D′ by the matrix

$$T = \begin{pmatrix} 0 & -1 \\ 1 & 0 \end{pmatrix}$$

     Draw and label ABCD and A′B′C′D′ on a graph.

   b A′B′C′D′ is transformed to A″B″C″D″ by the matrix

$$R = \begin{pmatrix} 3 & 0 \\ 0 & 3 \end{pmatrix}$$

     Draw and label A″B″C″D″ on the graph.

c Calculate **RT**.

d Describe geometrically the transformations represented by **T**, **R** and **RT**.

e Find the matrix which represents the transformation of A″B″C″D″ to ABCD.

6 Transformations are represented by the matrices:

$$P = \begin{pmatrix} -1 & 0 \\ 0 & 1 \end{pmatrix} \qquad R = \begin{pmatrix} -1 & 0 \\ 0 & -1 \end{pmatrix} \qquad T = \begin{pmatrix} 0 & 1 \\ 1 & 0 \end{pmatrix}$$

$$Q = \begin{pmatrix} 1 & 0 \\ 0 & -1 \end{pmatrix} \qquad S = \begin{pmatrix} 0 & -1 \\ 1 & 0 \end{pmatrix}$$

Describe geometrically the transformations represented by

a PQ   b QR   c SQ   d QS   e PT   f TQ.

*35 Iteration

An iterative process is one which is repeated several times, following exactly the same process on each repeat.

One iterative process for finding the square root of a number is:

  (i) take a guess at the square root i.e. make a first estimate.

 (ii) divide the number by your estimate then find the average of this number and your estimate.

(iii) this average is an improved estimate so repeat (ii) using this second estimate to find a third estimate.

(iv) repeat (ii) with each new estimate until a sufficiently accurate value has been found.

Algebraically, if the first estimate of the square root of a number A is called x_1, then the second estimate, x_2, is given by

$$x_2 = \frac{1}{2}\left(x_1 + \frac{A}{x_1}\right)$$

The third estimate is
$$x_3 = \frac{1}{2}\left(x_2 + \frac{A}{x_2}\right)$$

The fourth estimate is
$$x_4 = \frac{1}{2}\left(x_3 + \frac{A}{x_3}\right)$$

The $(n + 1)$th estimate is
$$x_{n+1} = \frac{1}{2}\left(x_n + \frac{A}{x_n}\right)$$

and this is an iterative formula for the square root of A.

Example

 a Using the iteration formula $x_{n+1} = \frac{1}{2}\left(x_n + \frac{A}{x_n}\right)$, find the square root of 85 correct to 5 decimal places.

 b By writing x for x_{n+1} and x_n in the iteration formula, show that your answer is a solution of the equation $x^2 = 85$.

 a
$$x_{n+1} = \frac{1}{2}\left(x_n + \frac{A}{x_n}\right)$$

One estimate of $\sqrt{85}$ is 9

$$x_1 = 9$$

$$x_2 = \frac{1}{2}\left(9 + \frac{85}{9}\right) = 9.2222222$$

(some calculators will display 10 digits)

Transfer $x_2 = 9.2222222$ into the calculator memory and repeat the calculation with this value using the following sequence of operations:

9.2222222 $\boxed{\text{Min}}$ 85 $\boxed{\div}$ $\boxed{\text{MR}}$ $\boxed{=}$ $\boxed{+}$ $\boxed{\text{MR}}$ $\boxed{=}$ $\boxed{\div}$ $\boxed{2}$ $\boxed{=}$

The display reads $\boxed{\textbf{9.2195448}}$

$$\therefore x_3 = 9.2195448$$

Transfer 9.2195448 into the memory using $\boxed{\text{Min}}$ and repeat the sequence of operations to obtain

$$x_4 = 9.2195445$$

The digits up to the sixth decimal place are the same for x_4 and x_3. Therefore the square root is correct to 5 d.p. That is:

$$\sqrt{85} = 9.21954 \text{ to 5 d.p.}$$

(In general, when $x_{n+1} = x_n$ to the desired degree of accuracy, then the solution x is given by $x = x_n = x_{n+1}$.)

b Substituting x for x_{n+1} and x_n in the iteration formula gives

$$x = \frac{1}{2}\left(x + \frac{85}{x}\right)$$

$$2x = x + \frac{85}{x}$$

$$2x^2 = x^2 + 85$$

$$x^2 = 85$$

This method of finding square roots was used before 1000BC by the Babylonians.

Iterative methods are also used for solving equations.

1 Use the iterative formula $x_{n+1} = \frac{1}{2}\left(x_n + \frac{A}{x_n}\right)$

to find the square root of 60, correct to 5 d.p.

2 One root of the equation $x^2 - 8x + 2 = 0$ can be solved by using the iterative formula

$$x_{n+1} = 8 - \frac{2}{x_n}$$

a Take any value for x_1 and use the formula to find a solution to the equation correct to 3 d.p.

b Use the quadratic formula to solve the equation.

3 **a** The second root of the equation

$$x^2 - 8x + 2 = 0$$

can be found by using the iteration

$$x_{n+1} = \frac{x_n^2 + 2}{8}$$

(i) Take $x_1 = 2$, (ii) take $x_1 = 0.2$ and calculate:

x_2, x_3, x_4, x_5, x_6 and x_7

Hence find the second root of the equation correct to 3 d.p.

b How many iterations are required to find the root, correct to 3 d.p., when:
(i) $x_1 = 2$ (ii) $x_1 = 0.2$?

4 An iterative formula for solving a cubic is

$$x_{n+1} = \frac{3}{x_n^2} - 4$$

a Take $x_1 = 4$ and calculate $x_2, x_3, x_4, x_5, x_6, x_7$ and x_8.
For each iteration, write down the first five digits.

b What is the solution, correct to 3 d.p.?

c How many iterations are required to find this solution?

d By replacing x_{n+1} and x_n with x, show that this value is a solution of the equation

$$x^3 + 4x^2 - 3 = 0.$$

5 A sequence of numbers $x_1, x_2, x_3, \ldots$ is defined by

the relationship $x_{n+1} = 4.5 + \dfrac{5}{2x_n}$ and by $x_1 = 4$.

a Find x_2, x_3, x_4, giving all the figures on your calculator each time.

b Suggest the limit that the sequence appears to be approaching.

c (i) Replace x_{n+1} and x_n in the relationship by x and show that the result is equivalent to the quadratic equation

$$2x^2 - 9x - 5 = 0.$$

(ii) Show that the limit suggested in **b** is a solution of this equation.

d (i) Factorise $2x^2 - 9x - 5$.

(ii) What is the complete solution of the equation $2x^2 - 9x - 5 = 0$?
(SEG, Spring 1988)

6 The equation $3x^2 - 11x - 4 = 0$ can be rearranged as

$$x = \frac{1}{3}\left(11 + \frac{4}{x}\right) \quad \text{or as} \quad x = \frac{3x^2 - 4}{11}$$

a Write down two possible iterative formulae for solving this equation.

b Take $x_1 = 3$ and calculate x_2, x_3, and x_4 for each of your formulae.

c What can you deduce from your results?

d Take $x_1 = 0.3$ and calculate x_2, x_3, x_4, x_5 and x_6 for each of your formulae.

e What do you deduce are the roots of the equation $3x^2 - 11x - 4 = 0$?

f Verify your answer to **e** by solving the equation $3x^2 - 11x - 4 = 0$ by factorising.

g Rearrange the equation $2x^2 - 4x + 1 = 0$ to find an iterative formula which will locate the solution to the equation which is greater than 1.

36 Making a Small Survey

Land surveying is a branch of mathematics which deals with measuring and recording the size and shape of any portion of the earth's surface and showing this on a map or plan.

Surveying the earth's surface is usually concerned with the measurement of lengths over the ground.

This measurement may be made with a **chain**, or a **tape measure**, or the circumference of a **trundle wheel**, though nowadays, aerial surveys are also made by means of photographs.

In a **chain survey**, the measurements of length are carried out using only a chain.

A chain is more likely to retain its original length than a tape and hence the survey should be more accurate.

36.1 The field book

When making a small survey, you must record the measurements obtained.

It is usual to use a 'field book', which can be any form of notepad.

You should start by drawing an 'artist's impression' of the area you are going to survey, and mark on this all the major landmarks.

All the measurements you obtain, together with a clear identification of which lengths you have measured, must be inserted into the field book.

36.2 Triangulation

Triangulation is a method of surveying using triangles.
The area to be surveyed is divided into a number of triangles and, by measuring all three sides, you can construct each triangle and draw the overall shape accurately.

Suppose you wanted to survey the area ABCDE. First of all, join BD and BE. This creates three triangles: ABE, BDE and BCD.

By measuring all the sides shown, you can construct triangle ABE. Side BE can then be used to construct triangle BDE and then triangle BCD is constructed on side BD.

It is usual to measure another diagonal, for example AC, as a check length.

SURVEYING

Draw the following polygons accurately and make the specified measurement. If the polygon has more than four sides, you may find it helpful to draw a rough sketch first.

The vertices of polygons are lettered in a cyclic order which may be either clockwise or anticlockwise.

1 AB = 5 cm,   BC = 6 cm,   CA = 4 cm.   Measure angle B.

2 AB = 6 cm,   BC = 8 cm,   CA = 7 cm.   Measure angle A.

3 PQ = 2.5 in,   QR = 3.7 in,   RP = 2.8 in.   Measure angle P.

4 RS = 3.2 in,   ST = 1.9 in,   TR = 2.7 in.   Measure angle T.

5 AB = 3.8 cm,   BC = 4.2 cm,   CA = 3.9 cm.   Measure angle C.

6 AB = 3.9 cm,   BC = 4.7 cm,   Angle A = 48°. Measure AC.

7 PQ = 4.9 cm,   QR = 3.1 cm,   Angle Q = 71°. Measure PR.

8 AB = 5 cm,   BC = 7 cm,   CA = 8 cm,   CD = 8 cm,   DA = 5 cm.
   Measure BD.

9 PQ = 3 in,   QR = 4 in,   RS = 2.5 in,   SP = 2 in,   PR = 3 in.
   Measure SQ.

10 KL = 12 cm,   KN = 13 cm,   LM = 3 cm,   MN = 8 cm,
    LN = 6 cm,   KP = 6 cm,   LP = 10 cm,   KQ = 4 cm,   LQ = 12 cm.
    Measure KM, MP and NQ.

36.3 Offsets

So far, we have been dealing with shapes that have straight edges. In practice, a surveyor is more likely to measure areas which have irregular or curved edges.

It is necessary, therefore, to find an accurate method of plotting the shape of boundaries. Look at the area here.

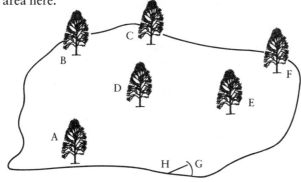

Initially, permanent, prominent features are chosen which are on, or near, the edges of the area to be surveyed.

In the above diagram, trees A, B, C, F and the gatepost G are suitable points.

By using the method of triangulation, the polygon ABCFG can be drawn.

To record the position of the actual edge of the field, the technique of taking offsets is used.

A **survey line** is a line, on the ground, connecting two points which can be clearly identified. These two points normally form part of a triangulation.

An **offset** is a line at right angles to the survey line. To plot the shape of a curved boundary, you set out a survey line joining two points on the edge, or near the edge, of the area and then use offsets to measure the boundary.

An example is shown below in the area on the right.

The survey line AB has been divided into ten equal parts, each 1 m in length. Offsets are measured to the curved boundary.

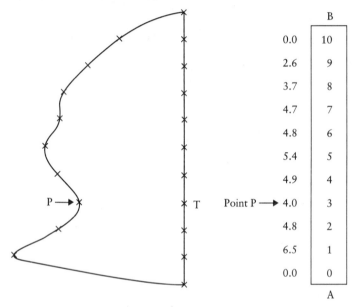

B	
0.0	10
2.6	9
3.7	8
4.7	7
4.8	6
5.4	5
4.9	4
Point P → 4.0	3
4.8	2
6.5	1
0.0	0
A	

To identify the boundary at point P, you record the distance AT along the survey line (3 m) and the distance TP perpendicular to the survey line (4.0 m).

All these measurements should be written down in a field book.

In this case, the offsets are to the left of the line when you proceed from A to B, and so they are recorded in the field book on the left of the survey line AB.

In practice, it is most unlikely that you would divide AB into equal parts.

Most boundaries are either a combination of straight edges or a combination of basic curves.

In the shape shown, you would only need to take offsets to points P, Q, R with extra measurements between P and Q.

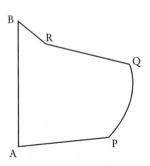

The more measurements you take, the greater the accuracy. However, the extra time spent, reflected in the number of offsets used, should be balanced against the greater accuracy.

Example 1

Draw a diagram (in cm) to correspond to the following offsets.

Q	
8	0
7	3
4	8
2	8
1	6
0	5
P	

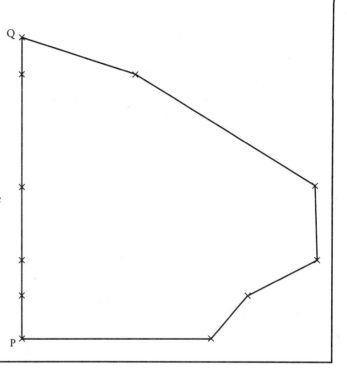

1 Draw PQ 8 cm in length.

2 Mark points along PQ at distances 1 cm, 2 cm, 4 cm, 7 cm from P.

3 Draw lines at right angles to PQ, and to the right of PQ, at all these points.

4 Draw a line at right angles to PQ at P since the offset at P is not zero.

5 Mark a point along each offset at the required distance.

6 Join these points with straight lines.

Example 2

XY is a line across a rock garden. Offsets from XY to the edges of the rock garden are shown opposite. Draw a plan of the rock garden.

The offsets written on the left of XY represent offsets on the left of the survey line. Similarly for those on the right of XY.

The offsets not found usually indicate that the edge of the boundary is straight between the two neighbouring points.

There is no left offset at *. Hence the edge is straight between the offsets A and B.

	Y	
4	10	1
6	8	3
	6	6
B 2	4	
*	3	4
A 3	1	2
1	0	0
	X	

Exercise 195

Draw diagrams which correspond to the following offsets.
Use straight edges to join up the offsets.

1 B
10	1
8	4
6	2
0	0

A

3 S
0	12
3	8
4	6
4	4
1	0

R

5 B
1	16	0
5	12	
4.1	9	2.8
	8	4.1
3.3	7	3.4
2.2	5	1
0	0	2

A

7 K
0.8	8.6	1.3
	7.4	2.8
	5.2	4.7
4.1	4.2	5.2
3.4	2.6	
2.1	1.2	3.4
1.1	0	2

J

2 Q
6	2
5	5
3	1
0	0

P

4 Y
0	16	1
2	14	2
6	12	1
5	8	2
2	5	3
1	4	2
0	0	1

X

6 G
1.3	11	0
2.6	9.5	
4.1	8	2.4
3	7	
	6	1.8
2	3	1.4
1	0	2

F

8 T
1	11	1.2
2.9	9.2	2.7
2.8	8.5	3.4
3.9	8.1	2.1
3.4	7.2	1.7
2.8	6.5	2.1
2	6	2.4
1	5	1.6
0	0	1

S

9 AB is a garden fence of length 11.1 m. Offsets are taken to four shrubs. Draw the line AB (in cm) and mark in the position of the four shrubs.

B
11.1	1.1
9.4	2.4
7.2	3.2
5.0	2.8

A

AB = 15

In questions 10 and 11, identify the positions of the shrubs found by the offsets.

10 D
108	51
65	28
50	46
20	31
	20

C

CD = 160

11 Q
29	115
11	98
34	94
27	58
41	34
18	22
12	18

P

PQ = 120

12 Draw triangle ABC (clockwise) and then draw the offsets to each side. Join all the points with a smooth curve.

B
0.2	7
2.1	5
3	3
2.8	2
1	0

A

C
0.7	8
1.3	7
2.0	4
1.1	2
0.4	0

B

A
0.2	6
1.3	5
1.8	3
2.1	1
0.1	0

C

13 Draw triangle XYZ (anticlockwise) and then draw the offsets. Join with a smooth curve.

Y
9	0.2
4.8	1.1
3	1.6
2	1.5
0	0.1

X

Z
7	0.4
5.6	1.1
4	1.2
2	1.1
1	0.6
0	0.3

Y

X
8	0.3
6.1	0.6
4.7	0.4
3.9	1.1
2.1	0.7
0	0.1

Z

37 Plane Areas, including Circles

Students should read pp. 197–209 before commencing this section.

37.1 The area of a trapezium

The area of a trapezium is:

$\frac{1}{2}$ × **Sum of parallel sides** × **Perpendicular height**

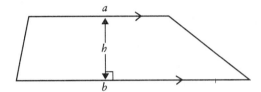

$$\text{Area} = \frac{1}{2}(a + b)h$$

Example

Trapezium PQRS has PQ parallel to SR. PQ = 8 cm, SR = 13 cm, and the perpendicular distance from S to PQ is 9 cm.

Find area PQRS.

$\text{Area} = \frac{1}{2}(13 + 8) \times 9$

$\phantom{\text{Area}} = \frac{1}{2} \times 21 \times 9$ (evaluate inside of bracket first)

$\phantom{\text{Area}} = 94.5 \text{ cm}^2$

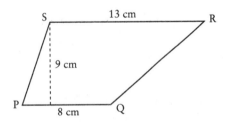

Exercise 196

1 Find the areas of the parallelograms with the following measurements:

   a base = 10 cm, height = 12 cm   **c** base = 34.7 in, height = 13 in

   b base = 12.6 cm, height = 6 cm   **d** base = 1.3 m, height = 26 cm.

2 Find the areas of the following trapeziums.

   a

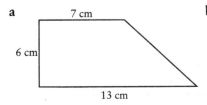

   b

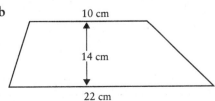

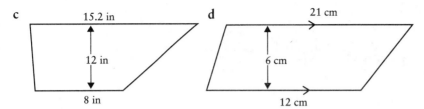

c   15.2 in   d   21 cm

12 in   6 cm

8 in   12 cm

3   Find the height of a parallelogram which has:

   a   area $32\,cm^2$, base $4\,cm$ **c**   area $42\,in^2$, base $8\,in$.

   b   area $36\,cm^2$, base $9\,cm$

4   Find the base of a parallelogram which has:

   a   area $70\,cm^2$, height $10\,cm$ **b**   area $20\,cm^2$, height $11.1\,cm$.

37.2   The areas of other polygons

The area of other polygons may be obtained by dividing the polygon into a number of simpler shapes (triangles, rectangles, etc.). Then the areas of each of these shapes can be worked out and the total added up.

Example

Find the area of a regular hexagon with sides 20 cm.

Draw the hexagon and add in extra lines as shown:

Exterior angle of a regular polygon is $\dfrac{360°}{6} = 60°$.

$\therefore$ Interior angle (e.g. AFE) = 120°.

Hence $\angle$ FAE = $\angle$ AEF = 30° (angles of $\triangle$ AFE add up to 180°).

$\therefore \angle$ BAE = 90°

From $\triangle$ AFP, AP = AF cos FAP = 20 cos 30° = 17.3 cm.

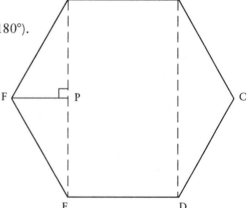

Area of rectangle ABCD = AB × AE
$$= 20 \times (17.32 \times 2)$$
$$= 693\,cm^2$$

Area of $\triangle$ AEF = $\frac{1}{2}$ × AE × FP
$$= \tfrac{1}{2} \times 34.64 \times 10$$
$$= 173\,cm^2$$

$\therefore$ Area of polygon ABCDEF = Area ABCD + Area $\triangle$ AEF + Area $\triangle$ BCD
$$= 692 + 173 + 173$$
$$= 1038$$

The area of polygon ABCDEF is $1040\,cm^2$.

Exercise 197

1 Find area ABCDE in the shape shown.

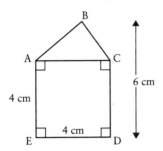

2 Find the area of a regular octagon of side 8 cm.

3 Find the area of the hexagon ABCDEF.

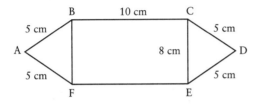

4 Find the area of ABCD.

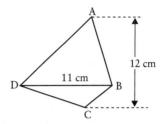

5 Find the area of PQRS.

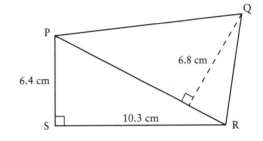

37.3 The area of a sector

The area of sector OPQ is proportional to the angle at O. If the angle is 30°, ($\frac{1}{12}$ of the whole angle at O which is 360°), the area of the sector is $\frac{1}{12}$ of the whole area of the circle.

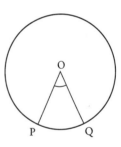

In the general case:

$$\frac{\text{Area of sector}}{\text{Area of circle}} = \frac{\text{Angle of sector at O}}{360°}$$

> ## *Example*
>
> Find the area of a sector of a circle radius 8 cm subtended by an angle of 50° at the centre of the circle shown below.
>
> $$\frac{\text{Area of sector}}{\text{Area of circle}} = \frac{50°}{360°}$$
>
> $$\text{Area of sector} = \frac{50}{360} \times \text{Area of circle}$$
>
> $$= \frac{50}{360} \times \pi \times 8^2 \qquad (\text{Area of circle} = \pi r^2)$$
>
> $$= 27.93\,\text{cm}^2$$
>
> $\therefore$ The area is $27.9\,\text{cm}^2$.

Exercise 198

In questions 1 – 3, find the areas of the sectors given:

1

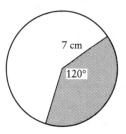

2

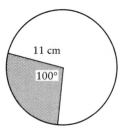

3

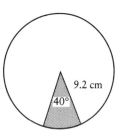

4   The area of the sector is $70 \, \text{cm}^2$. Find the radius of the circle.

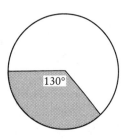

5   The area of the sector is $32 \, \text{in}^2$. Find the radius of the circle.

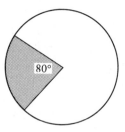

6   Find the angle x if the area of the sector is $30 \, \text{cm}^2$.

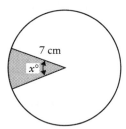

7   Find the angle y if the area of the sector is $87 \, \text{cm}^2$.

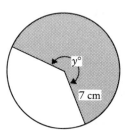

*38 Volumes

38.1 The volume of a pyramid

The volume of a pyramid is given by:

$$\text{Volume} = \frac{1}{3} \times \text{Base area} \times \text{Height}$$

$$= \frac{1}{3} Ah$$

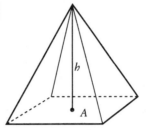

> **Example**
>
> A pyramid has a square base of side 8 in. The height of the pyramid is 6 in. What is its volume?
>
> $$\text{Base area} = 8 \times 8$$
> $$= 64 \, \text{in}^2$$
>
> $$\therefore \text{Volume} = \frac{1}{3} \times 64 \times 6$$
> $$= 128 \, \text{in}^3$$
>
> $\therefore$ The volume is $128 \, \text{in}^3$.

38.2 The volume of a cone

The volume of a cone is given by:

$$\text{Volume} = \frac{1}{3} \pi r^2 h$$

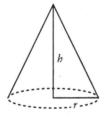

> **Example 1**
>
> Find the volume of a cone of height 12 cm and base radius 8 cm.
>
> $$\text{Volume} = \frac{1}{3}\pi r^2 h$$
>
> $$= \frac{1}{3} \times \pi \times 8^2 \times 12$$
>
> $$= \frac{1}{3} \times \pi \times 64 \times 12$$
>
> $$= 804.2 \, \text{cm}^3$$
>
> $\therefore$ The volume is $804 \, \text{cm}^3$.

Example 2

Find the volume of a cone of slant height 20 cm, and radius 12 cm.

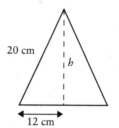

Since you are given the slant height, you must first find the vertical height h.
Pythagoras' theorem is used to do this.

$$h^2 = 20^2 - 12^2$$

$$= 400 - 144$$

$$= 256$$

$$h = 16$$

$$\therefore \text{Height} = 16\,\text{cm}$$

Hence $\quad \text{Volume} = \dfrac{1}{3} \times \pi \times 12^2 \times 16$

$$= 2413\,\text{cm}^3$$

$\therefore$ The volume is $2410\,\text{cm}^3$.

---

*38.3 The volume of a sphere

The volume of a sphere of radius r is given by:

$$\textbf{Volume} = \frac{4}{3}\pi r^3$$

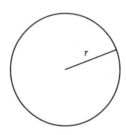

Example 1

Find the volume of a sphere of radius 6 cm.

$$\text{Volume} = \frac{4}{3}\pi r^3$$

$$= \frac{4}{3} \times \pi \times 6^3$$

$$= \frac{4}{3} \times \pi \times 216$$

$$= 904.8\,\text{cm}^3.$$

$\therefore$ The volume is $905\,\text{cm}^3$.

┌───┐
Example 2

The volume of a sphere is $280 \, \text{cm}^3$. What is its radius?

$$\frac{4}{3}\pi r^3 = 280$$

$$\therefore 4\pi r^3 = 3 \times 280$$

$$\therefore \pi r^3 = \frac{3 \times 280}{4}$$

$$= 210$$

$$\therefore r^3 = \frac{210}{\pi}$$

$$\therefore r = \sqrt[3]{\frac{210}{\pi}}$$

$$= 4.058 \, \text{cm}$$

$\therefore$ The radius is $4.06 \, \text{cm}$.
└───┘

Exercise 199

1 Find the volume of a sphere of radius:

   a 6 cm   **b** 3.8 cm   **c** 5.2 in

2 Find the volume of a pyramid of height 12 cm on a square base of side 4 cm.

3 Find the volume of a pyramid of height 6 cm on a rectangular base 8 cm by 10 cm.

4 Find the volume of a pyramid of height 9 cm on a rectangular base 6 cm by 8 cm.

5 Find the volume of a sphere of diameter:

   a 8 cm   **b** 9.4 cm   **c** 3.4 in

6 Find the volume of a cone:

   a height 9 cm, base radius 8 cm,

   b height 7 cm, base diameter 11 cm,

   c height 2.8 in, base radius 6.4 in.

7 The volume of a sphere is $252 \, \text{cm}^3$. Find its radius.

8 The volume of a sphere is $412 \, \text{cm}^3$. Find its diameter.

9 The volume of a cone is $280 \, \text{in}^3$. Its height is 21 in. Find its base radius.

10 The slant height of a cone is 25 cm. The base radius is 7 cm. Find its volume.

39 The Surface Areas of Bodies

39.1   The surface areas of bodies with straight edges

When a body has a polygon for each face (e.g. a cube
or cuboid) the surface area is obtained by adding
together all the areas of the faces. The faces are all
shapes identified on pp. 195 and 198.

Example 1

Find the surface area of a cuboid of length 12 cm, width 8 cm and height 6 cm.

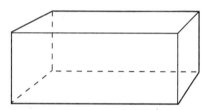

$$
\begin{aligned}
\text{Surface area} &= (\text{Top} + \text{Bottom}) + \text{Two sides} + \text{Two ends} \\
&= (12 \times 8 + 12 \times 8) + 2(12 \times 6) + 2(8 \times 6) \\
&= (96 + 96) + 144 + 96 \\
&= 432 \, \text{cm}^2
\end{aligned}
$$

∴ The area is $432 \, \text{cm}^2$.

Example 2

Find the surface area of a pyramid which has a square base of side 6 cm
and for which each sloping edge is 10 cm long.

The diagram shows a sloping face.

To find the area, you first need to find the length VP.

By Pythagoras' theorem

$$
VP^2 = 10^2 - 3^2
$$

$$
\therefore VP = \sqrt{91}
$$

$$
= 9.539 \, \text{cm}
$$

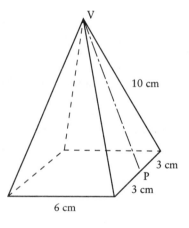

∴ The area of a sloping face $= \dfrac{1}{2} \times 6 \times 9.539 \, \text{cm}^2$

$$
= 28.62 \, \text{cm}^2
$$

The area of the base is $6 \times 6 = 36 \, \text{cm}^2$.

There are four sloping faces.

$$
\begin{aligned}
\therefore \text{Total surface area} &= \text{Area of base} + 4 \times 28.62 \\
&= 36 + (4 \times 28.62) \\
&= 150.48 \, \text{cm}^2
\end{aligned}
$$

∴ The total area is $150 \, \text{cm}^2$ (to 3 s.f.).

1   Find the surface area of a cube with edges:

   a 5 cm   **b** 6 cm   **c** 8.1 cm.

2   Find the surface area of a cuboid with dimensions:

   a 6 cm   3 cm   8 cm

   b 4.1 cm   5.2 cm   3.8 cm

   c 8 in   6 in   3.5 in

3   The volume of a cube is 64 cm³. Find the length of each edge, and hence the surface area of the cube.

4   The volume of a cube is 28 cm³. Find the length of each edge, and hence the surface area of the cube.

5   The volume of a cuboid is 32 cm³. Its length and width are 8 cm and 5 cm. Find the height and surface area of the cuboid.

6   A pyramid with slant edge 10 cm is on a square of side 8 cm. Find the surface area of the pyramid.

7   A pyramid on a triangular base (i.e. a tetrahedron) has every edge 12 cm. Find its surface area.

8   A tetrahedron (see question 7) has an equilateral base of side 20 cm. The slant edges are of length 26 cm. Find the surface area of the tetrahedron.

39.2   The surface area of a cylinder

The curved area of a cylinder of height h and radius r is $2\pi rh$.

Closed cylinders also have two circular ends, each of which is of radius r.

Example 1

A tin of baked beans has height h cm, and has a radius of r cm. A label is wrapped around the tin as shown.

What area of paper is covered by the label?

The label can be removed from the tin and opened out to form a rectangle:

$$\text{Length of label} = \text{Circumference of tin}$$
$$= 2\pi r \text{ cm}$$
$$\text{Breadth of label} = \text{Height of tin}$$
$$= h \text{ cm}$$
$$\text{Area of label} = \text{Area of rectangle}$$
$$= \text{Length} \times \text{Breadth}$$
$$= 2\pi r \times h$$
$$= 2\pi rh \text{ cm}^2$$

This example verifies the formula for the curved surface of a cylinder:

Curved surface area = $2\pi rh$

Example 2

Find the curved surface area of a cylinder, radius 12 cm and height 10 cm.

$$\text{Curved surface area} = 2\pi rh$$
$$= 2 \times \pi \times 12 \times 10$$
$$= 754 \text{ cm}^2$$

∴ The curved surface area is 754 cm^2.

Example 3

A drinking glass is in the shape of a cylinder, radius 4 cm and height 8 cm. Find its outer surface area.

$$\text{Curved surface area} = 2\pi rh$$
$$= 2 \times \pi \times 4 \times 8$$
$$= 201.1 \text{ cm}^2$$

$$\text{Area of one end (the other is open!)} = \pi r^2$$
$$= \pi \times 4^2$$
$$= 50.26 \text{ cm}^2$$

$$\therefore \text{Total surface area} = 201.1 + 50.26$$
$$= 251.4 \text{ cm}^2$$

∴ The total surface area is 251 cm^2.

Exercise 201

1 Kitchen roll is wrapped around a tube of cardboard of length 23 cm and diameter 4.2 cm. What area of cardboard is used in making the tube?

2 A cylindrical block of wood is 25 cm high and has a diameter of 14 cm.
Calculate the area of the curved surface.

3 A garden cloche is made by covering three wires, each bent into a semicircle, with polythene.

a If the height of the cloche is 30 cm, what length of wire is required?

b The length of each cloche is 60 cm.
What area of polythene is required to cover the wires? (The ends of the cloche are left open.)

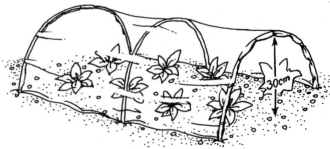

*39.3   The surface area of a sphere

Surface area of a sphere $= 4\pi r^2$

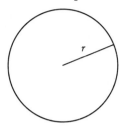

Example 1

Find the surface area of a sphere of radius 5 cm.

$$\text{Surface area} = 4\pi r^2$$
$$= 4 \times \pi \times 5^2$$
$$= 4 \times \pi \times 25$$
$$= 314 \, \text{cm}^2$$

$\therefore$ The surface area is $314 \, \text{cm}^2$.

Example 2

A sphere has curved surface area $220 \, \text{cm}^2$.
Find its radius.

$$4\pi r^2 = 220$$
$$r^2 = \frac{220}{4\pi}$$
$$= 17.51$$
$$r = \sqrt{17.51}$$
$$= 4.184$$

$\therefore$ The radius is $4.18 \, \text{cm}$.

Exercise 202

1   Find the surface area of a sphere of radius:

   a 5 cm   **b** 8 cm   **c** 6 in.

2   Find the surface area of a sphere of diameter:

   a 8 cm   **b** 20 in   **c** 12.1 cm

3   Find the radius of a sphere with surface area:

   a $28 \, \text{cm}^2$,   **b** $28.4 \, \text{in}^2$.

4   Find the volume of a sphere with surface area:

   a $20 \, \text{cm}^2$ **b** $36 \, \text{cm}^2$.

40 Areas and Volumes of Irregular Shapes

It is more likely that a surveyor will be working with shapes which have irregular, curved edges. There are two main methods of finding the approximate area of such shapes.

These are the **trapezium rule** and **Simpson's rule**.

As examples of these two rules, we will use the same area so that we can compare the answers. It is sensible to convert from the dimensions of the scale diagram to the dimensions of the actual area as soon as possible, definitely before any area is found.

40.1   The trapezium rule

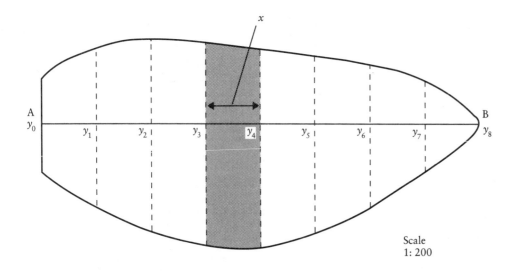

Scale
1: 200

Consider the shape shown. The main survey line, AB, has been divided into eight equal parts. If a main survey line had not been given, any line which passed through the extremities of the shape could be used.

From A, B and each division mark, offsets are drawn to the curved boundary on both sides of the survey line shown by dotted lines on the diagram, and the lengths of these lines are measured. They are the lengths $y_0, y_1, y_2, \ldots, y_8$. Note that the numbering of these lengths always starts with y_0.

These lengths are the whole lengths of the lines from one curved boundary to the other curved boundary. By doing this, we have a set of strips, a typical one of which has been shaded in the above diagram. Each strip is approximately a trapezium, and so its area can be found and hence the total area can be calculated.

This gives the **trapezium rule** which states:

$$\text{Area} \simeq \frac{x}{2}\{(y_0 + y_n) + 2(y_1 + y_2 + \ldots + y_{n-1})\}$$

for n strips, of width x, and lengths $y_0, y_1, \ldots y_n$, or
in other words:

$$\text{Area} \simeq \tfrac{1}{2}\text{ Width (First + Last + 2} \times \text{Middles)}$$

Example

Use the trapezium rule to find the area of the shape given on p. 386.

The scale is 1:200, hence we multiply every measurement by 200 to convert from the scale diagram (in cm) to the actual length (in cm).

Since the scale of 1:200 is also 1 cm to 2 m, we multiply the length (in cm) on the scale diagram by 2, and convert this length to m. This gives the actual length.

Length on scale diagram	Actual length
$y_0 = 2.5$ cm	5 m
$y_1 = 4.2$ cm	8.4 m
$y_2 = 5.1$ cm	10.2 m
$y_3 = 5.4$ cm	10.8 m
$y_4 = 5.3$ cm	10.6 m
$y_5 = 4.7$ cm	9.4 m
$y_6 = 3.7$ cm	7.4 m
$y_7 = 2.4$ cm	4.8 m
$y_8 = 0$ cm	0 m

The width of each strip is 1.5 cm, which represents 3 m. Substituting in the trapezium rule with $n = 8$ (AB was divided into 8 strips), we have:

$$\text{Area} \simeq \frac{x}{2}\{(y_0 + y_n) + 2(y_1 + y_2 + \ldots + y_{n-1})\}$$

$$= \frac{3}{2}\{(5 + 0) + 2(8.4 + 10.2 + 10.8 + 10.6 + 9.4 + 7.4 + 4.8)\}$$

$$= \frac{3}{2}(5 + 2 \times 61.6)$$

$$= \frac{3}{2}(5 + 123.2)$$

$$= \frac{3}{2} \times 128.2$$

$$= 192.3 \text{ m}^2$$

$\therefore$ The area is 192 m^2.

Note. The main survey line does not have to be divided into 8 equal parts. If the survey line were 21 m long, say, it would be sensible to divide it into 7 equal parts.

In questions 1 – 8, use the trapezium rule to find the area of the shapes. Use the survey line shown, and the number of strips indicated.

1 6 strips
Scale 1 : 2000.

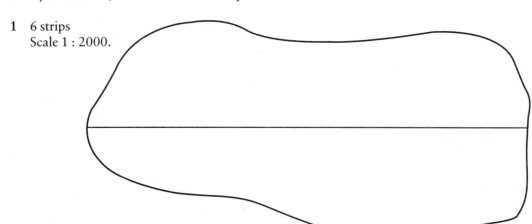

2 8 strips
Scale 1 : 2000.

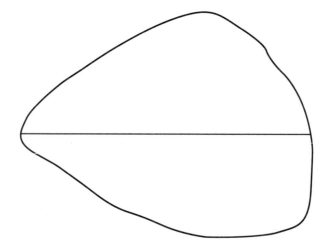

3 10 strips
Scale 1 : 3000.

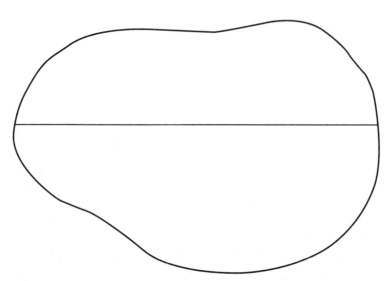

4 7 strips
Scale 1 : 4000.

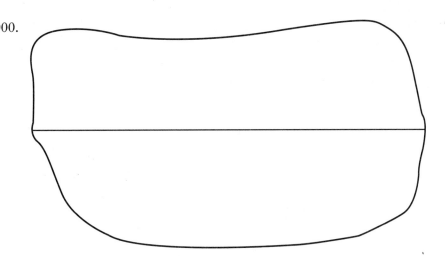

5 5 strips
Scale 1 : 5000.

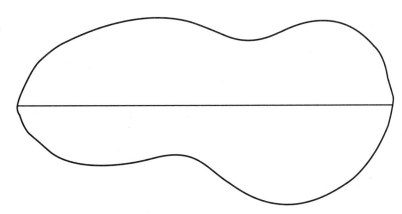

6 8 strips
Scale 1 : 4000.

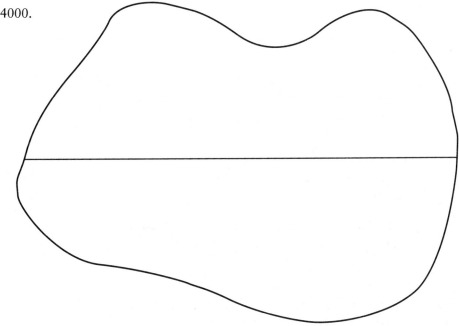

7   7 strips
Scale 1 : 3000.

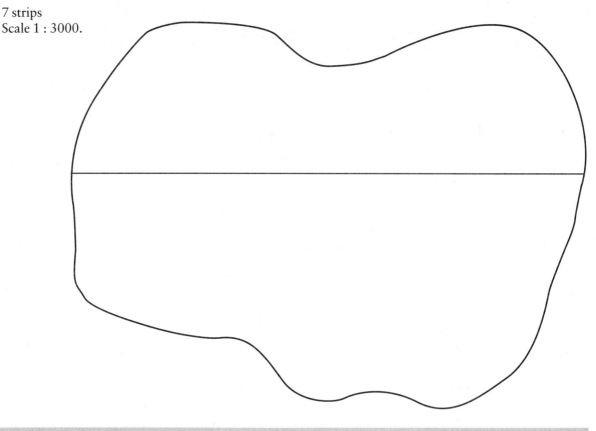

40.2   Simpson's rule

A more accurate value is obtained by using Simpson's rule, which considers pairs of strips as having curved edges.

$$\text{Area} \simeq \frac{1}{3}\,\text{Width (First + Last + 2 × Evens + 4 × Odds)}$$

or

$$\textbf{Area} \simeq \frac{x}{3}\{(y_0 + y_n) + 2(y_2 + y_4 + \ldots + y_{n-2}) + 4(y_1 + y_3 + \ldots + y_{n-1})\}$$

for n strips (n must be an even number), of width x, and lengths $y_0, y_1, \ldots, y_n$.

Note. With Simpson's rule, n must be *even*, since the rule uses pairs of strips, i.e., there must be an even number of strips.

When $n = 8$, the formula gives:

$$\text{Area} \simeq \frac{x}{3}\{(y_0 + y_8) + 2(y_2 + y_4 + y_6) + 4(y_1 + y_3 + y_5 + y_7)\}$$

┌─ *Example* ─────────────────────────────────────

Use Simpson's rule to find the area of the shape given on p. 386.

We will use the lengths found on p. 387. Once again, we must remember to convert to actual lengths before using the formula. We must also remember to identify the first and the last values of y and separate the odd y lengths (y_1, y_3, y_5, etc.) from the even y lengths (y_2, y_4, etc.)

For Simpson's rule	Length on scale diagram	Actual length Even y	Actual length Odd y
First	$y_0 = 2.5\,\text{cm}$	5 m	
Odd	$y_1 = 4.2\,\text{cm}$		8.4 m
Even	$y_2 = 5.1\,\text{cm}$	10.2 m	
Odd	$y_3 = 5.4\,\text{cm}$		10.8 m
Even	$y_4 = 5.3\,\text{cm}$	10.6 m	
Odd	$y_5 = 4.7\,\text{cm}$		9.4 m
Even	$y_6 = 3.7\,\text{cm}$	7.4 m	
Odd	$y_7 = 2.4\,\text{cm}$		4.8 m
Last	$y_8 = 0\quad\text{cm}$	0 m	

$y_2 + y_4 + \ldots + y_{n-2}$ or sum of evens $= 28.2\,\text{m}$

$y_1 + y_3 + \ldots + y_{n-1}$ or sum of odds $= 33.4\,\text{m}$

The width of each strip is 3 m.

Substituting into Simpson's rule with $n = 8$ (we divided AB into 8 strips), we have:

$$\text{Area} = \frac{3}{3}\{5 + (2 \times 28.2) + (4 \times 33.4)\}\ \text{m}^2$$

first + last   sum of evens   sum of odds

$$= \frac{3}{3}(5 + 56.4 + 133.6)$$

$$= \frac{3}{3} \times 195.0$$

$$= 195\,\text{m}^2$$

$\therefore$ The area is $195\,\text{m}^2$.

This answer is similar to that found by the trapezium rule. The diagrams below show that for each pair of strips the trapezium rule gives too small an approximation for a convex shape and too large an approximation for a concave shape.

Simpson's rule uses a quadratic curve as an approximation and hence Simpson's rule usually gives a more accurate approximation.

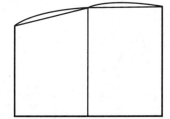
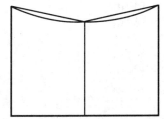

Exercise 204

Use Simpson's rule to find the areas of the shapes given in Exercise 203 on pp. 388–9. In questions 4, 5 and 7, use 6 strips, otherwise use the number of strips indicated in section 40.1.

40.3   Volumes using the trapezium rule and Simpson's rule

The volume of a solid having a uniform cross-section is given by:

Cross-sectional area × Length of the solid

This can be used to find the volumes of valleys, or mounds of earth etc.

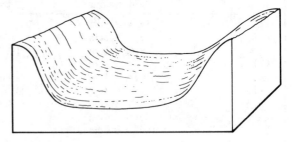

The diagram shown represents part of a glaciated river-bed for which the cross-sectional area is constant for 400 m.

The cross-section is drawn accurately below. The area can be found using the trapezium rule, and from this we can find the approximate volume of water in the valley.

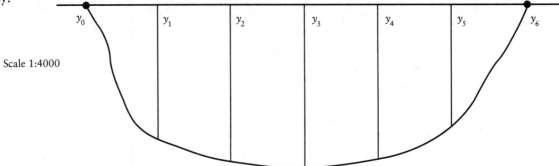

Scale 1:4000

From the cross-sectional diagram, we can measure the lengths $y_0, y_1, \ldots, y_6$ (we are now using 6 strips), and convert these to actual lengths.

	Length in scale diagrams	Actual length
y_0	0	0
y_1	3.5 cm	140 m
y_2	4.1 cm	164 m
y_3	4.3 cm	172 m
y_4	4.1 cm	164 m
y_5	3.2 cm	128 m
y_6	0	0

Sum of ends

$y_0 = 0\,\text{m}$
$\underline{y_6 = 0\,\text{m}}$
$0\,\text{m}$

Others

$y_1 = 140\,\text{m}$
$y_2 = 164\,\text{m}$
$y_3 = 172\,\text{m}$
$y_4 = 164\,\text{m}$
$y_5 = \underline{128\,\text{m}}$
Total $\underline{768\,\text{m}}$

The width of each strip is 80 m (2 cm on the scale diagram).
By the trapezium rule, the area of the cross-section is:

$$\frac{x}{2}\{(y_0 + y_6) + 2(y_1 + y_2 + y_3 + y_4 + y_5)\}$$

(We are using 6 strips, so $n = 6$.)

$$\text{Area} \quad = \frac{80}{2}(2 \times 768)$$

$$= \frac{80}{2} \times 1536$$

$$= 40 \times 1536$$

$$= 61\,440\,\text{m}^2$$

∴ The area is $61\,440\,\text{m}^2$.

Volume = Area of cross-section × Length of valley

$$= 61\,440 \times 400$$

$$= 2\,458\,000$$

Hence the volume is $2\,460\,000\,\text{m}^3$.

1   Find the approximate volume of earth in a mound whose constant
    cross-sectional area is shown below, and whose length is 50 m.

    Scale 1 : 200
    (use 6 strips)

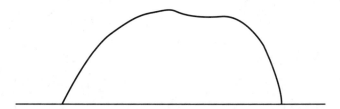

2   Find the approximate volume of water in a valley whose cross-sectional
    area is shown below, and whose length is 1300 m.

    Scale 1 : 1000
    (use 8 strips)

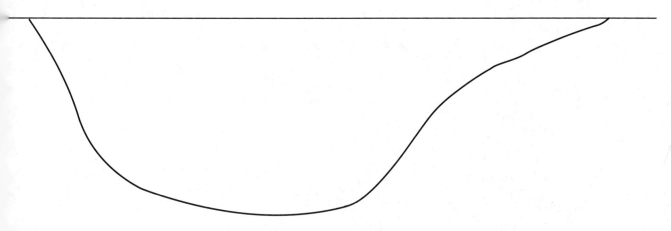

3 Find the approximate volume of earth in a mound whose constant cross-sectional area is shown below, and whose length is 50 m.

Scale 1 : 3000
(use 6 strips)

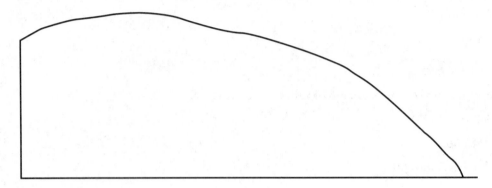

4 The diagram, which is not drawn to scale, shows Mr Blake's garden which has been infected by fungus. He decides to remove the soil to a depth of 2 feet. In order to estimate the amount of soil to be removed, Mr Blake measures offsets from the straight edge AB to the curved edge. The results (in feet) are given in the table below.

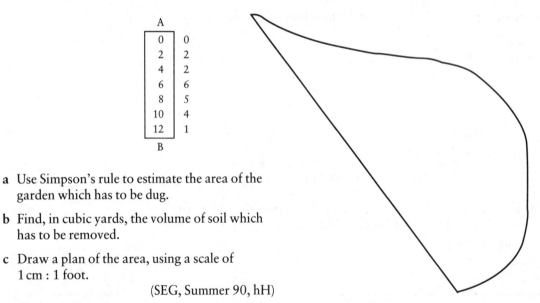

A

0	0
2	2
4	2
6	6
8	5
10	4
12	1

B

a Use Simpson's rule to estimate the area of the garden which has to be dug.

b Find, in cubic yards, the volume of soil which has to be removed.

c Draw a plan of the area, using a scale of 1 cm : 1 foot.

(SEG, Summer 90, hH)

41 Ordnance Survey Maps

41.1 Triangulation points

An Ordnance Survey map contains a great deal of information about the area it shows. It can inform us both about the natural landscape of the area (e.g. hills and rivers) and non-natural features (e.g. roads and churches).

In particular, an Ordnance Survey map can show us the features of the landscape – for example, whether the area is low or high land, and whether it is flat or sloping.

The heights of particular points may be shown as triangulation points $\triangle$ 195 m, where a block on the ground shows the true height, or as a spot height 195, where the height is only shown on the map. Ordnance Survey maps show many spot heights, particularly at the tops of hills.

The height of the land is measured above sea level (stated more precisely in the form 'above mean sea level at Newlyn'). Triangulation points are the most accurately surveyed heights in the country.

41.2 Contour lines

Contour lines are lines joining places of equal height on a map. For example, any point on a contour line marked 100 (metres) is 100 m above sea level. These lines are not, of course, shown on the ground.

Contour lines tell us about the shape of the ground. For example, the closer the contour lines are to each other, the steeper the slope, and an absence of contour lines indicates that the land is almost flat.

On modern scale Ordnance Survey maps, contour lines are shown at intervals of 10 metres, and every fifth contour line is darker so that the map is easier to read.

On older Ordnance Survey maps (either the imperial maps or the original metre maps), contour lines are shown at intervals of 50 ft, with every 250 ft emphasised.

In each case, the larger-scale maps show more frequent contour lines.

Contour lines are regularly identified on maps so that particular heights and the frequency with which contour lines are marked can be readily found.

41.3 Gradients

The **gradient** of a slope is a measure of its steepness.

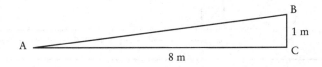

The gradient of 1 : 8 means that for every 8 metres (km or feet, etc.) of **horizontal** distance, the land rises or falls 1 m (km or feet, etc.)

(*Note.* The actual distance AB is more than 8 metres. The map identifies the horizontal distance AC.)

A gradient is calculated by measuring the horizontal distance between two points and dividing this by the vertical difference in height between these points, both measurements being in the same units.

The gradients of roads are now being converted to percentages:

$$1 : 25 \text{ becomes } \frac{1}{25} \times 100 = 4\%$$

The **average gradient** between two points is the gradient of the imaginary straight line joining these points.

Gradient = Vertical difference in height: Horizontal distance

which is expressed as a fraction $\dfrac{1}{x}$ or 1 in x.

Example 1

On a map, scale 1 : 40 000, the points A and B are 8 cm apart. Calculate the average gradient of AB when A is on the contour line 200 metres, and B is on the contour 280 metres.

Construct a diagram:

$$\text{Horizontal map distance} = 8 \text{ cm}$$

$$\text{Actual horizontal distance} = 8 \text{ cm} \times 40\,000$$

$$= 3.2 \text{ km}$$

$$\text{Vertical difference in height} = 280 - 200 \text{ metres}$$

$$= 80 \text{ metres}$$

$$\text{Gradient} = \frac{\text{Vertical difference in height}}{\text{Horizontal distance}}$$

$$= \frac{80 \text{ m}}{3.2 \text{ km}}$$

$$= \frac{80 \text{ m}}{3.2 \times 1000 \text{ m}}$$

$$= \frac{80}{3200}$$

$$= \frac{1}{40}$$

$\therefore$ Gradient = 1 in 40 or $\dfrac{1}{40}$ or 2.5%.

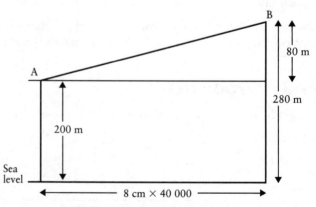

Example

Calculate the average gradient between the two points A and B.

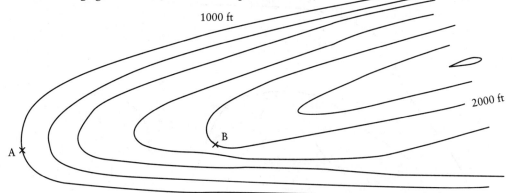

The distance AB on the map is 2.1 inches. A is on the contour 1000 ft, and B is on contour 2000 ft. The scale of the map is 1 inch : 1 mile.

$$\text{Actual horizontal distance} = 2.1 \text{ miles}$$

$$= 2.1 \times 5280\,\text{ft}$$

$$= 11\,088\,\text{ft}$$

$$\text{Vertical difference in height} = 2000\,\text{ft} - 1000\,\text{ft}$$

$$= 1000\,\text{ft}$$

$$\frac{\text{Horizontal distance}}{\text{Vertical difference}} = \frac{11\,088\,\text{ft}}{1000\,\text{ft}}$$

$$= 11.1$$

∴ Gradient = 1 in 11.1 or 1 : 11.1

Exercise 206

Find the average gradient of the lines stated.

1 a AB, **b** CD   (Scale 1 : 10 000)

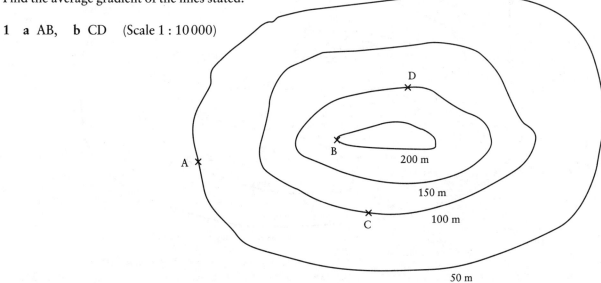

2 **a** AC, **b** BE, **c** CD

(Scale 1 : 30 000)

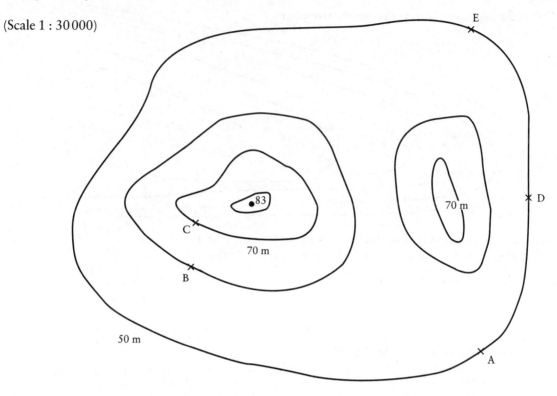

3 **a** AB, **b** AC, **c** BD, **d** CE, **e** BE

(Scale 1 : 25 000)

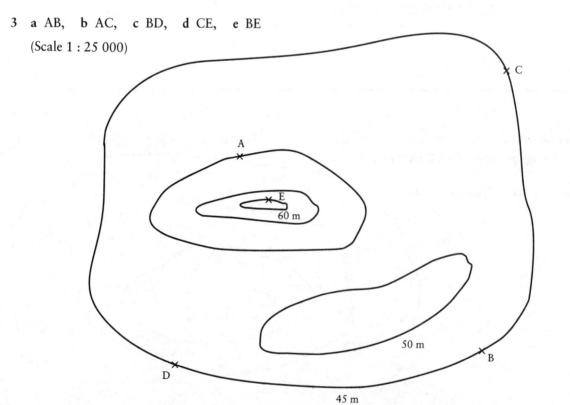

41.4 Cross-sections

If you walked in a straight line from one point to another across country, you would usually go up and down slopes. The graph which shows these slopes is called a cross-section.

The graph shows the shape of the land along this line.

To draw a cross-section from the map below along the line A to B:

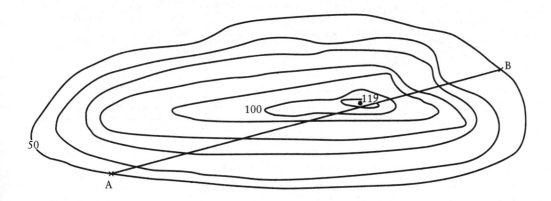

- Line up the edge of a piece of paper along AB and mark the positions of A and B.

- Mark on the edge of the piece of paper the position of each contour line and write the heights of each of the contours.

- Place the edge of the paper along an axis on a graph.

- Mark off the position of each contour height on this axis.

- Draw a vertical scale to give the heights, and plot the points appropriately.

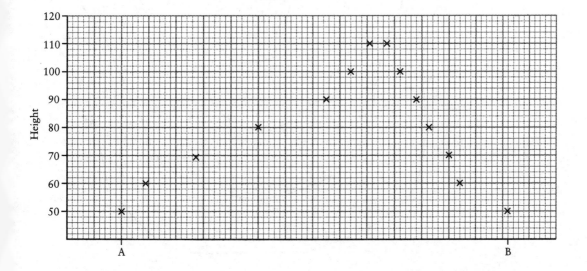

● Join these points with a smooth curve as shown, remembering that the valleys and hills cannot cross the next contour line.

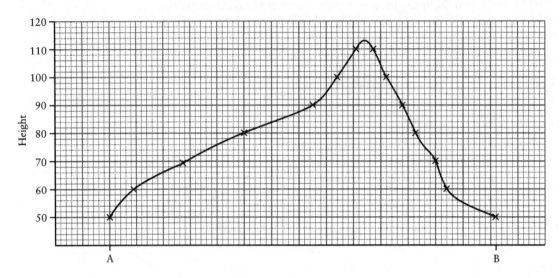

Exercise 207

In questions 1 – 3, find the cross-sections of the lines stated. The maps relate to those in Exercise 206.

1 CD **2** BE, CD **3** AB, CD

4 The map shows the contour lines of a hill. The map is drawn to a scale of 1 cm : 40 m, and the contour lines are plotted at intervals of 5 m.

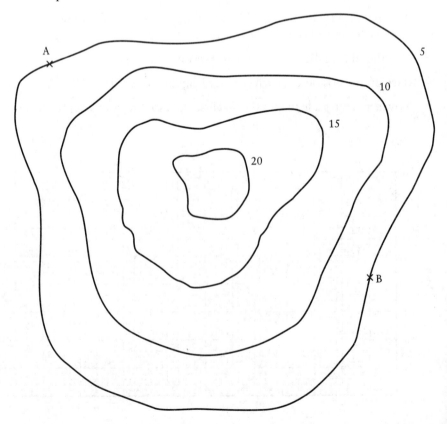

a Using the grid below, draw a cross-section of the hill from A to B.

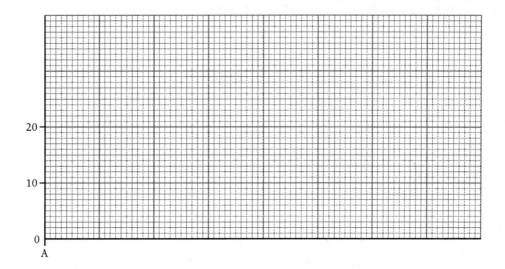

b Using your cross-section, estimate the average gradient from the point A
to the highest point of the hill, giving your answer as a percentage,
correct to the nearest whole number.

(SEG, Summer 89 hH)

Solutions to Exercises

Note. Solutions are given for exercises with a 'closed' numerical answer. They are not provided for 'open-ended' questions and for some questions leading purely to an illustration (e.g. a pie-chart or a graph).

Exercise 1

1 a £1.38 b £318.62 c £22.12 d £13.91 e £5.17
  f £8.73 g £97.20 h £44.20

Exercise 2

1 a £10.12 b £4.07 c £5.50 d £10.79 e £10.04
2 a £108.18 b £1.05 c £792.00 d £233.75
  e £156.00 f £21.69

Exercise 3

1 a £24.00 b £8.40 c £14.08 d 36p e £10.39
2 a £49.77 b 59p c £51.20 d £23.76 e £59.43
  f £1.09

Exercise 4

1 a 80.0% b 3.45% c 175% d 50%
  e 54.5% f 130%
2 a +23.1% b +20.0% c +375% d −20.0%
  e −45.5% f +464%

Exercise 5

1 £2.86   2 29.6%
3 10% if earnings greater than £100, £10 if earnings less than £100
4 6195   5 9.18%   6 £946.20   7 12%

Exercise 6

1 5:6   2 4:5:10   3 20:9   4 a 5:4 b £10 c £15
5 a 9:10:6:4 b 2.25:2.5:1.5:1
  c *Star*, 40; *Sun*, 45; *Daily Mirror*, 35

Exercise 7

1 a £2.70 b £150 c £75
2 a £2535 b £1780 c £4402.50   3 £16.80

Exercise 8

1 £260, £390   2 £1500, £4500, £6000
3 £48, £32, £24   4 £1750, £1250, £1000
5 £375, £625   6 a £75, £50
7 a £6318, £4212 b £15760 c £9924
8 £26154, £34872, £17436, £26154

Exercise 9

1 £172.20   2 £479.67   3 £6.43   4 £11582.40
5 a £259.70 b £155.75 c £441

Exercise 10

1 £249.90   2 £344.10
3 Andrews, £163.18; Collins, £214.92; Hammond, £210.94; Jali, £254.72; Longman, £246.76
4 £259.89
5 a £3.50 b 4 hours

Exercise 11

1 £1560   2 £24.72   3 £304   4 £1456
5 a Firm B b Firm A, £250

Exercise 12

1 £59.04   2 £125.44   3 £52.38   4 £152.22
5 a Mrs English, £520; Mrs Beckett, £432
  b Mrs English, £530; Mrs Beckett, £424
  c Increase for Mrs English, decrease for Mrs Beckett; the new scheme favours the more productive worker.

Exercise 13

1 a (i) £3295   (ii) £14205   (iii) £3551.25
  b (i) £5015   (ii) £7525   (iii) £1881.25
  c (i) £3295   (ii) £33   (iii) £8.25
2 £3760   3 £295.98   4 a £4295 b £3865
5 £284.10, £150.50, 66p

Exercise 14

1 a (i) £5927 (ii) £493.92
  b (i) £6559 (ii) £546.58
  c (i) £8869 (ii) £739.08
2 a £6439 b £391.50   3 £12206
4 a £8687 b Yes

Exercise 15

1 a £6.50 per week b £36.96 per month
  c £56.67 per month d £18.70 per month
  e £7.20 per week
2 a £54.08   b £2649.92
3 a £184 b £12.92 c £141.83
4 a £107.00 b £1284 c £5848.75 d £8812.75
  e £1782.27

Exercise 16

1 £26.60   2 £40.95   3 £1.66   4 25p   5 £3.36
6 £105.16
7 A, £14.81; B, £14.30; C, £14.97; B offers the best deal
8 £97.49   9 £7.53, £95.88
10 a 59p, £5.54 b £27.20, £367.20 c £2.60, £31.50
   d £46.92, £322.92 e £1.92, £15.66

Exercise 17

1 £1.22   2 £9114.89   3 £17.98   4 £898.23
5 £106.55   6 a £134.03 b £765.87

7 a £3.11, £25.88 b £1.93, £16.06 c £5.89, £49.10
  d £21.43, £178.56 e £35.36, £294.63
  f £12.32, £102.67 g £64.28, £535.71
  h £1.28, £10.71
8 $4.73

Exercise 18

1 a (i) £1   (ii) 25%
  b (i) £7   (ii) 35%
  c (i) 9p   (ii) 12%
  d (i) £18   (ii) 20%
  e (i) £100   (ii) 19%
  f (i) 3.99   (ii) 11.1%
2 33.3% 3 a £4.32 b 12% 4 £9719.80
5 20% 6 22.4% 7 a 95.1% b 1.46%

Exercise 19

1 a £35 b 17.9% 2 a £1174.50 b £225 c 23.7%
3 a £28.20 b 24.5% 4 a £3.25 b 18.1%
5 a £1200 b £1518.30 c £219.70 d £1738
  e £538; 44.8%

Exercise 20

1 a £252 b £82.50 c 2 years d 5 years
2 £2295, £9045 3 £231.25 4 $7\frac{1}{2}$%

Exercise 21

1 £7986 2 £2315.25 3 £7504.67, £984.67
4 £4655.45; £1551.82 5 a £69.16 b 6.92%
6 The second paying 8.75% per annum, ease of access to
  the account/services available/extras e.g. cheque book,
  credit card.

Exercise 22

1 £1986 2 £1800 3 £14 532 4 £8428.96
5 a £69 540 b £90 402 c £226 127

Exercise 23

2 a Tax is deducted from the interest earned at the current
    rate.
  b e.g. National Savings bonds, NSB Investment
    Account, special building society accounts
  c non-taxpayers
5 a Save as you earn
  b at a bank, post office or employer's pay office
6 a Tax exempt special savings account
  b (i) maximum of £9000 to be invested over 5 years
      (£1800 per year or £500 per month or £3000 in
      first year and £600 in year 5).
    (ii) only interest after tax may be withdrawn during
      the 5 years, otherwise income tax must be paid.
7 (i) the type of account
  (ii) the amount of money invested

Exercise 24

1 (i) Year is incorrect – should be 1992.
  (ii) Alteration should be initialled.
  (iii) Amount in figures should read £43.
  (iv) It is usual to add 'only' to an amount in whole
    pounds and/or draw a line through the remaining
    space to prevent additions to the amount.

2 (i) Insufficient funds in the account to cover the
    cheque.
  (ii) The cheque is unsigned.
  (iii) The cheque was completed incorrectly.
  (iv) The cheque is out of date.
3 The cheque has 'bounced' and the payee must go back to
  the drawer (writer of the cheque) for payment.
4 Crossed cheques must be paid into a bank account, an
  open cheque can be exchanged for cash.
5 The cheque must be paid into the bank account of the
  payee.
6 The card guarantees that the bank will honour cheques
  up to the amount on the card.
7 In the case of theft, one cannot be used by the thief
  without the other.

Exercise 25

1 a Southford b 0123456 c 30/3/89 d £444.01
  e £29.65 f £215.00 g Direct debit h £225
  i The account is in credit. j Cheque numbers

Exercise 26

1 Statement 2 Traveller's cheque 3 Access
4 Not negotiable 5 Direct debit 6 Investment
7 Night safe 8 Giro 9 Overdrawn 10 Repayment
11 Deposit account 12 Eurocheque
13 Rate   Initial letters spell STANDING ORDER.

Exercise 27

1 a £43.00 b £38.70 2 £26.25
3 a £223.44 b £33.45 c 8.8% per annum d 17.6%
4 a £5309.74 b 19.7%
5 a £1849.75 b £5549.25 c £8809.75 d 25.4%
  e 25.4%
6 a £11 138.70, £12 111.90 b 17.6%, 15.6%
7 A because APR values are 28% and 34.9%.

Exercise 28

1 a (i) £1111.32   (ii) £111.32   (iii) 11.1%
  b (i) £1220.88   (ii) £220.88   (iii) 11.0%
  c (i) £1337.04   (ii) £337.04   (iii) 11.2%
  d (i) £372   (ii) £72   (iii) 12%
  e (i) £3209.40   (ii) £809.40   (iii) 11.2%
  f (i) £2075.76   (ii) £375.76   (iii) 11.1%
2 a Something of value which the borrower possesses
    (e.g. a house) which can become the property of the
    lender if the borrower is unable to repay the loan.
  b A loan which is not covered by insurance. The
    insurance company would repay the loan if the
    borrower died.
3 a (i) £7429.80 b (i) £8871 c (i) £7953.60
4 a £1 b 96p, 89p
5 a (i) £67.78 (ii) £1.48
  b (i) £79.74 (ii) £11.96 (iii) £35.3%

Exercise 29

7 a £12.50 b £237.50 c £5.23
8 a £237.50 b £4.51 c Mrs Magee should change.
9 a (i) £203.50 (ii) 33.9% b (i) £104.41 (ii) 26%
  c The catalogue gives the better deal.

Exercise 30

1 a (i) £1.52; £1.34 (ii) 500 g size b 250 g size
2 a (i) 4.49 g (ii) 4.65 g b 600 g jar
  c possible wastage with larger sizes/sell by date/taste
3 a E15: £5.49; E10: £5.68½; E3: £6.10
  b (i) E15: more economical for a family
    (ii) E3: less to pay out of pension, not heavy to carry
    (iii) E10: good compromise, but an argument could
      be made for any size
    (iv) E3: small size for limited storage space.
5 a (i) A, £14.51 (ii) £14.01

Exercise 31

1 a 235.8 Sch b 24 920 Pts c $161.16
2 a £19.90 b £5.03 c £2.44
3 £16 4 a £57.09 b 28.5p
5 a 1678.80 DM b 6023.78 Sch
6 a 1 370 250 Lira b 320 250 Lira c £70.93 d £7.68
7 a 945 Fr b £94.03 c £5.97

Exercise 32

1 a £1 b £1 c £7.97 2 a £3 b £6
3 a 304.98 DM; 243.34 SFr; 2338.35 Sch
  b (i) £570 (ii) £5.70 c £226.43

Exercise 33

1 6.15 am; 0615 2 Ten past eleven in the evening, 2310
3 Half past nine in the morning, 9.30 am
4 Twenty to two in the morning; 0140
5 Ten to two in the afternoon, 1.50 pm
6 Twenty past ten in the evening; 10.20 pm
7 9.50 pm; 2150
8 Quarter to eleven in the morning; 1045
9 12.25 am; 0025
10 Four minutes to five in the afternoon; 4.56 pm.

Exercise 34

1 14 minutes 2 0655 3 Nine thirty 4 0942
5 a 1623 b 28 minutes 6 1 hour 45 minutes
7 a 1945 b 2030

Exercise 35

1 a 20 minutes b 4 c 39 minutes
  d 1451, 1543 (or 1551)
2 a 4 b 0130 Belgian time c 25 minutes d 0930
  e 1 hour f 1335

Exercise 36

1 £67 500, £11 000 2 £95 400, £14 600
3 £59 375, £1625 4 £113 600, £33 400
5 £160 000, £100 000

Exercise 37

1 £51 000 2 £78 750 3 £56 250
4 less than £16 667 5 £83 250

Exercise 38

1 a (i) £138.90 (ii) £33 336
  b (i) £103.68 (ii) £37 324.80

c (i) £260.78 (ii) £78 234
d (i) £304.20 (ii) £54 756
2 a £240.45 b £72 134(5)
3 a £28 025 b £242.14 c £87 170.40 d £2975
  e £90 145.40
4 a (i) £524.70 (ii) £157 410
  b (i) £433.44 (ii) £104 025.60
  c (i) £419.58 (ii) £75 524.40
  d (i) £583.95 (ii) £210 222

Exercise 39

1 £6240 2 £4440 3 £36 4 £175 p.c.m.
5 a £5640 b (i) £6100 (ii) £5797.44
  c Yes – they will own their own house eventually at not
    too much extra cost. Also because rent would rise in
    due course.
6 a Repairs and maintenance are usually carried out by
    the landlord.
    Freedom to change jobs and move around the country
    or abroad.
  b You own the property which, over a period of time,
    will appreciate in value. Freedom to do what you wish
    with the house and garden.

Exercise 40

1 a £577.24 b £57.72 2 £52.80 3 £1005.44
4 £3333.35 5 a £12.61 b 3.49%
6 £224 7 £7.20 per month

Exercise 41

1 £277.38 2 £83.77, £83.78 3 £67.51, £6.75
4 19.9p 5 a £101.75 b £67.54 c £85.65 d £93.44
6 £191 7 Yes

Exercise 42

1 a 66 865 b 42 735 c 68 138 d 94 238
2 a £87.64 b £126.15 c £91.92 d £97.06
  e £30.87
3 a £21.38 4 a 902 b 65 210

Exercise 43

1 a £168.30 b £118.15 c £164.72 d £190.19
2 388.5 3 £41.49
4 a 148 b 155.4 c £66.82 d £76.22
5 a (i) 5839 (ii) 211
  b (i) 221.55 (ii) £88.18 (iii) £96.88

Exercise 44

1 £32.21 2 £31.41 3 a £29.29 b £34.42
4 a £17.05 b 169 c £7.44 d £24.49 e £4.29
  f £28.78

Exercise 45

1 2.01 litres
2 a 1 litre b 2.5 litres c (5 + 2.5) litres d 5 litres
  e 2 × 1 litres f (10 + 1) litres
3 a £7.98 b £15.97 c £11.98
4 a 3.28 litres, 2.72 litres, 2.2 litres b 8.2 litres
  c £25.95 d £20.97

Exercise 46

1 a 4 rolls b 4 rolls c 2 rolls d 4 rolls e 3 rolls
2 £13.50 3 a 7 rolls b £43.40
4 a 5 rolls b £37.85

Exercise 47

1 a 6 rolls b 8 rolls c 6 rolls d 5 rolls
2 a 9 rolls b 6 rolls c 5 rolls 3 20 rolls
4 a £146 b £59.80 c £24.84 d £84.64

Exercise 48

1 117 tiles 2 a 120 b 112
3 a 486 tiles b 15.4 cm c 442
4 a £28.40 b £113.40 c £71.00 d £10.40
  e £201.00 f £138.60
5 a 224 b (i) 12 boxes (ii) £201.60

Exercise 49

1 a 5.4 metres of 4 m width b 4.9 metres of 4 m width
  c 3.4 metres of 3 m width d 2.5 metres of 4 m width
  e 7.4 metres of 3 m width and 7.4 metres of 4 m width
  f 12.4 metres of 4 m width
2 a £302.18 b £117.40 c £77.42 d £29.90
  e £916.34 f £620
3 a £346.45 b £67.28 4 £606.94
5 a (i) £38.31 (ii) £263.78 b £5.20 c £313.99
  d £332.28 e £24.92

Exercise 50

1 a £209.24 b £69.75
2 a (i) A: size 1, 25; size 3, 50. B: size 2, 66; size 3, 18.
      C: size 2, 50; size 3, 50
  (ii) A: £144.25; B: £149.76; C: £164
3 a (i) 2.4 kg (ii) £9.62 b (i) 96 (ii) £44
4 a (i) 48 (ii) 3 litres b £42.15

Exercise 51

1 £56 2 £84 3 a £36 b £114
4 a £375 b £31.25 5 a £179.28 b £14.94

Exercise 52

1 £67.64 2 £137.92 3 £14.78 4 £36; 23.1%
5 £290.75; £30.28

Exercise 53

1 a £19.95 b £30.35 c £3.49 d £120.80 e £40.46
2 £5042 3 £10.80 4 £77.52; £18604.80
5 Insurance is taken out against an event which *might*
  happen e.g. an accident; assurance is taken out against
  an event which *will* happen, i.e. death.
6 It increases because you are likely to have a shorter time
  in which to pay the premiums.

Exercise 54

1 a £457 b £702 c £892 d £1589
2 a £631.80 b £568.62 c £284.31
3 £155.38 4 a £960 b £864 c £604.80
5 £262.40 6 £192.60

Exercise 55

1 a £25 b £54 c £1477.33 d £222 e £45
2 a £60 b £34 c £1.01 d £1200 e £85
3 £180 4 £215.20 5 £7180 6 £2.66

Exercise 56

1 50% 2 21% 3 $10\frac{1}{2}$% 4 9.6% 5 6.5%

Exercise 57

1 £2304 2 £630 3 £328.05 4 £2414.45

Exercise 58

(D = discrete, C = continuous, QL = qualitative,
QN = quantitative)
1 QN,D 2 QN,C 3 QN,C 4 QN,D 5 QL
6 QN,C 7 QN,D 8 QN,D 9 QN,D 10 QL
11 QN,C 12 QL

Exercise 59

1 a total population
  b people asked at a particular time in a particular
    shopping street
  c Specify the intended catchment area of the new
    shopping centre. Use the electoral role for this area to
    select a random sample.
2 a all the students in the College
  b She only asks people in the common room, and she
    only asks girls.
  c Find students' enrolment numbers, and select 50
    random numbers in the range of numbers given.
3 a all cars in the UK
  b The police only check cars between 5pm and 6pm.
  c Sample at different times of day on different days of
    the week on different roads.
4 a the soil in his garden
  b only one sample. Soil beyond the range of his throw
    cannot be sampled.
  c Draw a plan of the garden and divide into numbered
    squares. Use random sampling. Choose several
    squares and take samples from these areas.
5 a All teenagers
  b Carol only asked people entering a tobacconist's. The
    sample is biased towards smokers. Carol should only
    ask teenagers.
  c Obtain a list of pupils in the school; select a random
    number from the list.
6 a any homes in the telephone area
  b The salesgirl only asks those with a telephone. The
    sample includes only those who are at home when the
    salesgirl rings.
  c The salesgirl needs to know the number of homes in
    the area. She needs to use the rating register (used in
    water rates) to identify homes; use their reference
    numbers to obtain a random sample.
7 a all cars owned by people in an area
  b Peter only asked people at home during one
    afternoon. The responses came from a small part of
    the town.
  c Use the DVLC in Swansea to produce a list of cars
    registered as being owned by people in the required
    area. Use random numbers to obtain a sample of the
    required size.

8 a all people who travel to work in John's area
 b People at the station would be biased towards those using a train to go to work.
 c Use the electoral role and ask a random sample from this how they travel to work. If they do not work, delete them from the sample. Continue with a random selection until you have a suitable number who do work.

Exercise 61

1 1:5 times. 2:4 times. 3:4 times. 4:5 times. 5:6 times. 6:4 times
2 0–10: 9 times. 11–20: 7 times. 21–30: 10 times. 31–40: 6 times. 41–50: 5 times. 51–60: 6 times.
3 0–4: 9 times. 5–9: 11 times. 10–14: 8 times. 15–19: 11 times. 20–24: 5 times. 25–29: 8 times. 30–34: 12 times. 35–39: 11 times. 40–44: 5 times.
4 A: 19. E: 29. I: 12. O:26. U:4.

Exercise 67

1 a 160 **b** 120 **c** 220 **d** 140
2 a 325 **b** 290 **c** 210, 87.8%
3 a 240 **b** 210 **c** 210 **d** 120
4 Conservative by 960 votes
5 40 476, 20 238, 85 714
6 a 298; **b** 46; **c** 306
7 a 72.2% **b** 458 300 acres
8 a 75 **b** 65 **c** 40

9

Cost of lunch (£)	Frequency
0–1.99	2
2.00–3.99	4
4.00–5.99	12
6.00–7.99	16
8.00–9.99	16
10.00–11.99	8
12.00–13.99	6
14.00–15.99	10

10

Distance (m)	Number of lobsters
0–5	30
5–10	10
10–15	20
15–20	30
20–25	30
25–30	30
30–35	40
35–40	10

11 140, 56%

12

Number of peas	Frequency
100–199	15
200–249	20
250–299	30
300–349	30
350–399	22
400–499	20

13

Ages	Frequency
11–13	160
13–14	160
14–15	190
15–16	190
16–18	140
Total:	840

14 81
15 56

Exercise 68

1 Missed: 1988, 1989. No scale on sales.
2 Different widths of bars. Non–linear vertical scale.
3 Vertical scale does not start at origin. No time interval on horizontal scale.
4 Different width bars. No vertical scale. No indication of meaning of 'goodness'. Food not identified.
5 Other drinks not identified. Different width bars. No vertical scale. Probably picked one constituent of milk not in other drinks.
6 No vertical scale. Different width bars. 'Typical other drink' meaningless.
7 No vertical scale. 'Pollution' not defined.

Exercise 69

1 £130   **2** 69.06 mph   **3** 68.27 mph   **4** 38.58
5 The ages are given to the last completed year. A person aged 17 is between 17 years 0 days and 17 years 364 days old. Therefore, the true mean is probably above $28\frac{3}{4}$, hence John is probably correct.
6 a £1710 **b** £1820 **c** £182   **7** 1.848 m
8 46.5 people

Exercise 70

1 4   **2** 44p   **3** 8–10m   **4** 110–114   **5** red

Exercise 71

1 34   **2** 12.5   **3** 4   **4** 50p   **5** 6.35 kg

Exercise 72

1 1   **2** 6   **3** 26.60 min (= 20 min 36 s)
4 7.79 words   **5** £167.64

Exercise 73

1 b 6-point moving average
 c Th–F, $28\frac{5}{6}$ F–S, $29\frac{1}{6}$ S–Su, 29 Su–M, $28\frac{5}{6}$ M–T, $29\frac{1}{6}$ T–Th, $29\frac{2}{3}$ Th–F, $30\frac{1}{6}$
 e Steady at beginning of the two weeks, then an improvement in trade
2 b 3-point moving average
 c Summer, 43 Autumn, $42\frac{1}{3}$ Spring, 45 Summer, $45\frac{1}{3}$ Autumn, $46\frac{1}{3}$ Spring, $46\frac{2}{3}$ Summer, $47\frac{1}{3}$
 e Gradual increases in absences. You need to know the total number of students in each year.
3 b 4-point moving average
 c W–F, 450 F–S, 442.5 S–T, 435, T–W, 427.5 W–F, 427.5 F–S, 420 S–T, 422.5 T–W, 415 W–F, 407.5

e The graph shows a gradual drop in the attendance.
4 b 4-point moving average
 c Su–A, 75.75 A–W, 76.5 W–S, 77.75 S–S, 76.5
   S–A, 77.75
 e Rise in late 1989 to 1990. Fall in early 1990. Rises
   again in middle of 1990.
5 b 5-point moving average
 c Th, 13.8 F, 13.2 S, 14.2 M, 14.6 T, 15
   Th, 14.4 F, 15.2 S, 14.8 M, 15 T, 15.2 Th, 15.6
 e There is an increase in the number of sausages sold.
6 b 7-point moving average.
 c Th, $17\frac{1}{7}$ F, $17\frac{2}{7}$ S, $17\frac{1}{7}$ S, $17\frac{2}{7}$ M, $17\frac{3}{7}$ T, $17\frac{4}{7}$
   W, 18 Th, $18\frac{2}{7}$ F, $18\frac{3}{7}$ S, $18\frac{2}{7}$ S, $18\frac{1}{7}$ M, 18 T, $17\frac{6}{7}$
   W, $17\frac{5}{7}$ Th, $17\frac{5}{7}$
 e The average increases before decreasing.

Exercise 74

1 John 8.6, Sarah 5.631.
2 Jane 54%, Helen 67%. Helen gets the better result.
3 £1.50 per litre with wine D. £1.31 per litre without D.
4 8.82%

Exercise 75

1 a 32 b 28 c 35 d 7 e 8
2 a 71 cm b 66 cm c 79 cm d 13 cm e (i) 4 (ii) 45
3 a £12700 b £11200 c £13800 d £2600
 e £14100
4 a 58 b 44 c 71 d 27 e (i) 84 (ii) 83 (iii) 80
5 b (i) 190 g (ii) 74 c (i) 34 kg (ii) 21 kg

Exercise 76

1 34 2 11 3 86 4 8.9

Exercise 77

1 31, 15.7 2 15, 1.14
3 459, 5.44. Minimum at 1 lb. 4 16, 2.30.
5 a Town 23 min, 4.9 min; by-pass 23 min, 2.14 min
 b The by-pass. There is less variation in the times, i.e. it
   is more reliable.
6 69.8; 11.9. 58.8; 8.62.
 Saturday's cars are travelling slower, and at a more
 uniform speed.

Exercise 78

1 $\frac{5}{8}$ 2 0 3 $\frac{1}{3}$ 4 0

Exercise 79

1 $\frac{3}{10}$ 2 $\frac{1}{6}$ 3 $\frac{1}{2}$ 4 a $\frac{1}{2}$ b $\frac{1}{13}$ c $\frac{1}{52}$
5 $\frac{6}{11}$ 6 a $\frac{7}{20}$ b $\frac{1}{10}$ 7 $\frac{17}{20}$

Exercise 80

1 a $\frac{1}{9}$ b $\frac{1}{18}$ c $\frac{1}{18}$ 2 a $\frac{1}{4}$ b $\frac{1}{2}$ 3 a $\frac{1}{4}$ b $\frac{3}{16}$
4 a $\frac{1}{6}$ b $\frac{0}{0}$ 5 a $\frac{1}{3}$ b $\frac{1}{8}$

Exercise 81

1 $\frac{2}{13}$ 2 a $\frac{3}{10}$ b $\frac{3}{20}$ 3 $\frac{4}{9}$ 4 $\frac{16}{25}$ 5 $\frac{1}{2}$ 6 $\frac{4}{11}$ 7 $\frac{7}{10}$

Exercise 82

1 a $\frac{9}{49}$ b $\frac{16}{49}$ c $\frac{24}{49}$ 2 $\frac{17}{30}$ 3 $\frac{53}{100}$

4 a $\frac{1}{36}$ b $\frac{5}{18}$ 5 a $\frac{1}{4}$ b $\frac{31}{50}$ 6 a $\frac{4}{13}$ b $\frac{48}{91}$
7 b $\frac{2}{9}$ c $\frac{5}{9}$ d $\frac{1}{12}$

Exercise 83

1 a 24 b 4, 49 c 17 or 41 d 4 e 41
2 a −243, 729, multiply preceding term by −3
 b 9, 4, subtract 5 from preceding term
 c 125, 216, cubes of the positive integers
 d 26, 37, add 1 to the preceding square of an integer
   (or add successive odd numbers to successive terms)
 e 13, 21, add the two preceding terms 3 16
4 a 601300
 b six hundred and one thousand three hundred
5 17 6 a 24 b 1
7 three thousand one hundred and eighty three
8 a −4 b 5 c 4 d 3 e 4
9 a $\sqrt{19}$, $\sqrt{7}$ b $\sqrt{16}$, $\sqrt{81}$ c $\sqrt{16}$ d $\sqrt{49}$, $\sqrt{81}$
 e $\sqrt{49}$
10 62 11 17 12 417
13 a −8, 4, 19 b 4, 19 c $\sqrt{5}$, π d 19 e −8, 4
14 18043 15 21, 28 16 one hundred and thirty
 thousand four hundred and two
17 1, 2, 3, 4, 6, 8, 12, 24
18 a 2, 19 b Any two of $\sqrt{16}$, 16, 9 c $\sqrt{101}$
 d 15 or 2 e 45 or 9
19 9.34, 3^2, 2.9^2, $\sqrt{65}$, $\sqrt{63}$
20 a 10.5 b 0.105 c 0.03 d 0.006 e 67030 f 0
 g 4570 h ∞ i 120

Exercise 84

1 a 2 b −4 c −10 d 4 e 6 f 40 g 0 h −49
2 a −1 b 17 c −13 d 20 e −3 f 0 g 50 h 36
3 a 12 b 637 c 5295 d −1186 e −3462 f 1766

Exercise 85

1 a −18 b −3 c 28 d $1\frac{3}{4}$ e −15 f −27 g 8
 h 36 i −1 j −10
2 a $-5\frac{1}{4}$ b 3431 c −390 d 168 e 1.2 f $-\frac{1}{2}$
3 a $-\frac{1}{4}$ b $-\frac{3}{4}$ c $-\frac{3}{4}$

Exercise 86

1 a 13 b −16 c 44 d −14
2 a 21.1°C b −2.2°C c −20°C d −23.3°C
3 a 36 b −125 c −9.96

Exercise 87

1 1, 4, 9, 16, 25, 36, 49, 64, 81, 100
2 1, 8, 27, 64, 125, 216
3 a 32 b 729 c 25 d $\frac{1}{8}$ e−64
4 a 32 b 2187 c 4 d 9
5 a 5 b 11 c 25 d 15 e 14
6 a (i) 3 (ii) −4 (iii) 9 b 2 c (i) 3 (ii) 5
7 a 6 b $\frac{4}{3}$ c 8 d $\frac{4}{15}$ e $\frac{2}{19}$
8 a $\frac{3}{4}$ b $1\frac{1}{3}$ 9 1 + 3 + 5 + 7 + 9
10 a $\frac{1}{121}$ b $\frac{1}{12}$ c $\frac{9}{8}$ d −27 e 1
11 8 12 27 13 81 14 125
15 a 32 b 81 c $\frac{1}{16}$ d −8 e 0.25
16 a 2 b 3 c 6 d $\frac{1}{2}$ e −4
17 3 18 a $\frac{1}{4}$ b 9 c 2 d $\frac{2}{7}$ e $\frac{2}{17}$ 19 100

Exercise 88

1 a 1×15 b 1×27 c 1×36 d 1×64
$\quad$ 3×5 $\quad$ 3×9 $\quad$ 2×18 $\quad$ 2×32
$\quad$ $\quad$ $\quad$ $\quad$ $\quad$ 3×12 $\quad$ 4×16
$\quad$ $\quad$ $\quad$ $\quad$ $\quad$ 4×9 $\quad$ 8×8
$\quad$ $\quad$ $\quad$ $\quad$ $\quad$ 6×6

$\quad$ e 1×100
$\quad\quad$ 2×50
$\quad\quad$ 4×25
$\quad\quad$ 5×20
$\quad\quad$ 10×10

2 a (i) 1, 2, 3, 4, 6, 8, 12, 16, 24, 48
$\quad$ (ii) 1, 2, 3, 4, 6, 8, 9, 12, 18, 24, 36, 72
$\quad$ b 1, 2, 3, 4, 6, 8, 12, 24
3 64, 144, 116, 2620
4 a 5, 10, 20, 25 b 5, 10, 25 c 5, 10 d 5, 25
$\quad$ e 5, 10, 20
5 a 240, 87, 96, 255 b 240, 96 c 240, 255

Exercise 89

1 a $2^2 \times 3 \times 5$ b $3^2 \times 5 \times 7$ c $3^2 \times 5^2$
$\quad$ d $2^2 \times 3^2 \times 5 \times 7^2$ e $2 \times 7 \times 11^2$
2 a (i) $2^2 \times 3^2$; $2 \times 3^2 \times 5$ (ii) 2×3^2
$\quad$ b (i) $2^3 \times 3 \times 7$; $2^3 \times 3^2$ (ii) $2^3 \times 3$
$\quad$ c (i) $2^3 \times 3 \times 13$; $2^2 \times 5 \times 13$ (ii) $2^2 \times 13$
$\quad$ d (i) $2 \times 7 \times 11$; $2^2 \times 5 \times 11$ (ii) 2×11
$\quad$ e (i) $2 \times 3^3 \times 17$, $2^2 \times 3^2 \times 13$ (ii) 2×3^2
$\quad$ f (i) $3 \times 5 \times 7^2$, $2 \times 3 \times 5^2 \times 7$ (ii) $3 \times 5 \times 7$
3 a (i) 14, 28, 42 (ii) 30, 60, 90 (iii) 24, 48, 72
$\quad$ (iv) 30, 60, 90
$\quad$ b (i) multiples of 14 (ii) multiples of 30
$\quad\quad$ (iii) multiples of 24 (iv) multiples of 30

Exercise 90

1 a 32 b 15 c 28 d 14 e 115 f 22
2 a 150 b 144 c 156 d 12600 e 252 f 144
3 a 12 b $\frac{13}{18}$ 4 a 225 b $\frac{4}{13}$ 5 72
6 a 5 b 4 c 40 metres 7 1.24 am
8 a 252 b Artemis, 14; Khons, 9; Soma, 7
9 a 250 b 15, 28 10 18

Exercise 91

1 9.74 2 0.36 3 147.5 4 29 5 0.53 6 4.20
7 1250 8 0.004 9 270 10 460.0

Exercise 92

1 a 3500 b 240 c 1200 d 300 e 250 f 10 g 40
$\quad$ h 5
2 a 9 b 0.06 c 24
3 Incorrect calculations are c, d, e, f, h, j.

Exercise 93

1 a 7.9×10^5 b 4.6×10^{-3} c 3.13×10^4
$\quad$ d 9.41×10^{-5} e 1×10^5 f 2.82×10^{-4}
$\quad$ g 1.57×10^4 h 4.7×10^7 i 3.4×10^{-2}
$\quad$ j 2.7×10^{-6} k 5×10^{-1} l 1.4×10^{-7}
2 a 7530 b 240 c 0.0019 d 0.837 e 0.0451
$\quad$ f 40420 g 3192000 h 0.000974 i 0.0000068
3 a 6×10^4 b $5.18 \times 10^\circ$ c 4×10^1
$\quad$ d 1.35×10^4 e 2.174×10^3 f 4×10^{-2}

$\quad$ g 4.32×10^{-5} h 2.84×10^4 i -5×10^{-4}
$\quad$ j 6×10^1
4 390
5 8 minutes 20 seconds

Exercise 94

1 11.8 2 73700 3 2.70 4 0.137 5 0.874
6 29.0 7 21.4 8 882.65 9 0.818 10 171.39
11 7.564 12 a Leo's b shoes 13 Lise
14 d 5.828 15 c 2.2361
16 (i) 1.7321 (ii) 3.1623
17 2.2361; 1.7321; 3.1632 The method used in questions
$\quad$ 15 and 16 calculates the square root of a given number.

Exercise 95

1 a 8 b 8 c 18 d 9 e 7 f 9 g 5 h 16
2 a $\frac{2}{3}$ b $\frac{3}{4}$ c $\frac{2}{9}$ d $\frac{1}{3}$ e $\frac{3}{4}$ f $\frac{2}{3}$ g $\frac{6}{13}$ h $\frac{2}{3}$
3 a $\frac{20}{9}$ b $\frac{31}{6}$ c $\frac{59}{12}$ d $\frac{111}{10}$ e $\frac{35}{4}$ f $\frac{38}{3}$
$\quad$ g $\frac{143}{20}$ h $\frac{143}{7}$
4 a $1\frac{2}{7}$ b $7\frac{1}{5}$ c $3\frac{1}{3}$ d $7\frac{5}{8}$ e $13\frac{10}{11}$ f $10\frac{7}{12}$
$\quad$ g $8\frac{4}{5}$ h $23\frac{1}{2}$

Exercise 96

1 $1\frac{3}{20}$ 2 $\frac{1}{24}$ 3 $\frac{2}{3}$ 4 $\frac{3}{4}$ 5 $\frac{1}{2}$ 6 $5\frac{1}{20}$ 7 2 8 $1\frac{11}{18}$
9 $1\frac{5}{9}$ 10 $\frac{1}{20}$ 11 $\frac{7}{8}$ 12 $1\frac{1}{4}$

Exercise 97

1 $\frac{7}{10}$ 2 $1\frac{1}{3}$ 3 12 4 $1\frac{1}{9}$ 5 $7\frac{3}{20}$ 6 $1\frac{5}{7}$ 7 8
8 $\frac{7}{8}$ 9 $1\frac{3}{17}$ 10 5 11 3

Exercise 98

1 a $\frac{8}{12} = \frac{2}{3}, 0.\dot{6}$ b $\frac{1}{4}, 0.25$ c $\frac{5}{8}, 0.625$
2 $0.\dot{1}$ 3 $0.\dot{5}$ 4 $0.7\dot{5}$ 5 1.45 6 $4.8\dot{4}$ 7 $2.8\dot{3}$
8 $7.\dot{4}$ 9 $2.\dot{8}$ 10 $0.8\dot{1}$ 11 $3.\dot{1}4285\dot{7}$

Exercise 99

1 a 20% b $12\frac{1}{2}$% c 70% d 65% e $66\frac{2}{3}$% f 36%
$\quad$ g 175% h 250%
2 a $\frac{3}{5}$ b $\frac{1}{4}$ c $\frac{1}{10}$ d $\frac{17}{20}$ e $\frac{3}{20}$ f $1\frac{3}{10}$ g $\frac{3}{8}$ h $\frac{1}{3}$

3
a	0.75	75 %
b	$\frac{1}{2}$	50 %
c	0.125	$12\frac{1}{2}$%
d	$\frac{1}{3}$ 0.3	
e	$\frac{3}{8}$	$37\frac{1}{2}$%
f	0.7	70 %
g	$\frac{7}{20}$ 0.35	
h	$\frac{2}{3}$	$66\frac{2}{3}$%
i	0.6	60 %
j	$\frac{5}{8}$ 0.625	

4 a 406 b 146 5 a 200 b 75 c $\frac{3}{10}$
6 a 30 b 8 7 a £162000 b $91\frac{2}{3}$% or 91.7%

Exercise 100

1 a 3 : 2 b 10 : 1 c 12 : 1 d 1 : 4 e 4 : 3 f 3 : 2
$\quad$ g 3 : 8 h 3 : 2 i 2 : 3
2 a 8 b 2 c 1.2 d 0.6 e 3.2 f 2.6
3 42 cm 4 a 13.5 cm b 8 inches
5 a 3.5 kg b 5.25 kg 6 410 and 1025

Exercise 101

1 2 acres 2 8 cm², 6 cm², 2 cm²
3 a 5 : 4 b £150 : £120
4 a (i) 5 : 3 (ii) £7.50, £4.50 b £7.33, £4.67
5 flour, 12 oz; butter, 8 oz; sugar, 4 oz
6 1.36 kg; 0.24 kg (or 1360 g and 240 g)

Exercise 102

1 £3.18 2 £2.66 3 £1.26 4 £1.84 5 £1.79
6 4 hours 7 88p
8 a 516.5 g b 34p c no (≈ 17% more)

Exercise 103

1 £200 2 3 hours 3 a 9.6 days b 12 people
4 36 minutes 5 10 extra 6 £10625

Exercise 104

1 81p 2 £374.50
3 a 90 miles b 7 gallons c approximately 8 miles/litre
4 7.14% 5 £17.92 6 9.7 kg 7 a 8p b £42.24

Exercise 105

1 a 54 mph b 8.6 m/s (31.0 km/h) c 88 km/h
2 a 204 km b 2760 miles c 98.125 km
3 a 3 hours b 7 hours 39 minutes
c 3 hours 48 minutes
4 $7\frac{1}{2}$ miles 5 800 m
6 a 3 hours 52 minutes b 6 km/h

Exercise 106

1 24 km/h 2 a 45 mph b 12 minutes c 49.6 mph
3 a 10 miles b $3\frac{1}{2}$ mph c 3 mph
4 a 85 km b 135 km c 1 hour 40 minutes
d 75.4 km/h

Exercise 107

1 a 7.6 cm by 16 cm b 6.4 cm by 4.0 cm c 1.6 cm
2 a (i) 7 cm (ii) 14 cm × 12 cm (iii) 8 mm
b (i) 93 cm (ii) $13\frac{1}{2}$ cm (iii) 69 cm × 60 cm
3 a 9 ft 2 in b 39.7 cm
4 a 20.4 m b 68 feet
5 a 17 mm b (i) 5 ft 9 in (ii) 172.5 cm

Exercise 108

1 a 1 cm b 4.8 cm c 1.1 m d 28.8 cm
2 a 20 m b 124 m c 168 m d 9000 in (250 yd)
e 9000 in
3 a 1 : 10000 b 1 : 50000 c 1 : 25000
d 1 : 12500 e 1 : 1000000
4 a 250 m b (i) 1850 m (ii) 4.4 km (iii) 500 m
(iv) 155 m
c (i) 20 cm (ii) 7.2 cm (iii) 19 cm (iv) 42.4 cm
5 a 2 cm
b (i) 2 km (ii) 24.4 km (iii) 8.15 km (iv) 18.45 km
c (i) 54 cm (ii) 21.2 cm (iii) 1.16 mm (iv) 1.188 m
6 a 2.5 km b 6.25 km² c 3.75 km² d 4 cm²
7 a (ii) 1050 m b (i) 700 m (ii) 950 m

Exercise 109

1 a 1.232 km b 32 mm c 0.626 kg d 73.1 cl
e 16.2 ml f 12700 m g 0.0591 litres h 3400 kg
i 13.9 mg j 85 kg
2 230 g 3 200 bags
4 25000 mg 5 a 2.93 m b 293 cm
6 a 28.4 g b 28400 mg c 2.84 × 10⁴ 7 1.2 litres
8 27.7 litres

Exercise 110

1 a 3.25 ft b 2 lb 8 oz c 5.5 yards d 27 inches
e 55 oz f 2.5 gal g 28 pints h 14 fl oz
2 40 bottles 3 $\frac{3}{4}$ pint 4 1.2 tons
5 $8\frac{3}{4}$ oz 6 a 46.5 ft b 15.5 yards
7 $6\frac{1}{8}$ (6.125) oz 8 5 bottles

Exercise 111

1 a 15 oz b 15.5 oz c 14.6 oz
2 a 454 g b 5.3 oz c 2.3 kg d 336 g e 0.5 oz
3 a 2 lb 3 oz b 14 oz c 3 lb 5 oz d 4 lb 10 oz
4 a 0.219 gal b 4.4 gal 5 8.8 mph
6 a 75 miles b 202 miles c 268 miles d 35 miles
e 231 miles
7 Yes (≈ 15 cm wide)

Exercise 112

1 a 22 cm b 28 m c 24 in d 32 cm e 43.8 cm
2 a 114 b 28 c 10 d 28 e 68 f 66.6
3 132 cm (1.32 m)

Exercise 113

1 a 12 cm² b 13.5 cm² c 15.75 cm² d 70 cm²
e 42.75 cm² f 151.29 cm²
2 a 60 cm² b 300 cm² c 75 cm² d 43.75 cm²
e 264.48 cm² f 240.25 cm²
3 a 120 b 378 c 31.35 d 2
4 13700 cm² 5 a 30 m² b £299.70 6 26 m²
7 a 144 cm² b 256 cm² 8 a 12 m b 50 m
9 30.25 cm²
10 a 50000 cm² b 864 square inches c 1260 mm²

Exercise 114

1 a 11.25 cm² b 32.76 m² c 41.31 in²
2 a 88 cm² b 245 mm² c 63 in² d 19.32 m²
3 a 10 m b 16 cm
4 a 16 cm b 8 cm c 12 mm d 12.24 m² e 23.65 in²
5 a 72.7 cm² b 12.7 cm² c 122.2 m² d 130.4 m²
e 33.6 cm² f 12.2 cm²

Exercise 115

1 a 120 cm² b 75.6 cm² c 451.1 m²
d 3380 cm² (0.338 m²) e 1978 mm² (19.78 cm²)
2 a 2.4 cm b 0.33 m c 13.4 cm² d 4 m e 2.4 yd
3 a 94.6 b 54.9 c 378

Exercise 116

1 a 5 cm, 31.4 cm b 8.6 in, 4.3 in c 7 mm, 22.0 mm
d 16.2 cm, 8.1 cm e 1.9 m, 1.0 m
2 9.42 cm 3 3.46 cm 4 10.68 m 5 11.0 m
6 a 45.5 b 59.1 c 25.5

Exercise 117

1 a $158\,\text{cm}^2$ b $2730\,\text{in}^2$ c $145\,\text{mm}^2$ d $37.7\,\text{cm}^2$
   e $33.8\,\text{m}^2$
2 a $3\,\text{cm}$ b $7\,\text{in}$ c $6\,\text{cm}$ d $4\,\text{m}$ e $10\,\text{ft}$
3 a $2.93\,\text{cm}$ b $6.91\,\text{in}$ c $5.86\,\text{cm}$ d $3.91\,\text{m}$ e $9.77\,\text{ft}$
4 $8.0\,\text{m}$ 5 £7.35   6 $56\,\text{cm}$

Exercise 118

1 a $74.3\,\text{m}$ b $11\,000\,\text{m}^2$
2 a (i) $104\,\text{cm}$ (ii) $745\,\text{cm}^2$
   b (i) $123\,\text{cm}$ (ii) $914\,\text{cm}^2$
   c (i) $85.7\,\text{cm}$ (ii) $442\,\text{cm}^2$
   d (i) $99.4\,\text{cm}$ (ii) $547\,\text{cm}^2$
3 a $773\,\text{m}^2$ b $514\,\text{cm}^2$ c $1500\,\text{cm}^2$ d $373\,\text{m}^2$
4 a $9.82\,\text{cm}^2$, $6.28\,\text{cm}^2$, $3.53\,\text{cm}^2$ b $9.82\,\text{cm}^2$
   c The area of the large semicircle equals the sum of the
     areas of the other two semicircles.
   d $25.6\,\text{cm}^2$
5 a πR^2 b πr^2 c $\pi R^2 - \pi r^2$ d $221\,\text{cm}^2$
6 $88.0\,\text{m}^2$

Exercise 119

1 a $90\,\text{cm}^3$ b $51.3\,\text{cm}^3$ c $0.336\,\text{m}^2$ d $1450\,\text{cm}^3$
   e $1.2\,\text{m}^3$
2 $3.33\,\text{cm}$
3 a $512\,\text{cm}^3$ b $1331\,\text{cm}^3$ c $68.9\,\text{cm}^3$
4 $4.0\,\text{cm}$
5 a $27\,\text{in}^3$ b $3\,\text{in}$ 6 a $96\,\text{cu ft}$ b $2.27\,\text{ft}$
7 a $788\,\text{cm}^3$ b $136\,\text{cm}^3$ c $657\,\text{cm}^3$
8 $313\,\text{m}^3$
9 a $603\,\text{cm}^3$ b $3050\,\text{cm}^3$ c $1690\,\text{in}^3$ d $388\,\text{cm}^3$
10 a $1.43\,\text{cm}$ b $2.60\,\text{cm}$ c $1.72\,\text{in}$
11 $15\,800\,\text{m}^3$ 12 $79.7\,\text{cm}^3$

Exercise 120

1 $28\,\text{cm}^2$ 2 $950\,\text{cm}^2$ 3 $7.75\,\text{cm}^2$ 4 $8.5\,\text{cm}^2$ 5 2
6 4 7 $\frac{1}{2}$ 8 $3:4$

Exercise 121

1 a $192\,\text{in}^3$ b $8\,\text{in}$ 2 a $7\,\text{cm}^3$ b $35\,\text{cm}^2$
3 $15\,\text{cm}$ 4 a $3:2$ b $9:4$ c $42\,\text{cm}$
5 a $378\,\text{cm}^3$ b £2.02 (£2.03)

Exercise 122

1 $3a + 4b + 2c$ 2 $3a + 5c$ 3 $6f + 3s$

4 $7x + 5y$ 5 $2p - q$ or $q - 2p$ 6 $\dfrac{2a}{b}$

7 $5x - 8$ 8 $\dfrac{y}{4} + 6$ 9 $\dfrac{2n + 3}{8}$

10 $\dfrac{4n - 1}{2n}$

Exercise 123

1 $5x$ 2 $7n$ 3 $8a$ 4 $2y$ 5 $7a + 6b + 4c$
6 $x + 5y + 9z$ 7 $9p + 2q + r$
8 $5x - 3y + z$ 9 $-a - b + 4c$
10 $3x + 5y$

Exercise 124

1 a 11 b -1 c 13 d 6 e $\frac{1}{2}$ f 2 g -30 h -4
   i 0 j 17 k -1 l -2
2 a 1 b -9 c -5 d -6 e 0.3 f -14 g 6 h 8
   i 10 j -5 k 1 l -2
3 a 9 b -24 c 8 d $\frac{4}{3}$ e -5 f $-\frac{3}{4}$ g 38 h 7
   i 28 j 1 k 0 l 14
4 a $\frac{1}{4}$ b 1 c 2 d -4 e $2\frac{1}{2}$ f $-\frac{1}{4}$ g $8\frac{1}{2}$ h $\frac{3}{5}$ i 3
   j -1 k -8 l $4\frac{1}{2}$

Exercise 125

1 $5x$ 2 $3mn$ 3 $\dfrac{2y}{z}$ 4 abc

5 $6pqr$ 6 $\dfrac{4bc}{d}$ 7 $6xy$ 8 $\dfrac{-2p}{q}$

9 $\dfrac{9xy}{z}$ 10 a^3 11 a^2b^2 12 $3a^3$

13 $3a^3b^2$ 14 $-6a^2b^2c$ 15 $6a^2b^4c^2$ 16 $9a^2b$

17 $-18ab^3$ 18 $-2a$ 19 x^3y^3 20 $\dfrac{-9a^2b}{c}$

Exercise 126

1 x^8 2 x^6 3 x^3 4 x^{-1}

5 $x^0 = 1$ 6 $\dfrac{1}{x^2}$ 7 1 8 $\dfrac{3x^2}{2}$

9 $\dfrac{1}{2x^2}$ 10 $4x^2$ 11 $-18x^4$ 12 $-\frac{7}{2}$ (or $-3\frac{1}{2}$)

13 $10x^2y$ 14 x^2y^2 15 $\dfrac{2a^2b}{c^2}$ 16 $\dfrac{3xy}{z^2}$

17 $\dfrac{8z}{y}$ 18 $\dfrac{a}{b}$ 19 $4x$ 20 $\dfrac{xz}{2y}$

Exercise 127

1 $20x + 8y$ 2 $6a - 12b$ 3 $-6p + 12q$
4 $5x - 15y + 10z$ 5 $-a + b + c$
6 $3x^2 - xy + 2xz$ 7 $-a^2 - ab + ac$
8 $2x^3 + 3x^2 + 2x$ 9 $4y^3 - 12y^2 + 4y$
10 $-5x + 10x^2 - 5x^3$ 11 $5x + 2y$ 12 $y^2 + y - 12$
13 $9a + 17b$ 14 $8p - 15q$ 15 $x^2 - 11xy$
16 $-x^3y + x^3y^2 - 2x^2y^3$ 17 $5p - 4q + 7r$
18 $-2x + 4y - 4z$ 19 $3x + 5y$
20 $5x^2y + xy^2$

Exercise 128

1 $4(x + 3y)$ 2 $3(p - 2q)$ 3 $5(a + 2)$ 4 $5(2b - 1)$
5 $7(2m - 3n)$ 6 $4(3s + 5t)$ 7 $x(y + z)$ 8 $x(y + x)$
9 $y(y - 2)$ 10 $2y(y - 2)$ 11 $3x(2y + 1)$
12 $5x(1 - 2x)$ 13 $2(a + 4b - 2c)$ 14 $3(3x - y - 2z)$
15 $5(3x - y + 2)$ 16 $7(p - 2q + r)$ 17 $2a(a + 2b - 4)$
18 $3p(p + 2q - 3)$ 19 $xy(x + z + y)$ 20 $4pq(r - 3p)$

Exercise 129

1 $\dfrac{3x}{5}$ 2 a 3 $\dfrac{x}{4}$ 4 $\dfrac{7x}{4}$ 5 $\dfrac{7x}{12}$ 6 $\dfrac{x}{15}$ 7 $\dfrac{13x}{12}$ 8 $\dfrac{a}{4}$

9 $\dfrac{y}{3}$ 10 $\dfrac{y}{15}$ 11 $\dfrac{3}{x}$ 12 $\dfrac{1}{10x}$ 13 $\dfrac{3}{2y}$ 14 $\dfrac{1}{2y}$ 15 $\dfrac{3}{20a}$

16 $\dfrac{2x+7}{2x}$ 17 $\dfrac{x+8}{12}$ 18 $\dfrac{12x+1}{10}$

19 $\dfrac{2x+1}{2}$ 20 $\dfrac{3(x-5)}{10}$

Exercise 130

1 $n+5=12$ 2 $n-7=13$ 3 $2n+4=10$
4 $7n=21$ 5 $5n-6=29$ 6 $\frac{1}{2}n+10=22$
7 $2n-3=3$ 8 $\frac{1}{4}n-3=3$ 9 $2(w+2w)=18$
10 $2(w+w+4)=28$

Exercise 131

1 7 2 20 3 3 4 3 5 7 6 24 7 3 8 24
9 width = 3 10 length = 9 11 4 12 3 13 6
14 8 15 3 16 6 17 10 18 8 19 9 20 20

Exercise 132

1 4 2 3 3 2 4 $\frac{1}{3}$ 5 9 6 2 7 0 8 10
9 4 10 1 11 3 12 −5.2

Exercise 133

1 a (i) $(x+25)$ pence (ii) $12x$ pence (iii) $10(x+25)$
  b (i) $12x+10(x+25)=690$ (ii) $x=20$p (iii) 45p
2 a £$9x$ b £$5(20-x)$ c £$(4x+100)$
  d $4x+100=148$ e 12 matches
3 a £$(x-1.8)$ b $(18x-18)$ c £8.60, £6.80
4 55p 5 50, 47, 42 6 68 kg 7 £1.50

Exercise 134

1 a (i) 30 m/s (ii) 36 m/s b (i) 67.5 mph (ii) 81 mph
2 5 3 a 10 b 25 c −10 4 754 cm²
5 a 7.2 b 4.2 6 £1918.44 7 a 24 cm²
b 41.44 cm² 8 7 seconds

Exercise 135

1 a $B=\dfrac{P}{2}-L=\dfrac{P-2L}{2}$ b $h=\dfrac{S}{2\pi r}$

c $r=\sqrt{\dfrac{A}{\pi}}$ d $t=\dfrac{2s}{u+y}$ e $a=\dfrac{v-u}{t}$

2 a $x=\dfrac{y-c}{m}$ b (i) 2 (ii) 7

3 a 12 b $a=3M-b-c$ c 4
4 a $h=\dfrac{2A}{a+b}$, 5 b $a=\dfrac{2A}{h}-b=\dfrac{2A-bh}{h}$, 8

5 b 43 c $d=\dfrac{N-a}{n-1}$
  d (i) 3 (ii) 2, 5, 8, 11, 14
6 a £113 b $C=20\left(\dfrac{D-A}{N}\right)$ c £340

Exercise 136

1 a $5x+3y$ b $9p-2q-2r$ 2 a 4 b −4 c 8
3 a $12ab$ b $8a^2b$ c x^2 d $10a^6$ e 1 f $12p^2q^3r^3$

4 a x^3 b $\dfrac{1}{x^3}$ c 1 d $\dfrac{3a^3}{2}$ e $\dfrac{6}{x}$

5 a $9x-4$ b $9x+3$ c $-2x$
6 a $3a(2b+3c+1)$ b $xy(z-y)$
  c $2abc(c-3a+2b)$

7 a $\dfrac{x}{12}$ b $\dfrac{1}{2x}$ c $\dfrac{5x+1}{12}$

8 a 8 b 30 c 2 9 a 210 b 1275

10 a 2.5 b $x=\dfrac{2}{y-3}$ c (i) $-4, \frac{1}{2}$

Exercise 137

1 a 44° b 48° c 124° d 54°
2 a acute b reflex c obtuse d obtuse e reflex
  f acute
3 a 145°, 35° b 125°, 55°, 95° c 60°, 90°, 30°

Exercise 138

1 $a=52°$, $b=128°$, $c=128°$, $d=52°$
2 $p=65°$, $q=115°$, $r=65°$, $s=115°$, $t=65°$
3 $r=32°$, $s=148°$, $t=32°$, $u=148°$, $v=32°$,
  $w=148°$
4 $l=74°$, $m=74°$, $n=74°$, $p=106°$, $q=74°$
5 $a=42°$, $b=42°$, $c=48°$, $d=48°$
6 $a=120°$, $b=120°$, $c=30°$, $d=30°$

Exercise 139

1 a 1 b 3 c 5 d 2
2

A H X Z D

3 a   b   c

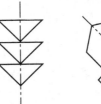

Exercise 140

1 H, 2; I, 2; N, 2; O, ∞; S, 2; X, 4; Z, 2.
2 a Shapes b, c and d b 3; 5; 2.
3 a (i) 0 (ii) 5 b (i) 0 (ii) 3 c (i) 0 (ii) 6
  d (i) 2 (ii) 2 e (i) 4 (ii) 4 f (i) 1 (ii) 1

Exercise 141

1 20°, 80° 2 A = B = $62\frac{1}{2}°$ 3 72°, 36°, 108°, 36°
4 70°, 40°, 140° f 60°, 50°

Exercise 142

1 △ ABC and △ RPQ   2 △ ABD and △ CBD or △ ABD and △ CDB
3 △ ABD and △ ACD
4 Not necessarily congruent
5 Not congruent   6 △ LMN and △ QOP

Exercise 143

1 3.2cm, 10.4cm
2 OP = 4.5 cm, OQ = 5.4 cm, PQ = 6.3 cm
3 a 3 b 8.1cm   4 a 4.5 cm b 5.0 cm c 6.25 cm

Exercise 144

1 Two pairs of adjacent sides are equal.
 The diagonals bisect each other at right angles.
 The diagonals bisect the angles between the equal pairs of sides.
2 ∠P = 136° = ∠ Q; ∠ R = 44°   3 45°
4 a 32° b 58°   5 a 57° b 81° c 28°

Exercise 145

1 a 45° b 135°   2 a 36° b 144°
3 18   4 9   5 540°   6 1440°
7 a 11 b 15

Exercise 146

1 a 53° b 118°   2   B = 52°,   D = 28°,   E = 100°
3 a 71° b 48°   4 a 32° b 106° c 42°
5 a 41° b 41°   6 a 65° b 25° c 70°
7   OBC = 21°   8 a 37° b 37° c 90°

Exercise 147

1 6.5 cm   2 3.6 cm   3 88.4°   4 61.3°   5 4.9 cm

Exercise 148

2 a 52° b 4.1cm   5 148 m   6 375 m   7 26.8 m

Exercise 149

1 b 37°, 53°
2 a 13cm b 2.67mm c 2.8in
3 no   4 a 7.67in b 8.94cm, 7.35cm c 10.3cm, 11.1cm d 13.4cm, 12.7cm, 12.0cm
5 24.0cm   6 129ft   7 145.5cm   8 134.5cm
9 170cm

Exercise 150

1 6.09cm   2 28.8°   3 6.36cm   4 11.0cm, 3.59cm
5 a 7.26cm b 26.9cm²   6 36.0°
7 a 5.49cm b 5.85cm

Exercise 151

1 a 4.84cm b 7.21cm   2 a 7.28cm b 9.10cm
3 a 22.2° b 16.1°   4 a 33.6° b 51.1°
5 18.0cm, 5.27cm   6 p = 13.8cm, q = 2.16cm
7 11.3in, 33.7°   8 34.4°

Exercise 152

1 13.1°   2 7.84cm   3 7.77cm   4 8.38cm, 5.50cm

5 1.44m   6 a 86.4cm b 85.1cm   7 48.0°
8 37.7°   9 a 118cm b 38.3°

Exercise 153

1 a 090° b 135° c 315° d 330° e 195°
2 a 113° b 074° c 312° d 237°   3 012°   4 241°
5 144°   6 216°   7 a 126° b 306°   8 2.30 miles

Exercise 154

1 229ft   2 15.6°   3 736ft   4 449mph   5 7.70m
6 12.5°   7 a 87.5° b 89.0°   8 5.00°   9 49.8m

Exercise 155

1 34.4in   2 7.92m   3 7.46ft, 23.7ft   4 2.54m
5 5.51in   6 56.3°   7 a 5.79cm b 46.4°
8 a 65.1m² b 782m³ c 35.4°
9 6.65°   10 25.3°   11 a 9.95m b 37.5°   12 21.5°

Exercise 156

1 a 1040 b 1927, 1951 and 1974 c 1940 d 1960
2 a £75 b £200
 c There are more points plotted around 1984 which fit the straight line
 d 1976
3 a Sarah b 0.6m and 1.4m c 6.5m d 3m
4 a December, October b June, June c Feb
 d South of France (Nice)

Exercise 157

1 a 10st 4lb (+ 1lb) b 10st 8lb
 c February 1990 and August 1991 d Yes
2 a 60cm² b 4.1cm² c 90cm²
3 a October and February b 14.8 and 15.8 hours
 c December to January and mid-May to mid-June
4 a 25mm b 0.7 minutes c 235mm
5 a 2.6s b −10cm c 0.45s d 1.75s

Exercise 158

1 a 22 miles,
 b Orleans, 66 miles; Tours, 153 miles; Poitiers, 219 miles; Bordeaux, 350 miles c 96km
2 a 32°F b 77°F c 28°C d −10°C
3 Milk, 6floz; sugar, 9floz; flour, 4floz; apples, 53floz; butter, 1floz
4 a 3m b 5m c 11m

Exercise 159

1 a 18 minutes b 60mph c 45 minutes d 8.45am
 e 35 miles f 20mph
2 a 8 miles b 11am c 12 noon
3 a Ali, 9.15am; Barry, 9.25am b 8mph
4 a 9.45am b Ali, 3 miles; Barry, 2 miles
5 a 88km/h b 58km/h c 80km/h

Exercise 160

7 a (i) The same coefficient of x (ii) the same gradient
 b The coefficient of x is the gradient of the line.
8 a All the lines pass through the origin.
 b A graph of the form $y = ax$ passes through the origin.

9 The point on the y-axis through which the line passes.
10 The line slopes downwards, i.e. the gradient of the line is negative.

Exercise 161

1 a 3 **b** 2 **c** −2 **d** $\frac{1}{2}$ **e** $\frac{2}{3}$ **f** −2 **g** −1 **h** −2
  i $-\frac{1}{3}$
2 a 2 **b** −5 **c** 4 **d** −1 **e** 2 **f** $\frac{1}{2}$ **g** −2 **h** −4 **i** 2
3 a 2 **b** 3 **c** $\frac{1}{2}$ **d** −1 **e** −1 **f** $\frac{3}{2}$ **g** −4 **h** $-\frac{1}{2}$
4 a $y = 2x + 1$ **b** $y = 3x - 1$ **c** $y = -x - 1$
  d $y = -\frac{1}{2}x + 2$
5 a (v) **b** (iii) **c** (vi) **d** (iv) **e** (i) **f** (ii)
6 b $L = 8W + 20$ or $y = 8x + 20$ **c** 42.4 cm **d** 20 cm
7 b $P = -0.6A + 132$ **c** (i) 99 (ii) 121

Exercise 162

1 b (i) 115 m (ii) 15.8 s (16 s) (iii) 52 m (iv) 5.2 m/s

2 a

x	1	2	3	4	5	6
A	0.43	1.73	3.90	6.93	10.83	15.59

  c (i) 5.30 (ii) 4.80

3 a

x	−5	−4	−3	−2	−1	1	2	3	4	5
y	−0.20	−0.25	−0.33	−0.50	−1	1	0.50	0.33	0.25	0.20

  c (i) 0.67 (ii) −0.29 (iii) −2.5 (iv) 1.2
4 a $L = \dfrac{A}{B}$

  b

B	10	12	14	16	18	20
L	36	30	25.7	22.5	20	18

  d (i) 23 m (ii) 19 m

5 a 36 **b**

N	4	6	8	10	12
T	9	6	4.5	3.6	3.0

  d 5.1 hours

Exercise 163

1 a $y = -20, -7, 0, 1, -4, -15$
  b It has a minimum value for y instead of a maximum value.
    The coefficient of x^2 is a negative number.
  c −0.75 **d** 2.6, −1.3
2 a $y = 15, 8, 3, 0, -1, 0, 3, 8, 15$
  c (i) 2.06 (ii) 6.2, −0.2 (iii) −1 (iv) $x = 3$
3 a $y = 0, 1.99, 3.96, 5.91, 7.84, 9.75$
  c (i) no (ii) yes **d** 2 m and 2.5 m

4 a

v	0	20	40	60	80	100
d	0	9	28	57	96	145

  c (i) 12 m (ii) 64 m **d** yes
5 a $A = 0, 144, 336, 576, 864, 1200$
  c (i) 426 cm^2 (ii) 7.6 cm **d** 584 cm^2 **e** 1.4%

Exercise 164

1 b 0 **c** 9 m **d** 6 m/s **2 b** (i) −7 (ii) 8 **3 b** 21 m/s
4 b (i) −13 (ii) 20
5 a $A = 12, 20, 32, 48, 68, 92$ **c** 5.5 cm^2 **d** 8 cm/s^2
6 b (i) 20 m (ii) 0.7 s and 3.3 s **c** (i) 10 m/s

(ii) −5 m/s. The negative gradient indicates that the stone is falling back to earth.

Exercise 165

1 $x = 2$ **2** $x = 9$ **3** $x = 7$ **4** $x = 6$
  $y = 4$ $y = 1$ $y = 5$ $y = 4\frac{1}{2}$

5 $x = 3$ **6** $x = 2$ **7** $x = 2$ **8** $x = 5$
  $y = 1$ $y = 1$ $y = 6$ $y = 0$

9 $x = 5$ **10** $x = 3$ **11** $x = 1\frac{1}{2}$ **12** $x = -2$
  $y = 6$ $y = -12$ $y = 2\frac{1}{2}$ $y = -1$

Exercise 166

1 64p, 44p **2** 7, 3 **3** 88 m^2 **4** £616 **5** 10, 3
6 24 years, 3 years **7** 12 sweets

Exercise 167

1 $x = 2$ **2** $x = 3$ **3** $x = 5$ **4** $x = 2$ **5** $x = 2\frac{1}{2}$
  $y = 3$ $y = 1$ $y = 4$ $y = 3$ $y = 2$

6 $x = 2$ **7** $x = -1$ **8** $x = -1$ **9** $x = 4$
  $y = 3$ $y = 2$ $y = -7$ $y = -2$

Exercise 168

1 $x^2 + 7x + 12$ **11** $4x^2 - 25$
2 $x^2 + x - 2$ **12** $9x^2 - 16$
3 $x^2 - 8x + 15$ **13** $x^2 + 4x + 4$
4 $x^2 + 5x + 4$ **14** $x^2 - 2x + 1$
5 $x^2 + 2x - 8$ **15** $9x^2 + 6x + 1$
6 $x^2 - 10x + 21$ **16** $4x^2 - 12x + 9$
7 $x^2 - x - 2$ **17** $2x^2 + 9x + 4$
8 $x^2 - 8x + 12$ **18** $3x^2 - 4x - 4$
9 $x^2 - 9$ **19** $6x^2 + 7x + 2$
10 $x^2 - 1$ **20** $12x^2 + x - 6$

Exercise 169

1 $(x + 1)(x + 7)$ **10** $(x + 12)(x - 1)$
2 $(x + 2)(x + 5)$ **11** $(x - 6)(x + 1)$
3 $(x - 2)(x + 3)$ **12** $(x - 8)(x + 2)$
4 $(x - 2)(x - 3)$ **13** $x(x - 7)$
5 $(x + 4)(x - 2)$ **14** $2x(x + 3)$
6 $(x + 3)(x - 5)$ **15** $3x(2x - 1)$
7 $(x - 3)(x - 3)$ **16** $x(1 - 2x)$
8 $(x - 6)(x + 2)$ **17** $(x + 6)(x + 6)$
9 $(x - 3)(x - 4)$ **18** $(x - 4)(x - 4)$

Exercise 170

1 $(2x + 1)(x + 2)$ **7** $(3x - 2)(x - 2)$
2 $(2x + 3)(x + 2)$ **8** $(3x + 2)(x - 5)$
3 $(2x - 1)(x + 5)$ **9** $(2x - 3)(x - 4)$
4 $(2x + 1)(x - 7)$ **10** $(3x + 2)(x + 6)$
5 $(3x + 5)(x + 1)$ **11** $(3x + 2)(x - 8)$
6 $3(x + 1)(x - 3)$ **12** $(2x - 7)(x + 2)$

Exercise 171

1 $(x-1)(x+1)$
2 $(x-4)(x+4)$
3 $(x-5)(x+5)$
4 $(2x-3)(2x+3)$
5 $(3x-1)(3x+1)$
6 $(4x-5)(4x+5)$

7 $(7-x)(7+x)$
8 $(6-5x)(6+5x)$
9 $(5-7x)(5+7x)$
10 $2(x-2)(x+2)$
11 $9(x-2)(x+2)$
12 $3(2x-5)(2x+5)$

Exercise 172

1 $x=4$ or 1
2 $x=-7$ or -4
3 $x=7$ or -7
4 $x=-3$ or -3
5 $x=-3$ or $-\frac{1}{2}$
6 $x=0$ or -3

7 $x=-\frac{4}{3}$ or 2
8 $x=0$ or $\frac{2}{5}$
9 $x=\frac{2}{3}$ or 6
10 $x=-\frac{7}{2}$ or 2
11 $x=5$ or -4
12 $x=\frac{1}{2}$ or -1

Exercise 173

1 $-1.82, 0.82$
2 $-0.29, -1.71$
3 $0.73, -2.73$
4 $0.77, -0.43$
5 $1.36, -7.36$
6 $5.30, 1.70$

7 $3.24, -1.24$
8 $1.58, 0.42$
9 $0.89, -3.39$
10 $2.72, 0.61$
11 $1.45, -3.45$
12 $2.62, 0.38$

Exercise 174

1 a $(x+3)$ by $(x-1)$ b $(x+3)(x-1)=21$
 c $4\,\text{m}$ by $4\,\text{m}$
2 a 14 b 8 3 a $\frac{1}{2}x(x-1)$ b $x^2-x-56=0$ c 8
4 a $(20-x)$ metres b $x(20-x)\,\text{m}^2$ c $8\,\text{m}$ by $12\,\text{m}$
 d $14.4\,\text{m}$ by $5.6\,\text{m}$ e $10\,\text{m}$ by $10\,\text{m}$
5 $(7-x)(4-x)=14, x=1.47$
6 $(5+x)(5+2x)=80, x=2.7, 10.4\,\text{cm} \times 7.7\,\text{cm}$
7 a 24 b 54 c $S=6(n-2)^2$ d (i) 4 (ii) 6

Exercise 175

1 a $v=\dfrac{2s-ut}{t}$ b 18 c 0; the vehicle has stopped

2 a $8\,\text{cm}$ b $u=\dfrac{fv}{v-f}$ c $4.3\,\text{cm}$

3 a $l=10\left(\dfrac{T}{2\pi}\right)^2$ b $25.3\,\text{cm}$

4 a $r=100\sqrt{\dfrac{A}{P}-1}$ b 22.9% c $A=PR^2$

 d $R=\sqrt{\dfrac{A}{P}}$ e (i) 1.051 (ii) 5.12%

5 a $S=\dfrac{100-v^2}{20}$
 b (i) $3.75\,\text{m}$; above its starting position
   (ii) $5\,\text{m}$; above its starting position, i.e. at the highest point
   (iii) $0\,\text{m}$; level with its starting position
   (iv) $-6.25\,\text{m}$; $6.25\,\text{m}$ below its starting position
6 a $P=2x+\pi x+4h$ b $x=\dfrac{P-4h}{2+\pi}$ c $1.5\,\text{m}$

7 a $P=2(x+y)$ b $A=xy$ c $xy=6(x+y)$

 d $x=\dfrac{6y}{y-6}$ e $15\,\text{cm}$

Exercise 176

1 a (i) $x=3$ (ii) $x=2\frac{1}{2}$ b $x=3$ 2 $x=50$
      $y=-2$ $y=1$ $y=5$ $y=24$
3 a $x^2+7x+10$ b x^2-49 c $2x^2-17x+8$
4 a $(x-9)(x+1)$ b $3x(x+4)$ c $3(x-2)(x+2)$
 d $3(x+3)(x+1)$ e $(2x-1)(x+6)$
5 a $5, -1$ b $2, 0$ c $\frac{4}{3}, -\frac{4}{3}$ d $-2, -\frac{3}{2}$
6 a $1.83, -3.83$ b $2.28, 0.22$
7 a $(x^2+2x)\,\text{cm}^2$ b $(4x+4)\,\text{cm}$ c $x^2-2x-4=0$
 d $3.24\,\text{cm}$

8 a Nx pence b $\dfrac{Ny}{10}$ pence c $P=\dfrac{Ny}{10}-Nx$ d £60

 e $N=\dfrac{10P}{y-10x}$ f 5000

Exercise 177

1 a infinite b empty c finite d finite e infinite
2 a $A=\{w,x,y,z\}$
 b $B=\{5,10,15,20,25,\dots\}$
 c $C=\{5,7,11,13,17,\dots\}$
 d $D=\{\text{Mon, Tue, Wed, Thu, Fri, Sat, Sun}\}$
 e $E=\{2,4,6,8,10,12,14\}$
 f $F=\{4,5,6,7,8,9,10,11\}$
3 a the continents b factors of 12 c set of triangles
 d square roots of 4 e first 5 triangular numbers
4 a true b false c false d true e false f true
5 $\{1\},\{2\},\{3\},\{1,2\},\{1,3\},\{2,3\}$
6 a Prime numbers (except 2) are not even numbers.
 b 4 is not an odd number.
 c 5 is a prime number and an odd number.
 d Prime numbers (except 2) are odd numbers.
 e 5 and 9 are odd numbers.

Exercise 178

1 a students in LVI who smoke or are female
 b boys in LVI who smoke c girls in LVI
 d boys or non-smokers (i.e. all boys and girls who don't smoke)
 e All the girls in LVI smoke.
 f boys or girls in LVI (i.e. all the LVI)
 g There are 10 students in LVI.
 h 5 boys are non-smokers.
2 a $3,6$ b $1,2,3,4,5,6,12,15,30,60$
 c $10,12,15,20,30,60$ d $10,20$ e $10,20$
 f $1,2,4,5,10,12,15,20,30,60$
 g $1,2,4,5,10,12,15,20,30,60$
 h $3,6,10,12,15,20,30,60$
3 a $\{\text{rhombuses}\} \subset P$ b $P \subset Q$ c $Q \not\subset P$ d $S \subset R \subset P$
 e $\{\text{trapeziums}\} \cap P = \{\}$ or $\emptyset$ f $\{\text{kites}\} \subset \{Q \cap P'\}$
4 a 28 b 7 c 0 d 16 e 28 f 44
5 a A b   c A d $\emptyset$ e A f
6 a There are 8 boys in the tutorial group; $n(B)=8$.
 b Five girls study mathematics; $n(B' \cap M)=5$.
 c There are 9 girls in the tutorial group; $n(B')=9$
  or there are 17 students in the tutorial group;
  $n(T)=17$.

Exercise 179

1

a
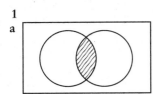

d
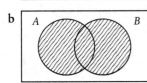

b
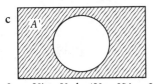

e
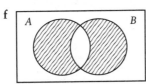

c
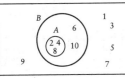

f
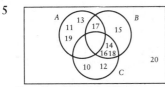

3 a $X' \cap Y$ **b** $(X \cup Y)'$ or $X' \cap Y'$ **c** $A \cap B \cap C$
 d $Q \cap R$ **e** $Y \cap Z \cap X'$ **f** $R \cap S' \cap T'$
4 $\mathscr{E} = \{1, 2, 3, 4, 5, 6, 7, 8, 9, 10\}$
 $A = \{2, 4, 8\}$
 $B = \{2, 4, 6, 8, 10\}$

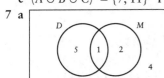

$A' = \{1, 3, 5, 6, 7, 9, 10\}$
$A \cap B = \{2, 4, 8\}$
$A \cap B' = \emptyset$
$(A \cup B)' = \{1, 3, 5, 7, 9\}$

5
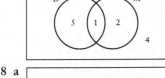

$B \cap C = \{14, 16, 18\}$
$A \cup B = \{11, 13, 14, 15, 16, 17, 18, 19\}$
$A \cap B \cap C = \emptyset$
$(A \cup B \cup C)' = \{20\}$
$A \cap B' \cap C' = \{11, 13, 19\}$
6 a $A \cap B \cap C = \{2, 4\}$ **b** $n(A \cap B \cap C) = 2$
 c $B \cup C = \{2, 4, 5, 6, 8, 9, 13, 14, 15\}$
 d $n(B \cup C) = 9$
 e $(A \cup B \cup C)' = \{7, 11\}$ **f** $n(A \cup B \cup C)' = 2$
7 a **b** 5
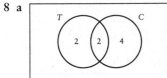

8 a

b (i) 2 (ii) 2 (iii) 4
 c (i) Two people drink both tea and coffee.
  (ii) Two people drink tea but not coffee.
  (iii) Four people drink coffee but not tea.

Exercise 180

1 a $x = 1$
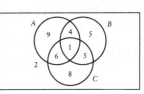

2 a 6 **b** $x = 6, y = 5, z = 3$
3 **b** $x = 3$ **c** 10
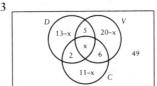

4 a $x = 2$ **b**
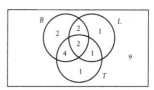

5 a $x = 3$ **b** 31%

Exercise 181

1 a $\begin{pmatrix} 6 \\ 3 \end{pmatrix}$; $6\mathbf{i} + 3\mathbf{j}$ **b** $\begin{pmatrix} 6 \\ -4 \end{pmatrix}$; $6\mathbf{i} - 4\mathbf{j}$ **c** $\begin{pmatrix} 6 \\ 3 \end{pmatrix}$; $6\mathbf{i} + 3\mathbf{j}$

 d $\begin{pmatrix} -7 \\ 2 \end{pmatrix}$; $-7\mathbf{i} + 2\mathbf{j}$

2 a $\begin{pmatrix} 0.5 \\ 2 \end{pmatrix}$ **b** $\begin{pmatrix} 7 \\ 0 \end{pmatrix}$ **c** $\begin{pmatrix} -4.5 \\ -2 \end{pmatrix}$ **d** $\begin{pmatrix} 5 \\ 4 \end{pmatrix}$ **e** $\begin{pmatrix} -6.5 \\ 2 \end{pmatrix}$

3 a (i) $\begin{pmatrix} 4 \\ 1 \end{pmatrix}$ (ii) $4\mathbf{i} + \mathbf{j}$ **b** (i) $\begin{pmatrix} 4 \\ -4 \end{pmatrix}$ (ii) $4\mathbf{i} - 4\mathbf{j}$

 c (i) $\begin{pmatrix} 9 \\ -1 \end{pmatrix}$ (ii) $9\mathbf{i} - \mathbf{j}$ **d** (i) $\begin{pmatrix} -10 \\ -5 \end{pmatrix}$ (ii) $-10\mathbf{i} - 5\mathbf{j}$

 e (i) $\begin{pmatrix} -7 \\ 7 \end{pmatrix}$ (ii) $-7\mathbf{i} + 7\mathbf{j}$ **f** (i) $\begin{pmatrix} -3 \\ 7 \end{pmatrix}$ (ii) $-3\mathbf{i} + 7\mathbf{j}$

Exercise 182

1 a 10 **b** 8.60 **2 a** 13 **b** 17 **c** 5 **3 a** 4.47 **b** 10

Exercise 183

1 (i) $\begin{pmatrix} 20 \\ -48 \end{pmatrix}$ (ii) $\begin{pmatrix} 2\frac{1}{2} \\ -6 \end{pmatrix}$ (iii) $\begin{pmatrix} -10 \\ 24 \end{pmatrix}$ (iv) $\begin{pmatrix} -5 \\ 12 \end{pmatrix}$

 (v) 13 (vi) 39
2 b, c and e

3 a parallel **b** equal magnitude **c** both **d** parallel
e equal

Exercise 184

1 a $\begin{pmatrix} 6 \\ 4 \end{pmatrix}$ (ii) $\begin{pmatrix} -2 \\ -6 \end{pmatrix}$ (iii) $\begin{pmatrix} 8 \\ 9\frac{1}{2} \end{pmatrix}$ (iv) $\begin{pmatrix} 8 \\ 17 \end{pmatrix}$

(v) $\begin{pmatrix} 1 \\ 0 \end{pmatrix}$ (vi) $\begin{pmatrix} 1 \\ 3 \end{pmatrix}$ (vii) $\begin{pmatrix} -\frac{1}{2} \\ -2 \end{pmatrix}$ (viii) $\begin{pmatrix} 0 \\ 6 \end{pmatrix}$

2 (i) $-5i + 3j$ (ii) 5.83 (iii) 11i (iv) 11 (v) $4\frac{1}{2}i - \frac{1}{2}j$
(vi) 4.53

3 (i) $\mathbf{p} = \begin{pmatrix} 17 \\ 19 \end{pmatrix}$ (ii) $\mathbf{q} = \begin{pmatrix} 17 \\ 19 \end{pmatrix}$

(iii) **p** and **q** are parallel and of equal lengths.

Exercise 185

1 (i) $\overrightarrow{PQ} = q - p$, $\overrightarrow{OR} = \frac{3}{4}p$, $\overrightarrow{OS} = \frac{3}{4}q$, $\overrightarrow{RS} = \frac{3}{4}(q - p)$
(ii) RS and PQ are parallel and $RS = \frac{3}{4}PQ$.
2 (i) $\overrightarrow{AP} = \frac{1}{2}b$, $\overrightarrow{CD} = (a - b - c)$, $\overrightarrow{CR} = \frac{1}{2}(a - b - c)$,
$\overrightarrow{PS} = \frac{1}{2}(a - b)$, $\overrightarrow{QR} = \frac{1}{2}(a - b)$,
$\overrightarrow{PQ} = \frac{1}{2}(b + c)$, $\overrightarrow{SR} = \frac{1}{2}(b + c)$

(ii) PQRS is a parallelogram.
3 $\overrightarrow{RS} = -p$, $\overrightarrow{ST} = -q$, $\overrightarrow{QT} = -2p$, $\overrightarrow{QR} = q - p$,
$\overrightarrow{PR} = 2q - p$, $\overrightarrow{OR} = 2q$
4 (i) $c + \frac{1}{2}a, \frac{1}{2}c + \frac{1}{4}a, -a, c - a, \frac{1}{2}(c - a), \frac{1}{2}(a + c), \frac{1}{4}a$
(ii) $XY = \frac{1}{4}OA$ and $XY \| OA$

5 (i) $\overrightarrow{AB} = b - a$, $\overrightarrow{AD} = \frac{2}{3}(b - a)$,
$\overrightarrow{CD} = \frac{1}{3}a + \frac{2}{3}b = \frac{1}{3}(a + 2b)$
$\overrightarrow{BE} = \frac{1}{2}a, \overrightarrow{DB} = \frac{1}{3}(b - a)$, $\overrightarrow{DE} = \frac{1}{6}a + \frac{1}{3}b = \frac{1}{6}(a + 2b)$

(ii) CD and DE are in the same straight line, CDE.
D divides CE in the ratio 2:1.

Exercise 186

1

	Large	Small
White	20	15
Wholemeal	30	25
Granary	24	20

2

	Rec	Cass	CD
Jazz	42	8	15
Class	21	12	35
Pop	19	30	10

3 a 3×2 **b** 3×1 **c** 2×2 **d** 2×3 **e** 1×2
f 3×3

4

	B	C	H	M	N
B	0	9	6	9	12
C	9	0	$4\frac{1}{2}$	$9\frac{1}{2}$	8
H	6	$4\frac{1}{2}$	0	5	6
M	9	$9\frac{1}{2}$	5	0	$8\frac{1}{2}$
N	12	8	6	$8\frac{1}{2}$	0

5

	3	6	9	12
3	1	1	1	1
6	0	1	0	1
9	0	0	1	0
12	0	0	0	1

Exercise 187

1 a $\begin{pmatrix} 4 & 9 \\ 4 & 9 \end{pmatrix}$ **b** $\begin{pmatrix} 1 & 3 & 0 \\ 6 & -1 & 4 \end{pmatrix}$ **c** incompatible

d $\begin{pmatrix} 2 & 2 & 4 \\ -2 & 4 & -2 \end{pmatrix}$ **e** $\begin{pmatrix} 1 & 1 & 4 \\ -2 & 5 & 2 \end{pmatrix}$ **f** $\begin{pmatrix} 3 & 6 & 4 \\ 7 & 3 & 6 \end{pmatrix}$

2 $\begin{pmatrix} 2 & 1 \\ 1 & 0 \\ 0 & 2 \\ 1 & 3 \end{pmatrix}$ **3** $\begin{pmatrix} 320 & 1000 & 2500 \\ 1020 & 3680 & 3000 \end{pmatrix}$ **4** $\begin{pmatrix} 12 & 7 & 7 \\ 15 & 5 & 6 \\ 16 & 7 & 3 \\ 7 & 8 & 10 \\ 6 & 5 & 15 \end{pmatrix}$

5 a $p = 4$, $q = 3$, $r = 3$ **b** $p = 7, q = -1, r = 2$
c $p = -2, q = -3, r = 2$

Exercise 188

1 a $\begin{pmatrix} 3 & 6 \\ 6 & -3 \end{pmatrix}$ **b** $\begin{pmatrix} -6 & 0 \\ 8 & 2 \end{pmatrix}$ **c** $\begin{pmatrix} -3 & 6 \\ 14 & -1 \end{pmatrix}$ **d** $\begin{pmatrix} 5 & 4 \\ 0 & -3 \end{pmatrix}$

e $\begin{pmatrix} 10 & 2 \\ -10 & -4 \end{pmatrix}$ **f** $\begin{pmatrix} -1 & 1 \\ 3 & 0 \end{pmatrix}$

2 a $\begin{pmatrix} 8.56 & 16.16 \\ 11.36 & 21.76 \end{pmatrix}$ **b** $\begin{pmatrix} 3.42 & 6.46 \\ 4.54 & 8.70 \end{pmatrix}$

3 a $\begin{pmatrix} 60 & 2.3 \\ 80 & 3.0 \\ 40 & 2.5 \end{pmatrix}$ **b** $\begin{pmatrix} 420 & 16.1 \\ 560 & 21.0 \\ 280 & 17.5 \end{pmatrix}$ **c** $\begin{pmatrix} 120 & 4.6 \\ 160 & 6.0 \\ 80 & 5.0 \end{pmatrix}$

4 a $\begin{pmatrix} 1 & 8 & 0 \\ 15 & -8 & -2 \end{pmatrix}$ **b** $\begin{pmatrix} 15 & 10 & -8 \\ -8 & -13 & 6 \end{pmatrix}$

c $\begin{pmatrix} -1 & 0 & 10 \\ 19 & -5 & -1 \end{pmatrix}$ **d** $\begin{pmatrix} -1 & 14 & -28 \\ -22 & 3 & -4 \end{pmatrix}$

e $\begin{pmatrix} 16 & -2 & 4 \\ -11 & -9 & 10 \end{pmatrix}$ **f** $\begin{pmatrix} -24 & 4 & -14 \\ 4 & 17 & -15 \end{pmatrix}$

Exercise 189

1 a (12) **b** (13) **c** (140) **d** (6)

2 a $\begin{pmatrix} 16 \\ 16 \end{pmatrix}$ **b** $\begin{pmatrix} 40 \\ 20 \\ 38 \end{pmatrix}$ **c** $\begin{pmatrix} 32 \\ 42 \end{pmatrix}$ **d** $\begin{pmatrix} 65 \\ 25 \\ 14 \end{pmatrix}$

3 a $\begin{pmatrix} 10 & 0 \\ 0 & 10 \\ 1 & 2 \end{pmatrix}$ b $\begin{pmatrix} 200 \\ 150 \\ 50 \end{pmatrix}$

4 $\begin{pmatrix} 12 & 3 & 4 & 0 \\ 8 & 4 & 0 & 10 \\ 8 & 3 & 3 & 5 \end{pmatrix}$ $\begin{matrix} 36 \\ 49 \\ 40 \\ 42 \end{matrix}$ $=\begin{pmatrix} 739 \\ 904 \\ 765 \end{pmatrix}$

5 $\begin{pmatrix} 1 & 1 & 0 & 2 & 9 \\ 0 & 0 & 0 & 2 & 11 \\ 2 & 1 & 1 & 0 & 9 \\ 1 & 2 & 3 & 0 & 7 \end{pmatrix}$; $\begin{pmatrix} 9 \\ 2 \\ 13 \\ 16 \end{pmatrix}$

6 $\begin{pmatrix} 31 \\ 35 \\ 39 \\ 22 \\ 17 \end{pmatrix}$; $\begin{pmatrix} \text{Croydon} \\ \text{Brunswick} \\ \text{Albion} \\ \text{Melbourne} \\ \text{Ringwood} \end{pmatrix}$

Exercise 190

1 a (i) $\begin{pmatrix} 9 & 8 \\ 7 & 4 \end{pmatrix}$ (ii) $\begin{pmatrix} 12 & 8 \\ 4 & 1 \end{pmatrix}$

b $AB \neq BA$
i.e. matrix multiplication is not commutative.

2 a $\begin{pmatrix} 2 & 3 \\ 1 & 1 \end{pmatrix}$ b $\begin{pmatrix} 4 & 1 \\ 2 & 3 \end{pmatrix}$ c $\begin{pmatrix} 1 & 0 \\ 0 & 1 \end{pmatrix}$ d $\begin{pmatrix} 0 & 1 \\ 1 & 0 \end{pmatrix}$

e $\begin{pmatrix} -1 & 0 \\ 0 & -1 \end{pmatrix}$ f $\begin{pmatrix} 1 & 0 \\ 0 & 1 \end{pmatrix}$ g $\begin{pmatrix} 1 & 0 \\ 0 & 1 \end{pmatrix}$ h $\begin{pmatrix} -1 & 0 \\ 0 & -1 \end{pmatrix}$

i $\begin{pmatrix} 1 & 0 \\ 0 & 1 \end{pmatrix}$

3 a AB, AC, BC, CA, CD, DA
b 2×3, 2×2, 3×2, 3×3, 3×2, 2×3

4 a $\begin{pmatrix} 12 & 24 & 22 \\ 18 & 12 & 31 \end{pmatrix}$ b $\begin{pmatrix} 15 & 7 & 10 \\ 18 & 9 & 16 \end{pmatrix}$

c $\begin{pmatrix} 12 & 7 \\ 11 & 5 \\ 14 & 9 \end{pmatrix}$ d $\begin{pmatrix} 42 & 38 & 42 \\ 54 & 30 & 63 \end{pmatrix}$

e $\begin{pmatrix} -15 & -6 \\ -1 & 1 \\ -1 & -8 \end{pmatrix}$

Exercise 191

1 a $\begin{pmatrix} -6 & -2 \\ -7 & -3 \end{pmatrix}$ b $\begin{pmatrix} -1 & -3 \\ 2 & -5 \end{pmatrix}$ c $\begin{pmatrix} -1 & 3 \\ 4 & -6 \end{pmatrix}$

2 a 1 b 2 c 0 d −1 e 0 f −2 g 0 h −4 i 1
3 c, e and g
4 a $\begin{pmatrix} 3 & -2 \\ -4 & 3 \end{pmatrix}$ b $\begin{pmatrix} 1 & -\frac{3}{2} \\ -1 & 2 \end{pmatrix}$ c no inverse

d $\begin{pmatrix} -3 & 4 \\ 4 & -5 \end{pmatrix}$ e no inverse f $\begin{pmatrix} -\frac{3}{2} & 2 \\ 1 & -1 \end{pmatrix}$

g no inverse h $\begin{pmatrix} -\frac{1}{2} & \frac{3}{4} \\ -\frac{1}{2} & \frac{5}{4} \end{pmatrix}$ i $\begin{pmatrix} 1 & 0 \\ 0 & 1 \end{pmatrix}$

5 a $\begin{pmatrix} -3 & 1 \\ 0 & -4 \end{pmatrix}$ b $\frac{1}{12}\begin{pmatrix} 4 & 1 \\ 0 & 3 \end{pmatrix}$

Exercise 192

1

a reflection in the x-axis
b enlargement by factor of 3
c rotation through 180°
d reflection in the y-axis
e reflection in $y = x$
f rotation through 270° (−90°)
g rotation through 90°
h reflection in $y = -x$

2 a A: reflection in $y = x$
B: rotation through 90°
C: reflection in the x-axis
D: rotation through 180°
E: reflection in $y = -x$
F: rotation through −90° (270°)
G: reflection in the y-axis

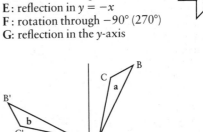

3

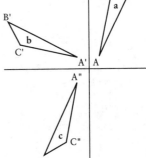

d $T^2 = \begin{pmatrix} -1 & 0 \\ 0 & -1 \end{pmatrix}$

e T^2 rotates triangle ABC through 180°C

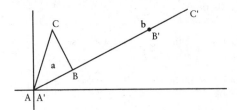

4 c The triangle has collapsed into a straight line.
d The determinant of the transformation matrix is 0, i.e. the transformation has no inverse – a line cannot be transformed into an area.

5

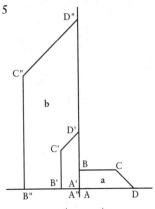

c $RT = \begin{pmatrix} 0 & -3 \\ 3 & 0 \end{pmatrix}$

d T: rotation through 90°
   R: enlargement by factor of 3
   RT: rotation through 90° and an enlargement by factor of 3

e $\begin{pmatrix} 0 & \frac{1}{3} \\ -\frac{1}{3} & 0 \end{pmatrix}$

6 a rotation through 180°
b reflection in the y-axis
c reflection in $y = x$
d reflection in $y = -x$
e rotation through 90°
f rotation through 90°

Exercise 193

1 7.74597   **2 a** 7.742 **b** 7.742, 0.258
3 a (i) and (ii) 0.258 **b** (i) 6 (ii) 4
4 a $x_2 = -3.8125, x_3 = -3.7936, x_4 = -3.7915,$
   $x_5 = -3.7913, x_6 = -3.7912, x_7 = -3.7912$
b $x = -3.791$ **c** 4
5 a $x_2 = 5.125$ **b** 5
   $x_3 = 4.9878048(78)$ **d** (i) $(2x + 1)(x - 5)$
   $x_4 = 5.0012224(94)$ (ii) $x = 5$ or -0.5

6 a $x_{n+1} = \frac{1}{3}\left(11 + \frac{4}{x_n}\right)$ and $x_{n+1} = \frac{3x_n^2 - 4}{11}$

b $x_2 = 4.1111111$ $x_2 = 2.0909091$
   $x_3 = 3.9909909$ $x_3 = 0.8287002$
   $x_4 = 4.0007524$ $x_4 = -0.1763425$

c The formula $x_{n+1} = \frac{1}{3}\left(11 + \frac{4}{x_n}\right)$ is approaching a positive root of value 4.
The formula $x_{n+1} = \frac{3x_n^2 - 4}{11}$ is approaching a negative root.

d $x_2 = 8.1111111$ $x_2 = -0.3390909$
   $x_3 = 3.8310502$ $x_3 = -0.3322774$
   $x_4 = 4.0147000$ $x_4 = -0.3335250$
   $x_5 = 3.9987795$ $x_5 = -0.3332984$
   $x_6 = 4.0001017$ $x_6 = -0.3333396$

e The roots are 4 and $-0.\dot{3}$ $(-\frac{1}{3})$

g $x_{n+1} = 2 - \frac{1}{2x_n}$

Exercise 194

1 41.4°
2 75.5°
3 88.4°
4 86.3°
5 55.8°
6 6.31 cm
7 4.87 cm
8 9.13 cm
9 4.70 in
10 KM = 14.5 cm, MP = 11.5 cm, NQ = 11.1 cm

Exercise 196

1 a 120 cm² **b** 75.6 cm² **c** 451 in² **d** 3380 cm²
2 a 60 cm² **b** 224 cm² **c** 139 in² **d** 99 cm²
3 a 8 cm **b** 4 cm **c** 5.25 in
4 a 7 cm **b** 1.80 cm

Exercise 197

1 20 cm² **2** 309 cm² **3** 104 cm² **4** 66 cm²
5 74.2 cm²

Exercise 198

1 51.3 cm² **2** 106 cm² **3** 29.5 cm² **4** 7.86 cm
5 6.77 in **6** 70.2° **7** 203°

Exercise 199

1 a 905 cm³ **b** 230 cm³ **c** 589 in³ **2** 64 cm³
3 160 cm³ **4** 144 cm³
5 a 268 cm³ **b** 435 cm³ **c** 20.6 in³
6 a 603 cm³ **b** 222 cm³ **c** 120 in³ **7** 3.92 cm
8 9.23 cm **9** 3.57 in **10** 1230 cm³

Exercise 200

1 a 150 cm² **b** 216 cm² **c** 394 cm²
2 a 180 cm² **b** 113 cm² **c** 194 in²
3 4 cm, 96 cm² **4** 3.04 cm, 55.3 cm²
5 0.8 cm, 101 cm² **6** 211 cm² **7** 249 cm²
8 893 cm²

Exercise 201

1 303 cm^2 **2** 1100 cm^2 **3 a** 283 cm **b** 5650 cm^2

Exercise 202

1 a 314 cm^2 **b** 804 cm^2 **c** 452 in^2
2 a 201 cm^2 **b** 1260 in^2 **c** 460 cm^2
3 a 1.49 cm **b** 1.50 in   **4 a** 8.41 cm^3 **b** 20.3 cm^3

Exercise 203

1 20 400 m^2 **2** 13 000 m^2 **3** 45 700 m^2
4 82 400 m^2 **5** 79 800 m^2 **6** 120 000 m^2
7 96 800 m^2

Exercise 204

1 21 200 m^2 **2** 13 200 m^2 **3** 46 600 m^2
4 85 400 m^2 **5** 87 100 m^2 **6** 122 000 m^2
7 99 700 m^2

Exercise 205

1 a 2200 m^3 **b** 2320 m^3
2 a 6 510 000 m^3
   b 6 660 000 m^3
3 a 1 800 000 m^3 **b** 1 820 000 m^3
4 a 42 square feet   **b** 3.11 cubic yards

Exercise 206

1 a 1 : 2.47 **b** 1 : 7 **2 a** 1 : 131 **b** 1 : 297 **c** 1 : 137
3 a 1 : 420 **b** 1 : 380 **c** nil **d** 1 : 122 **e** 1 : 118

Exercise 207

4 b 8%

Index

Accounts   30, 31, 33, 34
Acute angle   240
Acute–angled triangle   247
Additional law   144
Addition
    of algebraic terms   219
    of decimals   168
    of directed numbers   153
    of fractions   171, 227
    of matrices   350
    of probabilities   144
    of vectors   342
Algebra   218, 310
Algebraic fractions   227
Allowances   14
Alternate angles   242
Altitude   263
Angle between
    a line and a plane   281
    two planes   282
Angles   239
    in a circle   258
    of depression and
        elevation   278
    and parallel lines   242
Approximations   162
Annual percentage rate (APR)   36
Arcs   257
Areas   197, 207, 208, 214, 256,
    375, 382
Arithmetic mean   119
Arithmetic, four rules of   150
Assurance   77
Average
    moving   127
    weighted   129
Average gradient   396
Average speed   182, 184, 309
Axes   285, 289

Bank
    accounts   31
    giro   35
    loan   40
    statement   31, 32
Banking services   34
Bar charts   95
Basic algebra   218
Basic pay   10
Basic hourly rate   10
Bearings   276
Bias   89
Bonus   13
Bonus, no-claims   82
Brackets   224, 315
Budget account   34

Caculator, use of   162, 167, 367
Car insurance   80
Carpeting   71
Cartesian coordinates   298
Cash dispenser card   31, 32
Census   88
Chain survey   370
Charge card   43
Cheques   30, 34
Chord   257
Circles   195, 207, 257
Circumference   205, 257
Class
    boundaries   103
    frequency   103
    intervals   103
Coefficient   220, 317, 318
Commission   12
Common
    denominator   171, 227
    factor   160, 170, 226, 317
    multiple   159
Community charge   32, 58
Compass   276
Compatibility of matrices   358
Complementary
    angles   240
    sets   331, 334
Component bar charts   96
Composite areas   208
Compound
    interest   26
    interest tables   28
    percentages   85
Congruent triangles   250
Cone, volume of   379
Construction of triangles   261
Continuous variable   87
Contour lines   395
Convenience sampling   89
Conversion
    between metric and imperial
        units   194
    between fractions and
        decimals   174
    between fractions and
        percentages   175
Conversion graphs   293
Corresponding angles   242
Cosine ratio   272
Coursework   vi
Credit   38
Cross-section   211, 392, 399
Cube   210
Cuboid   210, 382
Cumulative frequency   131

Current account   31
Cyclic quadrilateral   259
Cylinder   211, 383

Data   87, 93
Decimal
    recurring   174
    terminating   174
Decimal place   162
Decimal system   150
Deductions from pay   14
Denominator   169
Deposit   38, 54
Deposit account   33
Depreciation   86
Determinant of a matrix   361
Diagonal   254
Diameter   205, 257
Difference   150
    of two squares   319
Direct
    debit   32
    proportion   7, 179
Directed numbers   150, 152, 222
Discounts   76, 82
Discrete variable   87
Dispersion   136
Distance – time graphs   296, 308
Division
    of algebraic terms   221
    of directed numbers   154
    of fractions   172
    in a given ratio   178
Dual bar chart   96

Electricity bills   61
Element of a set   328
Empty set   328
Endowment
    assurance   77
    mortgage   55
Enlargement   251, 364
Equations   229, 310, 320
Equilateral triangle   195, 247, 248
Equivalent fraction   169, 227
Estimation   163
Eurocheque   34
Even number   150
Exchange rates   49
Experimental
    data   87
    probability   141
Exterior angle
    of a polygon   256
    of a triangle   248

Factor   150, 158, 159, 170, 226, 316
Factorising   226, 317, 320
Field book   370
Finite set   328
Foreign currency   48
Formulae
    transformation of   236, 325
    substitution in   234
Fractions   150, 157, 169, 171, 174, 227
Frequency
    cumulative   131
    density   102
    distribution   93
    polygon   104
    table   93

Gas bills   63
Geometrical constructions   262
Geometry   239
    vector   345
Gradient   300, 308, 395, 396
Graphs   100, 285, 298, 303, 306, 313
Gross pay   10
Grouped distributions   102
    means of   123

Highest common factor   160
Hire purchase   38
Histogram   101
Household
    bills   59
    costs   54
Hypotenuse   265, 270, 272

Identity
    element   360
    matrix   360
Imperial units   191, 193
Improper fraction   157, 169
Income tax   14
Independent events   144
Index   156, 223
Infinite set   328
Insurance   74
    National   16
Integer   150
Intercept   300
Interest
    compound   26
    flat rate of   36
    simple   25
    true rate of   36
Interior angles
    of a polygon   256
    of a triangle   247
Interquartile range   133, 137
Intersection of sets   330, 333
Inverse
    matrix   360
    percentages   84
    proportion   180

Investments   25, 29
Irrational numbers   150
Isosceles triangle   247, 248
Iteration   367

Kite   253

Life assurance   77
Linear equations   229
    graphs of   298
Line
    graph   100
    segment   239
    of symmetry   243
Loans   40
Loss per cent   21
Lowest
    common multiple   160
    terms   7

Map scales   188
Matrices   348
Matrix transformations   362
Mean   119
    of grouped data   123
    weighted   129
Measures of rate   181
Median   121
Member of a set   328, 331
Meter dials   61
Metric units   191
Mixed numbers   169
Modal class   121
Mode   120
Modulus   341
Mortgages   54
Moving average   127
Multiple   150, 158
Multiplication
    of algebraic terms   221
    of directed numbers   154
    of fractions   172
    of matrices   354, 357
    by a scalar   342, 352
    law of probability   145
Mutually exclusive events   144

National Insurance   16
Natural numbers   150
Net pay   14
No-claims bonus   82
Null set   328
Number line   153

Obtuse angle   240
Obtuse-angled triangle   247
Odd number   150
Offsets   371
Ogive   131
Order
    of operations in arithmetic   155
    of symmetry   245
Ordnance Survey maps   395

Overdraft   40
Overtime rates   10

Painting   65
Pay   10, 14
Parallel lines   242
Parallelogram   195, 204, 253, 254
PAYE   14
Payment debit card   32
Percentages   1, 21, 84, 85, 175
Percentile   133
Perimeter   195
Periodic sampling   89
Personal
    allowance   14
    loan
Pictogram   94
Piecework   13
Pie chart   98
Perpendicular lines   240
Pilot survey   91
Poll tax   58
Polygons   195, 247, 256, 376
Population   89
Possibility space   142
Prime
    factor   159
    number   156
Prism   211
Probability   140
Product   150
Profit   21
Proportion   7, 179, 180
Proportional parts   8, 178
Pyramid   379, 382
Pythagoras' theorem   265, 281, 341

Quadratic
    equations   320
    expressions   316
    formula   321
    graphs   306
Quadrilaterals   253
Qualitative variable   87
Quantitative variable   87
Questionnaires   88
Quota sampling   89
Quotient   150, 159

Radius   206, 257
Range   136
Rate   181
    of exchange   49
    overtime   10
Rates, water   60
Ratio   6, 177
Rational numbers   150
Raw data   93
Reciprocal   157
Rectangle   198, 253, 254
Redecoration   65
Reflection   363
Reflex angle   241

Reflexive symmetry 243
Renting 57
Repairs 73
Repayment mortgage 55
Rhombus 195, 253
Right angle 240, 258
Right-angled triangle 247
Roots
    of numbers 156
    of quadratic equations 320
Rotational symmetry 245

Salary 10
Sample 88
Sampling methods 88
Savings 25, 29, 33, 34
Scalar 339
    multiple of a matrix 352
    multiple of a vector 342
Scalene triangle 247
Scales
    graphical 290
    map and model 187
Sectional bar chart 96
Sector of a circle 258, 377
Segment of a circle 258
Semi-interquartile range 133
Set notation 333
Sets 328
Significant figures 162
Similar
    areas 214
    triangles 251
    volumes 216
Simple interest 25
Simpson's rule 390
Simultaneous equations 310
    graphical solution of 313
Sine ratio 272
Singular matrix 361
Speed 182, 184, 308

Sphere 380, 385
Square 198, 253
Square matrix 348
Square numbers 150, 156
Square roots of numbers 156
Standard deviation 137
    form 164
Standing order 31
Statistics 87
Store card 43
Straight line 239
    equation of 298
Stratified random sample 89
Subset 328, 334
Substitution 220, 234
Subtraction
    of algebraic terms 219
    of directed numbers 153
    of fractions 171, 227
    of matrices 350
    of vectors 342
Supplementary angles 240
Surface areas 382
Survey 88, 91, 370
    line 372
Syllabus
Symmetry 243

Tally 92
    chart 93
Tangent to a circle 257
Tangent ratio 269
Taxable income 14
Tax 14, 18
Telephone bills 64
Three-dimensional
    trigonometry 281
Tiling 69
Time 50
Time-series graph 100
Timetables 52

Transformation of matrices 362
Transposing formulae 232, 236, 325
Trapezium 253, 375
    rule 386
Travel graphs 295
Traveller's cheques 34, 49
Tree diagram 146
Triangles 195, 201, 247, 250, 261
Triangulation 370
    points 395
Trigonometry 265
True interest rates 36

Union of sets 330, 333
Unit
    matrix 360
    vector 340
Units
    of area 198
    of volume 210
Universal set 328, 333

Value-added tax (VAT) 18
Variable 87, 220, 310
Variance 138
Vector geometry 346
Vectors 339
Venn diagrams 333
Vertically opposite angles 241
Volumes 210, 216, 379, 392

Wage 10
Wallpapering 67
Water rates 60
Weighted averages 129

Zero
    index 223
    matrix 360